Practical Engineering Geology

Steve Hencher

Spon Press
an imprint of Taylor & Francis

LONDON AND NEW YORK

First published 2012
by Spon Press
2 Park Square, Milton Park, Abingdon, Oxon OX14 4RN

Simultaneously published in the USA and Canada
by Spon Press
711 Third Avenue, New York, NY 10017

Spon Press is an imprint of the Taylor & Francis Group, an informa business

British Library Cataloguing in Publication Data
A catalogue record for this book is available from the British Library

Library of Congress Cataloging-in-Publication Data
Hencher, Steve (Stephen)
Practical engineering geology / Steve Hencher.
p. cm. - - (Applied geotechnics)
1. Engineering geology. I. Title.
TA705.H44 2012
624.1'51- -dc23
2011021261

ISBN: 978-0-415-46908-1 (hbk)
ISBN: 978-0-415-46909-8 (pbk)
ISBN: 978-0-203-89482-8 (ebk)

Typeset in Sabon
by Integra Software Services Pvt. Ltd, Pondicherry, India

MIX
Paper from
responsible sources
FSC   FSC® C013056
www.fsc.org

Printed and bound in Great Britain by
TJ International Ltd, Padstow, Cornwall

Practical Engineering Geology

This book presents a broad and fresh view on the importance of engineering geology to civil engineering projects.

Practical Engineering Geology provides an introduction into the way that projects are managed, designed and constructed and the ways that the engineering geologist can contribute to cost-effective and safe project achievement. The need for a holistic view of geological materials, from soil to rock, and of geological history is emphasised. Chapters address key aspects of

- geology for engineering and ground modelling
- site investigation and testing of geological materials
- geotechnical parameters
- design of slopes, tunnels, foundations and other engineering structures
- identifying hazards
- avoiding unexpected ground conditions.

The book is illustrated throughout with case examples and should prove useful to practising engineering geologists and geotechnical engineers and to MSc level students of engineering geology and other geotechnical subjects.

Steve Hencher is a Director of consulting engineers Halcrow and Research Professor of Engineering Geology at the University of Leeds.

Cover image *Am Buachaille* (The Herdsman), off Staffa in Scotland, is stunningly beautiful. It is also a succinct example of an engineering geological enigma so sits well on the front cover of this book. How were those curved columns formed and when in geological history? If we were to drill through (heaven forbid) would we find the same fractures that we can see at the surface? If we were to found a bridge on the island (again heaven forbid), how would we measure and characterise the rock? Could we simply use some rock mechanics classification to do the trick? Floating around the island, occasionally focusing on the distant horizon, one can ponder on such puzzles (see also page 78).

Applied Geotechnics

Titles currently in this series:

David Muir Wood *Geotechnical Modelling*
Hardback ISBN 978-0-415-34304-6
Paperback ISBN 978-0-419-23730-3

Alun Thomas *Sprayed Concrete Lined Tunnels*
Hardback ISBN 978-0-415-36864-3

David Chapman *et al. Introduction to Tunnel Construction*
Hardback ISBN 978-0-415-46841-1
Paperback ISBN 978-0-415-46842-8

Catherine O'Sullivan *Particulate Discrete Element Modelling*
Hardback ISBN 978-0-415-49036-8

Steve Hencher *Practical Engineering Geology*
Hardback ISBN 978-0-415-46908-1
Paperback ISBN 978-0-415-46909-8

Forthcoming:

Geoff Card *Landfill Engineering*
Hardback ISBN 978-0-415-37006-6

Martin Preene *et al. Groundwater Lowering in Construction*
Hardback ISBN 978-0-415-66837-8

Contents

Preface

The genesis of this book lies in a wet, miserable tomato field in Algeria. I was sitting on a wooden orange box, next to a large green Russian well-boring rig with a blunt bit. I was three weeks out of University. The Algerian driller hit the core barrel with a sledgehammer and a hot steaming black sausage of wet soil and rock wrapped itself around my hands. A Belgian contractor walked up and said to me (in French), 'What do you think? Four, six?' I looked at the steaming mass thoughtfully and said, 'Maybe about five.' He nodded approvingly. To this day I don't know what he was talking about or in what units.

I went to see the 'chef de zone' for this new steelworks, Roger Payne, who seemed totally in control and mature but was probably about twenty-eight, and suggested that we should write a book on engineering and geology. He, as a civil engineer, should write the geology bits and I should write the civil engineering bits as a geologist. That way we would see what we both considered important. We would edit each other's work. Well, we didn't do it but this book follows the blueprint. It includes aspects of geology that I consider most relevant to civil engineering, including many things that most earth science students will not have been taught in their undergraduate courses. It also provides an introduction into the parlance of civil engineering, which should help engineering geologists starting out. It is an attempt to set out the things that I wish I had known when I started my career.

Acknowledgements

Many have helped with this book mainly by reviewing parts, providing information and agreeing use of their data, figures and photos. These include: Des Andrews, Ian Askew, John Burland, Jonathan Choo, Chris Clayton, Gerry Daughton, Bill Dershowitz, Steve Doran, Francois Dudouit, Ilidio Ferreira, Chris Fletcher, John Gallerani, Graham Garrard, Robert Hack, Trevor Hardie, Roger Hart, Evert Hoek, Jean Hutchinson, Justyn Jagger, Jason Lau, Qui-Hong Liao, David Liu, Karim Khalaf, Mike King, Andy Malone, Dick Martin, Dennis McNicholl, David Norbury, Don Pan, Chris Parks, Andy Pickles, Malcolm Reeves, David Starr, Doug Stead, Nick Shirlaw, Kevin Styles, Nick Swannell, Leonard Tang, Len Threadgold, Roger Thompson and Derek Williams. I would also like to acknowledge the guidance of friends and mentors including Bob Courtier, Mike deFreitas, Richard Hart, Su Gon Lee, Keith Lovatt, Alastair Lumsden and Laurie Richards plus my research students whose work I have relied on throughout. Ada Li has drawn some of the figures and Jenny Fok has done some of the tricky bits of typing. Thanks to all.

Finally thanks to my long-suffering wife Marji – it has been a hard slog, glued to the computer and surrounded by piles of paper whilst the garden reverts to something resembling the Carboniferous rain forests. Sam Hencher has drawn some excellent cartoons and Kate and Jess have helped in their own sweet ways.

About the author

Steve Hencher is a Director of Halcrow China Ltd. (www.halcrow.com). He is also Research Professor of Engineering Geology at Leeds University, UK, and Honorary Professor in the Department of Earth Sciences at Hong Kong University.

He is a geologist by first degree and gained his PhD from Imperial College, London, on the shear strength of rock joints under dynamic loading. He then joined Sir WS Atkins & Partners where he was one of only nine geotechnical employees servicing what, even then, were the largest consultants in Europe. Atkins gave him wide experience in a very short term. This included the opportunity to investigate the ground for and supervise the construction and installation of piles at Drax Power Station, which provided a sharp insight into how large civil engineering projects work. Since then he has worked with the Hong Kong Government for five years, where he investigated major landslides, worked on shear strength of rock and first became involved in mapping and describing thick weathered profiles. Other major experience includes being part of the Bechtel design team for the High Speed Rail in Korea, working specifically on the design of very large span tunnels and underground stations. He taught the MSc in Engineering Geology at Leeds University full-time from 1984 to 1996 and supervised a large number of research students. Since 1997, he has headed geotechnics in the Hong Kong Office of Halcrow and was Regional Director of the Korean Office for seven years. He has worked and continues to work on various national and international committees in geotechnical engineering, in particular on weathered rocks, piling, landslides, rock slopes and rock mass characterisation. He has acted as an expert advisor and witness in several legal cases, including aspects of foundation design and construction, tunnelling, landslides and site formation.

1   Engineering geology

1.1   Introduction

Geology can be defined as the scientific study of the Earth and especially the rocks and soils that make up the Earth: their origins, nature and distribution, and the processes involved in their formation. Engineering geology then may be defined as the scientific study of geology as it relates to civil engineering projects such as the design of a bridge, construction of a dam or preventing a landslide. Engineering geologists need to identify the local rock and soil conditions at a site and anticipate natural hazards such as earthquakes so that structures can be designed, constructed and operated safely and economically. He (or she, throughout) needs to work with civil engineers and understand what they are trying to do and the constraints under which they work. His remit and responsibilities can be extensive, covering all of the Earth Sciences, including geophysics, geochemistry and geomorphology.

1.2   What do engineering geologists do?

Engineering geologists make up a high proportion of professional geologists throughout the world. Most of these work in civil engineering: in consulting (designing) or contracting (construction) companies with a team of engineers, some of whom will be specialised in the field of geotechnical engineering, which concerns the interface of structures with the ground.

One of the important tasks of an engineering geologist is to investigate the geological conditions at a site and to present these in a simplified ground model or series of models. Models should contain and characterise all the important elements of a site. Primary geological soil and rock units are usually further subdivided on the basis of factors such as degree of consolidation and strength, fracture spacing and style, hydrogeological conditions or some combination. Models must identify and account for all the natural hazards that might impact the site, as illustrated schematically in Figure 1.1 for a new high-rise

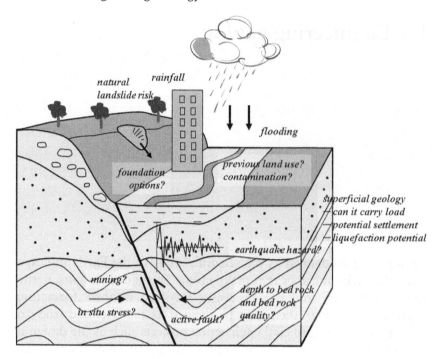

Figure 1.1 Site model for a new building, illustrating some of the factors and hazards that need to be addressed by the engineering geologist.

structure to be sited in a valley threatened by a nearby natural hillside. The ground model, integrated with the civil engineering structure, can be analysed numerically to ensure that the tolerance criteria for a project are achieved. For most structures, the design criteria will be that the structure does not fail and that any settlement or deformation will be tolerable; for a dam, the design criteria might include acceptable leakage from the impounded reservoir; for a nuclear waste repository, it would be to prevent the escape of contaminated fluids to the biosphere for many thousands of years.

1.3   What an engineering geologist needs to know

Many authors have attempted to define engineering geology as a subject separate to geology and to civil engineering (e.g. Morgenstern, 2000; Knill, 2002; Bock, 2006), but it is easier to define what a practising engineering geologist needs to know and this is set out in Table 1.1. Firstly, an engineering geologist needs to be fully familiar with geology to the level of a traditional earth sciences degree. He should be able to identify soil and rocks by visual examination and to interpret the geological history and structure of a site. He also needs to have knowledge of geomorphological processes, and be able to interpret terrain features and hydrogeological conditions. He must be familiar with ground investigation techniques so that a site can be

Table 1.1 Basic skills and knowledge for engineering geologists.

It is difficult to define engineering geology as a separate discipline but easier to define the subject areas with which an engineering geologist needs to be familiar. These include:

1. GEOLOGY

An in-depth knowledge of geology: the nature, formation and structure of soils and rocks. The ability to interpret the geological history of a site.

2. ENGINEERING GEOLOGY AND HYDROGEOLOGY

Aspects of geology and geological processes that are not normally covered well in an undergraduate geological degree syllabus need to be learned through advanced study (MSc and continuing education) or during employment. These include:

- Methods and techniques for sub-surface investigation.
- Properties of soil and rock, such as strength, permeability and deformability – how to measure these in the laboratory (material scale) and in the field and how to apply these at the large scale (mass scale) to geological models.
- Methods for soil and rock description and classification for engineering purposes.
- Weathering processes and the nature of weathered rocks.
- Quaternary history, deposits and sea level changes.
- Nature, origins and physical properties of discontinuities.
- Hydrogeology: infiltration of water, hydraulic conductivity and controlling factors. Water pressure in the ground, drainage techniques.
- Key factors that will affect engineering projects, such as forces and stresses, earthquakes, blast vibrations, chemical reactions and deterioration.
- Numerical characterisation, modelling and analysis.

These are dealt with primarily in Chapters 3, 4, 5 & 6.

3. GEOMORPHOLOGY

Most engineering projects are constructed close to the land surface and therefore geomorphology is very important. An engineer might consider a site in an analytical way, for example, using predicted 100-year rainfall and catchment analysis to predict flood levels and carrying out stability analysis to determine the hazard from natural slope landslides. This process can be partially shortcut and certainly enhanced through a proper interpretation of the relatively recent history of a site, as expressed by its current topography and the distribution of surface materials. For example, study of river terraces can help determine likely maximum flood levels and can also give some indication of earthquake history in active regions such as New Zealand. The recognition of past landslides through air photo interpretation is a fundamental part of desk study for many hilly sites. This is dealt with in Chapters 3 and 4.

4. CIVIL ENGINEERING DESIGN AND PRACTICE

An engineering geologist must be familiar with the principles of the design of structures and the options, say for founding a building or for constructing a tunnel. He/she must be able to work in a team of civil and structural engineers, providing adequate ground models that can be analysed to predict project performance, and this requires some considerable knowledge of engineering practice and terminology. The geological ground conditions need to be modelled mechanically and the engineering geologist needs to be aware of how this is done and, better still, able to do so himself. This is covered mainly in Chapters 2 and 6.

5. SOIL AND ROCK MECHANICS

Engineering geology requires quantification of geological models. Hoek (1999) described the process as 'putting numbers to geology'. That is not to say that pure geologists do not take a quantitative approach – they do, for example, in analysing sedimentary processes, in structural geology and in geochronology. However, a geologist is usually concerned with relatively slow processes and very high stress levels at great depths. The behaviour of soil and rock in the shorter term (days and months) and at relatively low stresses are the province of soil mechanics and rock mechanics. Knowledge of the principles and practice of soil and rock mechanics is important for the engineering geologist. This includes strength, compressibility and permeability at material and mass scales, the principle of effective stresses, strain-induced changes, critical states and dilation in rock masses.

characterised cost-effectively and thoroughly. Furthermore, he needs to understand the way that soils and rocks behave mechanically under load and in response to fluid pressures, how they behave chemically, and how to investigate their properties. To carry out his job properly, an engineering geologist also needs to know the fundamentals of how structures are designed, analysed and constructed, as introduced in Chapter 2 and presented in more detail in Chapter 6. Much of this will not be taught in an undergraduate degree and needs to be learnt through MSc studies or through Continuing Professional Development (CPD) including self study and from experience gained on the job.

The better trained and experienced the engineering geologist, the more he will be able to contribute to a project, as illustrated schematically in Figure 1.2. At the top of the central arrow, interpreting the geology at a site in terms of its geological history and distribution of strata is a job best done by a trained geologist. At the bottom end of the arrow, numerical analysis of the ground-structure interaction is usually the province of a geotechnical engineer – a trained civil engineer who has specialised in the area of ground engineering. There are,

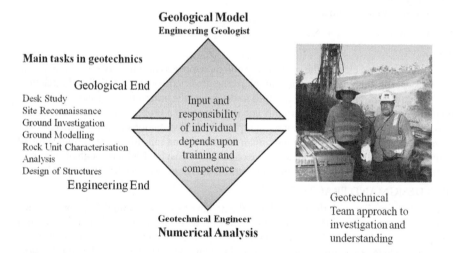

Geological Model
Engineering Geologist

Main tasks in geotechnics

Geological End

Desk Study
Site Reconnaissance
Ground Investigation
Ground Modelling
Rock Unit Characterisation
Analysis
Design of Structures

Engineering End

Input and responsibility of individual depends upon training and competence

Geotechnical Team approach to investigation and understanding

Geotechnical Engineer
Numerical Analysis

Figure 1.2 Roles of engineering geologists and geotechnical engineers. The prime responsibilities of the engineering geologist are 'getting the geology right' (according to Fookes, 1997) and 'assessing the adequacy of investigation and its reporting' (according to Knill, 2002), but an experienced engineering geologist with proper training can go much further, right through to the full design of geotechnical structures. Similarly, some geotechnical engineers become highly knowledgeable about geology and geological processes through training, study and experience and could truly call themselves engineering geologists. The photo shows David Starr and Benoit Wentzinger of Golder Associates, Australia, working in a team to investigate a major landslide west of Brisbane.

however, many other tasks, such as design of ground investigations and numerical modelling, that could be done by either an experienced engineering geologist or a geotechnical engineer. Many professional engineering geologists contribute in a major way to the detailed design and construction of prestigious projects such as dams, bridges and tunnels and have risen to positions of high responsibility within private companies and government agencies.

1.4   The role of an engineering geologist in a project

1.4.1   General

As discussed and illustrated later, some sites pose major challenges because of adverse and difficult geological conditions, but the majority do not. This leads to a quandary. If a 'one-size-fits-all' standardised approach is taken to site characterisation and more particularly to ground investigation (Chapter 4), then much time and money will be wasted on sites that do not need it but, where there are real hazards, then the same routine approach might not allow the problems to be identified and dealt with. This is when things can go seriously wrong. Civil engineering projects sometimes fail physically (such as the collapse of a dam, a landslide or unacceptable settlement of a building) or cost far more than they should because of time over-runs or litigation. Often, in hindsight, the root of the problem turns out to be essentially geological. It is also commonly found that whilst the difficult conditions were not particularly obvious, they were not unforeseeable or really unpredictable. It was the approach and management that was wrong (Baynes, 2007).

Engineering geologists can often make important contributions at the beginning of a project in outline planning and design of investigation for a site and in ensuring that contracts deal with the risks properly, as outlined in Chapter 2.

A skilful and experienced engineering geologist should be able to judge from early on what the crucial unknowns for a project are and how they should be investigated. Typical examples of the contributions that he might make are set out in Table 1.2.

1.4.2   Communication within the geotechnical team

The engineering geologist will almost always work in a team and needs to take responsibility for his role within that team. If there are geological unknowns and significant hazards, he needs to make himself heard using terminology that is understood by his engineering colleagues; the danger of not doing so is illustrated by the case example of a slope failure in Box 1-1.

Table 1.2 Particular contributions that an engineering geologist might bring to a project (not comprehensive).

1. Unravelling the geological history at a site. This will come initially from regional and local knowledge, examination of existing documents, including maps and aerial photographs, and the interpretation of exposed rock and geomorphologic expression. Geology should be the starting point of an adequate ground model for design.
2. Prediction of the changes and impacts that could occur in the engineering lifetime of a structure (perhaps 50–100 years). At some sites, severe deterioration can be anticipated due to exposure to the elements, with swelling, shrinkage and ravelling of materials. Sites may be subject to environmental hazards, including exceptional rainfall, earthquake, tsunami, subsidence, settlement, flooding, surface and sub-surface erosion and landsliding.
3. Recognising the influence of Quaternary geology, including recent glaciations and rises and falls in sea level; the potential for encountering buried channels beneath rivers and estuaries.
4. Identifying past weathering patterns and the likely locality and extent of weathered zones.
5. Ensuring appropriate and cost-effective investigation and testing that focuses on the important features that are specific to the site and project.
6. Preparation of adequate ground models, including groundwater conditions, to allow appropriate analysis and prediction of project performance.
7. An ability to recognise potential hazards and residual risks, even following high-quality ground investigation.
8. Identification of aggregates and other construction materials; safe disposal of wastes.
9. Regarding project management, he should be able to foresee the difficulties with inadequate contracts that do not allow flexibility to deal with poor ground conditions, if they are encountered.

Box 1-1   Case example of poor communication with engineers

The investigations into a rock slope failure are reported by Hencher (1983a), Hencher *et al.* (1985) and by Clover (1986). During site formation works of a large rock slope, behind some planned high-rise apartment blocks, almost 4,000m³ of rock slid during heavy rainfall on a well-defined and very persistent discontinuity dipping out of the slope at about 28 degrees. The failure scar is seen in Figure B1-2.1. The lateral continuity of the wavy feature is evident to the left of the photograph, beneath the shotcrete cover, marked by a slight depression and a line of seepage points. If the failure had occurred after construction, the debris would have hit the apartment blocks. A series of boreholes had been put down prior to excavation and the orientation of discontinuities had been measured using impression packers (Chapter 4). Statistical analysis of potential failure mechanisms involving the most frequent joint sets led to a design against shallow rock failures by installation of rock bolts and some drains. The proposed design was for a steep cutting, with the apartment blocks to be sited even closer to the slope face than would normally be allowed. Unfortunately, the standard method of discontinuity analysis had eliminated an infrequent series of discontinuities daylighting out of the slope and on one of which the failure eventually occurred. Pitfalls of stereographic analysis in rock slope design are addressed by Hencher (1985), a paper written following this near-disaster.

Examination of the failure surface showed it to be a major, persistent fault infilled with clay-bounded rock breccia about 700mm thick and dipping out of the slope (Figures B1-2.2 and B1-2.3). In the pre-failure borehole logs, the fault could be identified as zones of particularly poor core recovery; the rock in these zones was described as tectonically influenced at several locations. In hindsight, the fault had been overlooked for the design and this can be attributed to poor quality of ground investigation and

Figure B1-2.1 View of large rock slope failure in 1982, South Bay Close, Hong Kong.

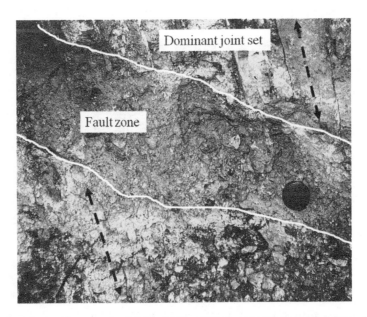

Figure B1-2.2 Exposure of brecciated and clay-infilled feature through mostly moderately and slightly weathered volcanic rock and with very different orientation to most other rock joints.

statistical elimination of rare but important discontinuities from analysis, as discussed earlier, but exacerbated by poor communication. The design engineers and checkers might not have been alerted by the unfamiliar terminology (tectonically influenced) used by the logging geologist; they should have been more concerned if they had been warned directly that there was an adversely oriented fault dipping out of the slope. The feature was identified during construction, but failure occurred before remedial

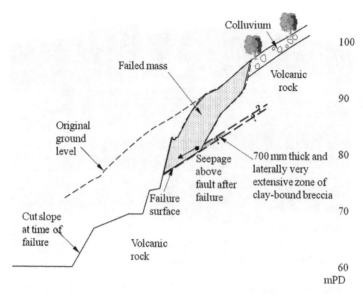

Figure B1-2.3 Cross section through slope showing original and cut slope profile at the time of failure. Geology is interpreted from mapping of the failure scar, but the main fault could be identified in boreholes put down before the failure occurred.

Figure B1-2.4 Slope in 2010 showing anchored concrete beams installed to prevent further failure in the trimmed-back slope above the apartment blocks.

measures could be designed (Clover, 1986). It was fortunate that the failure occurred before construction of the apartment blocks at the toe. The site as in 2010 is shown in Figure B1-2.4. The slope required extensive stabilisation with cutting back and installation of many ground anchors through concrete beams across the upper part of the slope and through the fault zones. These anchors will need to be monitored and maintained continuously for the lifetime of the apartments.

Inadequate site investigation that fails to identify the true nature of a site and its hazards can result in huge losses and failure of projects. Similarly, poorly directed or unfocused site investigation can be a total waste of time and money whilst allowing an unfounded complacency that a proper site investigation has been achieved (box ticked). The engineering geologist needs to work to avoid these occurrences. He needs to be able to communicate with the engineers and to do that he needs to understand the engineering priorities and risks associated with a project. Those risks include cost and time for completion. This book should help.

1.5 Rock and soil as engineering materials

In geology all naturally occurring assemblage of minerals are called rocks, whatever their state of consolidation, origins or degree of weathering (Whitten & Brooks, 1972). For civil engineering purposes it is very different. Geological materials are split into soil and rock, essentially on differences in strength and deformability. To make it more difficult, the definitions of what is soil and what is rock may vary according to the nature of the project. For many purposes, soil is defined as material that falls apart (disaggregates) in water or can be broken down by hand but, for a large earth-moving contract, materials may be split into soil and rock for payment purposes according to how easy or otherwise the material is to excavate; rock might be defined as material that needs to be blasted or that cannot be ripped using a heavy excavating machine. For engineering design, the distinctions are often pragmatic and there may be fundamental differences in approach for investigation and analysis. This is illustrated for slope stability assessment in Figure 1.3. In the left-hand diagram, the soil, which might be stiff clay or completely weathered rock, is taken as having isotropic strength (no preferential weakness directions), albeit that geological units are rarely so simple. To assess stability, the slope is searched numerically to find the critical potential slip surface, as explained in Chapter 6. In contrast the rock slope to the right is, by definition, made up of material that is too strong to fail through the intact material, given the geometry of the slope and stress levels. In this case, site investigation would be targeted at establishing the geometry and strength characteristics of any weak discontinuities (such as faults and joints) along which sliding might occur. If an adverse structure is identified then the failure mechanism is analysed directly. This conceptual split is fundamental to all branches of geotechnical engineering, including foundations, tunnels and slopes, and it is important that the engineering geologist is able to adapt quickly to seeing and describing rocks and soils in this way.

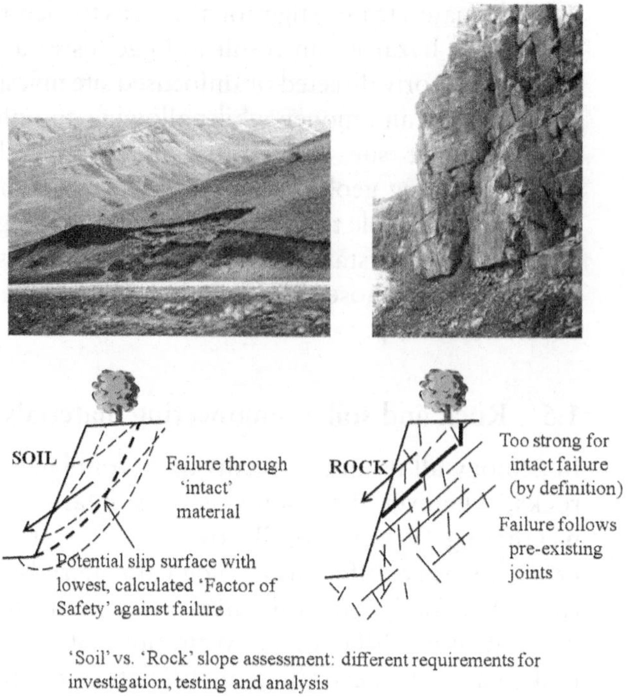

SOIL — Failure through 'intact' material

Potential slip surface with lowest, calculated 'Factor of Safety' against failure

ROCK — Too strong for intact failure (by definition)

Failure follows pre-existing joints

'Soil' vs. 'Rock' slope assessment: different requirements for investigation, testing and analysis

Figure 1.3 Distinction between soil and rock at a pragmatic level for slope stability analysis. Soil failure is near Erzincan, Turkey. Analysis involves searching for the slip plane that gives the lowest FoS for the given strength profile. The rock slope is in a limestone quarry, UK, and failure is totally controlled by pre-existing geological structure (bedding planes and joints).

The compartmentalisation of soil and rock mechanics is quite distinct in geotechnics, with separate international societies, which have their own memberships, their own publications and organise their own conferences. Details and links are given in Appendix A. Textbooks deal with soil mechanics or rock mechanics but not the two together. In reality, this is a false distinction and an unsatisfactory situation. Engineering geologists and geotechnical engineers need to appreciate that in nature there is a continuum from soil to rock and from rock to soil. Soil deposited as soft sediment in an estuary or offshore in a subsiding basin is gradually buried and becomes stronger as it is compressed by the weight of the overlying sediment, and strong bonds are formed by cementation, as illustrated in Figure 1.4. Conversely, igneous rock such as granite is strong in its fresh state but can be severely weakened by weathering to a soil-like condition, as illustrated in Figure 1.5, so that it might disintegrate on soaking and even flow into excavations below the water table.

An engineering geologist must be familiar with the full range of geological materials and understand the principles and methods of

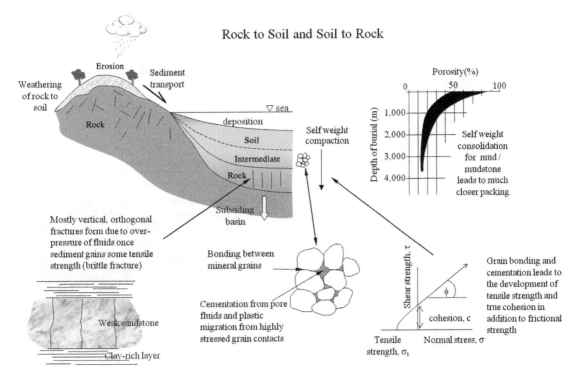

Figure 1.4 The cycle of rock to soil and soil to rock. Diagenetic and lithification processes cause soft sediment to transform into strong cemented rock during burial. Exposed rock breaks down to soil by weathering.

both soil and rock mechanics, which are tools to be adopted, as appropriate, within the engineering geological model.

1.6 Qualifications and training

Engineering geologists generally begin their careers as earth science graduates, later becoming engineering geologists through postgraduate training and experience. Within civil engineering, in many countries including the UK, Hong Kong and the USA, there is a career pathway that is measured through achievement of chartered status or registration as a professional, as summarised in Table 1.3. The aim is that engineering works should only be designed and supervised by competent persons who have received adequate training and experience. Chartered or registered status generally requires a recognised university degree followed by a period of training under the supervision of a senior person within a company. The practice of engineering is often legally defined and protected by government regulations. In some countries, only registered or chartered engineers or engineering geologists are permitted to use the title and to sign engineering documents (reports, drawings and calculations), thus taking legal responsibility. Details for career routes for various countries are set out in Appendix A, together with links to a number of learned societies and details of professional institutions that an engineering geologist might aspire to join.

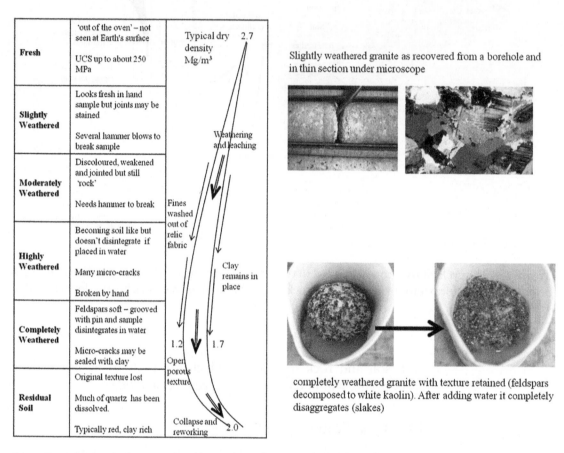

Figure 1.5 Typical stages of chemical weathering for an igneous rock.

Table 1.3 Typical routes for a career in geotechnical engineering (UK).

Engineering geologist	Geotechnical engineer
<ul><li>First degree geology or other earth sciences (BSc or MSc)</li><li>MSc in engineering geology</li><li>5+ years experience and training</li><li>Chartered Geologist (straight-forward route) – Geological Society of London</li><li>Chartered Engineer (more difficult route) – Institution of Civil Engineers or Institution of Mining, Metallurgy and Materials</li></ul>	<ul><li>First degree civil engineering (BEng or MEng)</li><li>MSc in geotechnical subject (e.g. soil mechanics or foundation engineering)</li><li>5+ years experience and training</li><li>Chartered Engineer (Institution of Civil Engineers)</li></ul>
Distinctive skills at early stage in career development	
<ul><li>Knowledge of the fabric and texture of geological materials and geological structures and how these will influence mechanical properties (more so for rock than soil)</li><li>Observation and mapping of geological data</li><li>Interpreting 3-D ground models from limited information following geological principles</li><li>Identifying critical geological features for a ground model</li></ul>	<ul><li>Numerate, with sound basis for analysis and the design of engineering structures</li><li>Good understanding of mechanics (more so for soil than rock)</li><li>Understanding of project management and business principles</li></ul>

2 Introduction to civil engineering projects

2.1 Management: parties and responsibilities

2.1.1 The owner/client/employer

All civil engineering projects have owners – otherwise known as the client or the employer, because the owner ultimately pays for all the works and employs the various parties involved in design and construction. The owner normally engages architectural and engineering companies to advise him and to manage, design and construct the project in a cost-effective manner. Most projects are designed by a consulting engineer and built by a contractor. Under such 'engineer's designs' the design responsibility rests with the project designer. Other projects are described as 'turnkey' or 'design and build', where a contractor is commissioned to deliver the whole project or part of a project as a complete package. Such arrangements – 'contractor's design' – often apply to specialist parts of projects such as a bored tunnel or piled foundations for a building. The typical relationships and tasks for a project designed by a consulting engineer are illustrated in Figure 2.1.

Sometimes the owner may have in-house technical expertise sufficient to overview the project (as in a government department or large energy company) but rarely will he have the staff or experience to design, construct and/or supervise all aspects of a large civil engineering project, which might require a huge range of skills – from site formation through numerical analysis to mechanical and electrical fitting out.

2.1.2 The architect and engineer

Engagement of an architect and engineer may be through competitive tender whereby several capable consulting companies are invited to make proposals for design and possibly supervision and for the cost control of construction and to give a price for carrying out this work. The owner will select and contract with one party or with a consortium of consultants known as a joint venture (JV), which might be a

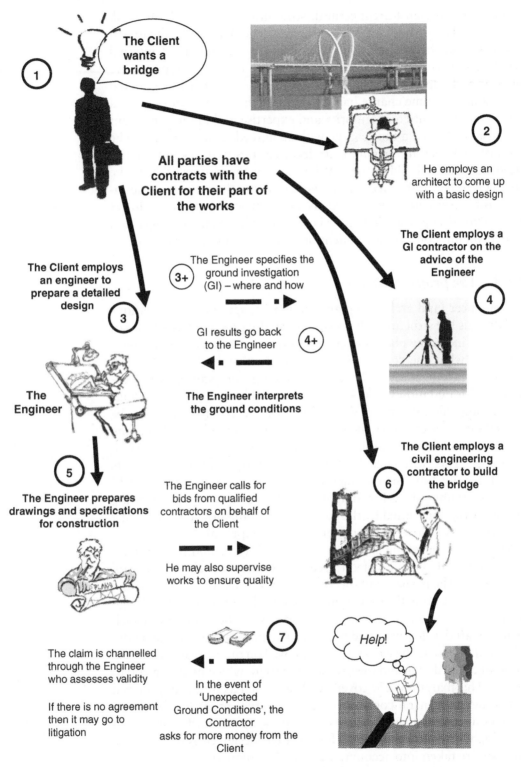

Figure 2.1 The client wants a bridge. This figure illustrates various contractual arrangements and relationships between the main parties in an engineered design – one where the project is designed by a specialist design engineer and built by a specialist contractor.

grouping of specialist architectural, structural, mechanical and civil/ geotechnical companies, which have joined together specifically to win and work on the project. The JV will need well-organised internal management to ensure that roles, responsibilities and payments are all clear and adhered to. The price paid by the owner may be a fixed lump sum on a time charge basis (usually with different rates quoted for engineers of different seniority and expertise within the consultant organisation) or on a time charge with an agreed ceiling estimate. The roles of architect and engineer are legal entities with responsibilities often defined by building regulations within the country where the project is to be constructed. An individual within the company responsible for design may be named as an approved person, architect or structural engineer and may be required to sign drawings and formal submissions to government or other checking organisations.

2.1.3   *The project design*

The engineer (and architect) plans the works, specifies investigations and designs the structure. The design is usually presented as a series of drawings, including plans and cross sections (elevations) to scale, with details of what the contractor is to construct and where. This will normally include an overall site plan showing, for example, the location of all foundation works – piles, pads or other features. Drawings are accompanied by specifications for how the construction is to be carried out – for example, the strength of concrete to be used and any restrictions such as prohibition on blasting because of proximity to buildings. This will later be supplemented by method statements (which set out how the contractor will carry out parts of the work) and programmes (dates for completion of the various activities making up the works) submitted by the contractor commissioned to construct the works (see below) to the designer for his approval.

Within the consulting engineers a project director and project manager will usually be appointed to see the project through to successful completion. The measures of success are not only delivery of the project to the satisfaction of the owner but also to make a profit for the design company and to meet internal requirements of the company, which include staff development and training. The price quoted to the owner when bidding to do the works is usually based on the estimated staff cost to produce the design and then adding a margin, which might be 100 to 200%. This margin would cover overheads such as office support and infrastructure, general company costs plus actual profit for the shareholders in the company. Whereas the mark-up on staff costs might seem high, actual profit margins for most UK design consultants, once all costs are taken into account, are often less than 10%.

The engineer is in a very responsible position, as he will plan any site investigation, seek tenders from contractors to carry out all tasks and works, and make recommendations to the owner regarding which contractors he should employ. He will take the site investigation data, design the works and probably supervise the works, although sometimes this is let as a separate contract or conducted in-house for consistency between separate sections of an ongoing project, as is the practice of the Mass Transit Railway Authority in Hong Kong, for example. During construction, the engineer will usually employ or nominate a resident engineer (RE) and other resident site staff who will deal with the construction on site, on a day-to-day basis. The site staff will refer any needs for design changes as the works progress back to the design office for resolution.

2.1.4 *The contractor*

Various contractors may be employed for the works. Contractors are usually invited to bid to carry out works, as set out in drawings, specifications and a bill of quantities (BOQ), which lists the works to be done and estimated amounts (e.g. volume of material to excavate). The contractor puts a price against each item in the BOQ and the sum of all the itemised costs will constitute his offer to the owner for completing the works. Generally, a specialist ground investigation contractor will be employed to carry out sub-surface investigation of the site following a specification for those works by the engineer. That specification will include locations and depths of sampling, types of testing and the equipment to be used (Chapters 3 and 4). Other contractors will be used to conduct and construct the various facets of a project.

Contractors, like engineers, need to ensure that they allow for some degree of profit. When the engineer assesses the various tenders, on behalf of the owner, he needs to be cautious that any particularly low bid is not unrealistic (which he would normally do by comparing with his own broad estimate of what the cost might be). A particularly low bid might mean that the contractor has misunderstood the scope of the works and whilst the low price might be attractive to the owner, quite often such situations end up in conflict or dispute, with the contractor desperately trying to compensate for his underestimation of the costs involved. Alternatively, the contractor might be trying to win or maintain market share at a time of high competition, so his bid has a deliberately low profit margin. A third possibility is that the contractor already has in mind ways to make claims for additional payment, especially if the contracts are not well drafted, as discussed later. The engineer may recommend that the owner does not accept the lowest tendered offer because of these various concerns, and some countries and governments have rules and methods in place for trying to eliminate unrealistic bids and ensuring that the most suitable contractor is employed.

Sometimes the contractor might identify some better or more cost-effective way of carrying out part or all of the works and can offer this as an alternative design to that presented in the tender documents (the conforming design); the owner might accept this proposal because of price, programme or quality reasons. The contractor (and his designer) might then take over responsibility for future design works and the owner may employ another engineer to check these designs.

The contractor may sub-contract parts of the works – for example, by employing a specialist piling sub-contractor to construct that element of the foundations. Whilst for a normal engineer-design project, the consulting engineer is responsible for overall design, the contractor may need to design temporary works necessary as intermediate measures in achieving the final design intent. For example, to construct a deep basement, the contractor may have to design some shoring system to support the excavation until the final walls and bracing slabs of the final structure have been completed. Temporary works should normally be designed to the approval of the engineer. In some instances, some of the permanent works are designed by the contractor or the temporary works somehow incorporated within the permanent works because to remove them might be too difficult or it is otherwise beneficial to do so.

Contractor's designs are sometimes adopted for parts of a project because of his local and specialist technical experience together with his knowledge of the costs of material, plant and labour. Another advantage is that there may be less ambiguity in terms of who is responsible for the performance of the works and in particular dealing with problems posed by difficult ground conditions. When it comes to foundations or tunnels, the contractor should be in a position to accept the risk of any unforeseeable ground conditions – providing he is allowed to design and conduct an adequate ground investigation to his own specification.

2.1.5   *Independent checking engineer*

For many large projects, an independent checker is employed by the owner to give added confidence that the design of permanent and temporary works is correct. The checker is usually a similar type of company to the design company, i.e. an engineering consultant. The check could be confined to a simple review of design assumptions and calculations but, in some instances, might involve a comprehensive and separate analysis of all aspects of a project.

2.2   Management: contracts

Civil engineering is a commercial business and the engineering geologist needs to understand how it works. The relations between all parties are governed by contracts. A contract is a legal document between the owner and each of the other parties involved with a project and defines the scope and specification of works, including payment schedules and responsibilities. Contracts also need to be made between consultants and specialist sub-consultants or JV partners, and between a contractor and specialist sub-contractors. It is very wise to use lawyers at this stage to ensure that contracts are well written to minimise the risk of later dispute, although standard forms of contract are often used and large companies tend to have internal documents. The experienced engineering geologist can help ensure that contracts are reasonable, realistic and fair with respect to their treatment of ground conditions, which is where many problems arise during construction. These problems need to be resolved in a pragmatic manner and quickly during construction, but there is often some dispute at a later stage over which party should pay for changes, additional costs and delays.

2.2.1　*Risk allocation for geotechnical conditions*

As discussed later, sites vary geotechnically from those that are extremely difficult to understand and characterise, to those that are simple and straightforward. In a similar fashion, site investigations vary in quality from focused, excellent and insightful, to downright useless, depending on the experience, capability and insight of the engineer and his team in planning and interpreting the investigation and the skill and quality of equipment of the ground investigation contractor. As a result, there are always risks involved in projects, especially where these involve substantial ground works, for example, in tunnelling or deep foundations. The risks need to be assigned under a contract and there are few mandatory rules. Each contract should state how variations are to be dealt with in the event of unforeseen ground conditions such as stronger or weaker rock (requiring different excavation techniques) or more water inflow to a tunnel (requiring additional ground treatment works) than had been anticipated. This is a large and important subject and guidance on how to identify critical ground conditions through a systematic approach for addressing hazards and risks, using focused site investigation, is presented in Chapters 4 and 6 and Appendix E. Chapter 7 takes this further and provides case examples of projects where things went wrong for some reason or other.

Some of the background and options for preparing a contract with respect to ground hazards are illustrated in Figure 2.2. Mostly, projects use standard contract forms such as the New Engineering Contract (NEC) (ICE, 2005) or Fédération Internationale Des Ingénieurs-Conseils (FIDIC) (discussed by Tottergill, 2006). Some contractual forms are suitable to engineer-design contracts and others to design and build situations.

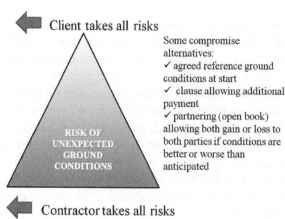

Contractor is paid the cost of completing works in full. Disadvantage is that there is no incentive for the contractor to resolve problems cost-effectively when they arise

Client takes all risks

Some compromise alternatives:
✓ agreed reference ground conditions at start
✓ clause allowing additional payment
✓ partnering (open book) allowing both gain or loss to both parties if conditions are better or worse than anticipated

RISK OF UNEXPECTED GROUND CONDITIONS

A disadvantage is that the contractor is unable (or unwilling) to price the risks with any certainty. Can go badly wrong (see Chapter 7)

Contractor takes all risks

Figure 2.2 The main options for forming a contract to deal with the risk of unexpectedly difficult ground conditions.

In some forms of contract, the owner accepts all the ground risks and that makes some sense in that it is his site, with all its inherent geological and environmental conditions. This kind of contract works quite well for simple sites and structures, for example, the cutting of a slope with the installation of soil nails, where the work done by the contractor is rather routine and can be simply re-measured against the provisional BOQ priced by the contractor when he tendered to do the work. If he excavates 2,300 m³ of soil and 52,050 m³ of rock during the contract, then that is what he will be paid for, at the prices he originally quoted for each type of excavation, although there might be some disagreement over the definition of soil and rock by the parties. Specialist engineers called quantity surveyors (QS) assess and recommend approval of such payments to the engineer and then on to the owner.

In an attempt to make it clear-cut where the responsibilities lie, some owners try to use contracts that place all the risks for ground conditions solely on the contractor, but this is inflexible and offers no way out when things go wrong. In practice, depending on commercial pressures, the contractor may take a serious gamble (sometimes without fully weighing up the risks) and it is then, when things start becoming difficult, such as when the ground conditions are worse than expected, that claims begin to be made and disputes can follow. Even where all the risk has been accepted by the contractor, when things become very difficult, he and his lawyers may try to use clauses in the contract, such as claiming that the works were physically or commercially impossible, or just give up on the project. The arguments can be long and extremely costly for all parties. Such contractual arrangements are rarely used these days for major projects.

For more complex projects and especially for constructions underground, usually some of the ground risks are accepted by the contractor. As Walton (2007) observes, the contractor, unlike the owner, is in the construction business, is a specialist in the particular type of works he is to undertake, and may be able to spread the risk over a number of contracts, to some degree. In order to get the contractor to accept some of the risk of encountering difficult conditions, however, the owner must expect to pay some additional sum to cover that insurance element through a higher contract price; if the risks do not materialise, he will have wasted money, but that is the nature of insurance.

In shared risk contracts, the contractor is expected to accept and cope with generally variable but predictable conditions, but is allowed to claim for additional money where something unpredictable and highly adverse is encountered. Despite the pressure release valve of old ICE Conditions of Contract Clause 12 (payment for unexpected ground conditions) and similar clauses in other standard forms of contract, it is in all parties' interests that all hazards and risks are foreseen and priced for by the contractor in terms of the extra work and delay which will occur if the risk materialises. This is definitely the province where the engineering geologist can play a major role and in particular by engineering geologists working within the engineer's consulting team, which is responsible for investigating the site and designing and specifying the works. There is a similarly important role for engineering geologists within the tendering contracting company, who must anticipate hazards and price the job sensibly.

Unfortunately, contractors sometimes fail to take account of all the perceived risks (even where aware) partly because they know that the owner (advised by the engineer) will be tempted to employ the contractor offering the lowest price. There are Machiavellian aspects to all this in that each party is trying to minimise its costs and risks whilst maximising profit. Contract writing and interpretation are key parts of this. For example, a contractor will try to predict where extra quantities might be required during construction, compared to the estimates by the engineer that will form part of the contract in the BOQ (for example, in the proportion or rock vs. soil to be excavated) and quote unit prices appropriately to maximise his profits. He might include high mobilisation charges, whilst trimming prices of other items on the bill to improve the payment schedule and his cash flow without jeopardising his chance of winning the contract in competition with other invited tendering contractors. This is all fair and above board but it does mean that the conduct of a civil engineering contract can be rather fraught at times.

2.2.2 *Reference ground conditions*

It is now common, for tunnelling works especially, to try to set out some reference ground conditions (presented in geotechnical baseline reports) that all parties buy into for contractual purposes before the works actually begin. For larger tunnelling contracts in the UK, and increasingly elsewhere, it is now mandatory that the hazards and risks are assessed and managed in a consistent manner (British Tunnelling Society, 2003). This is also the general case for some standard contracts (FIDIC). This was introduced largely because insurance companies were receiving an increasing number of claims due to tunnelling projects going seriously wrong and were threatening simply to withhold insurance on 'such risky, poorly investigated, poorly thought-through and mismanaged projects' (Muir Wood, 2000).

Unfortunately, in practice it is often not that simple to define engineering geological conditions in a distinct and unambiguous manner. If one tries to be very specific (say on the rock type to be encountered) then it would be relatively easy for the contractor to employ a specialist at a later stage to dispute the rock description in detail and then to allege that the slight difference in rock type caused all the difficulties that followed (excess wear, higher clay content etc., etc., plus delays and general loss of productivity). Drafters of reference conditions sometimes resort instead to broad characterisation, perhaps using rock mass classifications such as Q or Rock Mass Rating (RMR), as introduced in Chapters 4 and 5 and Appendix C. The problem there is that such classifications are made up of a range of parameters such as strength and fracture spacing, each of which can be disputed because geology is never that simple (or uniform). Furthermore, experienced persons can often draw very different conclusions from the same data set. Fookes (1997) reports an exercise where he asked two engineering geologists familiar with rock mass classifications to interpret the same sets of boreholes and exposures for a particular tunnel in terms of RMR and Q value. One came up with an RMR = 11 (extremely poor rock and danger of immediate collapse); the other RMR = 62 (fair rock and that no support is required). The Q value interpretations were similarly quite different (extremely poor vs. fair rock). In this particular case, the rock contained incipient cleavage (slate) and the different opinions on classifications mostly hinged upon whether that cleavage was considered a joint set or not – the standards and guidance documents do not help very much in this regard, as discussed in Chapters 3 and 4. The main point is that despite reference conditions being set out with good intentions of helping the contractor to price the job and avoiding dispute, there is no guarantee that this will be achieved.

It is the normal case that the extent of geological/geotechnical units and position and nature of faults, for example, are uncertain. The geotechnical baseline report should present the best interpretation of the ground conditions by the designers and state any limitations and reservations. In doing so, the rationale should not be, somehow, to outwit the contractor contractually, but to allow the contractor to select the right methods for construction, and to price and to programme his works adequately. Contractually, the reference conditions should be just that – something to refer to when considering whether some adverse ground was anticipated or anticipatable by an experienced contractor, given the available information. The contractor will have been expected to consider the site in a professional manner, which would include examining any relevant rock exposures, say in quarries adjacent to the route. Many contracts require the contractor to satisfy himself of the ground conditions at a site or along the route, but it is rarely practical for him to carry out his own ground investigation at tender stage (with no guarantee of winning the work) and often that constraint is accepted by an arbitrator in any subsequent dispute.

One point that follows is that it is very important for engineering geologists to keep good records throughout construction. These should be factual, with measurements, sketches and photographs, using standard terminology for description and classification, as introduced in Chapter 3. Quite often, especially for tunnels, the engineering geologist representing the contractor will prepare sketches of ground conditions encountered, together with engineering works installed (such as locations of rock bolts and instruments) and seek to get this agreed by the supervising team on a daily basis. This means that the basis for payment is clarified and, in the event of some contractual dispute later, there are clear records for all parties to review.

2.2.3 *Claims procedures*

Interestingly, when things become difficult during the works because of poor ground conditions, the contractor has to apply through the engineer for extra money (ultimately to be paid for by the owner). Now it is the engineer's responsibility to act impartially, within the terms of the contract, having regard to all the circumstances. In like manner, the engineer's representative on site and any person exercising delegated duties and authorities should also act impartially (ICE Conditions of Contract). In other standard contracts, in recognition that the engineer is employed by the owner, the engineer is expected to act reasonably rather than impartially, but nevertheless he is clearly expected to treat the contractor's claims in a proper manner with due regard to the contract and the actual situation. The engineer can, however, find himself in a position of conflicting interest, where the ground conditions that are causing the difficulty to the contractor might, and perhaps should, have been recognised and dealt with by the engineer's investigation, design and specification for the works (Dering, 2003). He might have to approve a claim by the contractor, in the knowledge that he himself is culpable because of poor ground investigation, modelling or design. Conversely, he might resist a claim that later proves valid following dispute resolution.

2.2.4   *Dispute resolution*

If a claim cannot be resolved between the contractor and the owner then the claim might be passed to a third party. The two parties can jointly appoint a technical expert to help resolve the issues through a process of adjudication. It is a far less formal process than going to court. The appointment of an adjudicator might be written into the original contract (as specified in the New Engineering Contract of ICE) and his decisions should be complied with. For larger projects, the parties might appoint an agreed panel of experts at the outset. The panel can be asked to adjudicate on the validity of any claim – whether conditions were different to those anticipated and whether they had the adverse consequences claimed by the contractor. This leaves the decisions in the hands of experienced professionals rather than lawyers whose knowledge of ground conditions and ground behaviour might be rather limited.

Mediation is an option where the parties to a dispute will plead their cases to an independent mediator (who might be a lawyer rather than a technical expert). He will try to get the parties to reach an agreement and will also provide an opinion as to the likely outcome if the matter is taken to the next, more expensive level. If a party (either the owner or contractor) is told by an independent mediator that their position over a claim is weak, then they may be more willing to reach an agreement with the other side. Arbitration is a higher-level process and is generally written into contracts as a way of having disputes resolved. Both parties agree at the outset that this should be so, and the location where any arbitration should be conducted. Arbitration takes place in a court and there may be up to three arbitrators – perhaps one agreed between both parties and the second and third chosen by each party independently. The cost, with lawyers (probably several on both sides), barristers, independent experts (see next section) and the court expenses, can be very high. In a recent case that the author was involved with, the final award to the winning party was essentially the same as had been previously offered in settlement, prior to arbitration, and was far exceeded by the cost of the legal proceedings.

Arbitration decisions are generally taken as final – however disgruntled one party might feel at the result. Arbitration reports and outcomes are generally kept confidential to the parties. Unfortunately, this means that the profession does not learn lessons, which is a great pity. The only cases that make their way into the literature as well-documented examples are those that are actually taken to court (public domain) or where there is some kind of forensic study in the case of a major collapse such as the collapse of the Heathrow Express tunnels and the failure of the Nicoll Highway excavations in Singapore. These and other case examples are presented in Chapter 7.

2.2.5 *Legal process and role of expert witness*

When disputes reach the stage of either arbitration or civil court, where one party sues another, it is usual for the parties to employ experts to advise them on the validity or otherwise of their case and, if they agree with their client's position, to make a report stating the reasons why. Because many ground condition claims are fundamentally linked to a poor appreciation of geology, engineering geologists often become involved in disputes as experts. Initially, the expert will be advising his client on the strength of the claim, outside any legal proceedings. If the expert disagrees with his client's position, he must tell him as soon as he recognises that situation. It will then be up to the client and his legal advisors to decide how to proceed.

If the expert thinks his client's case is valid and he writes a report that may be used in evidence, it is important that he recognises that his overriding duty is to the court rather than to the party that is paying for his services. The expert needs to understand that in complex, technical cases his evidence can often be pivotal. Questions put to an expert and his replies to them are treated as part of his evidence. Experts are required to make some statement of truth, that he believes that the facts stated in his report are true and that the opinions are correct and his own. He will also need to make an oath in court. The court will often request that the experts employed by the two sides hold meetings and prepare joint statements, identifying where matters are agreed and where matters are in disagreement. In principle, this sounds straightforward but sometimes instructing lawyers will prevent or limit the agreement of experts, partly because, whilst experts may agree broadly or compromise over some technical issue (as they would if they were working together on an engineering project), there may be subtleties in the legal considerations and case law that might hold sway. That said, technical experts must beware being led by lawyers in preparing their reports and must not attempt to argue points that they are not happy with or that fall outside their knowledge and expertise. In court, barristers, judges and arbitrators will question experts very thoroughly and a tenuous and weakly held position will usually be exposed for what it is.

Geotechnical experts need to recognise that they are not legal experts. The author was involved in a case where the contractor had accepted all ground risks. The situation appeared clear-cut and hopeless for the contractor to a layman, but a barrister educated me that the owner and his design engineers had made 'representations' which would affect things legally. Quite often a party is clearly at fault in some way but there may be some question over their legal responsibility. To resist a charge of negligence the engineer need not have done everything right – just to the same quality as his peers on average. I have heard an expert say in court, 'I have seen worse', as an excuse for poor practice. It is then up to the arbitrator or judge to decide whether that excuse is persuasive.

2.2.6   *Final word on contracts: attitudes of parties*

In practice, much depends upon the attitudes of the various parties. Even a poorly drafted contract can be made to work so that the owner gets his project constructed within his budget and the contractor makes a profit, but this requires co-operative and non-adversarial attitudes. To foster this attitude, formal partnering sessions are commonly used where everyone is asked to agree some set of rules of behaviour and professional dealings.

Whether or not this works is often down to individuals – especially the RE and the contractor's site agent. The author has experience of a large project involving several different contracts, where the RE had a high regard for one contractor but mistrusted another because of previous encounters on other projects. He was of the opinion that the second contractor had won the contract for an unrealistically low price and therefore would be out to make its profit through claims. The first contract went very well despite many technical problems, which were overcome in a pragmatic manner, working as a team. Reasonable claims were dealt with expediently and everything was completed on time, to the required technical standards and with the contractor leaving the site a happy man. The second contract was a direct contrast. All site supervisory staff were instructed by the RE not to give him any advice, help or site instructions, to avoid chinks in the contractual armour that the RE thought the contractor might exploit through spurious claims. The contract went badly wrong; there were technical difficulties, delays to all later works, financial losses and bad feelings all around.

The Mass Transit Railway Corporation (MTRC) in Hong Kong have run several very challenging projects with the construction of underground stations and tunnels in heavily congested urban areas with all sorts of problems to be overcome. As an MTRC spokesman put it verbally: 'conditions are tough enough anyway without contractual difficulties on top'. They therefore try to agree target-cost contracts on a 'cost-plus' basis for complex projects. The contractor does what he needs to do to construct the works and gets paid accordingly. If the contract is brought to completion below target price then the contractor receives a bonus, if not he 'shares the pain' with the MTRC. During one particularly challenging contract at Tsim Sha Tsui, more than 600 ideas for better working practices were presented during the works, together with 371 value-engineering proposals (ways to do things more cost-effectively). As a consequence, the programme was reduced by five months with significant savings in costs to everyone's benefit.

2.3 Design of structures: an introduction

The following section provides a brief illustration into how engineering projects are designed and constructed, so that the following chapters dealing with ground investigation and preparation of ground models can be better understood. The project types used for this introduction are a) foundations for a building and b) tunnels. Civil engineering design and construction is addressed in more detail in Chapter 6.

2.3.1 Foundations

2.3.1.1 Loading from a building

A building imposes a load on the ground. This will include the vertical dead weight of the building – its walls and fixed fittings and a live load including transient loads such as from snow, wind or earthquake loading, as discussed in detail in Chapter 6. The stress from a building, if placed directly on to flat, essentially isotropic ground, will decrease with depth and can be expressed as a bulb of pressure, as illustrated in Figure 2.3. At a depth of perhaps 1.5 to 2.0 times the diameter of a building, the stress level can be anticipated to reduce to 10% of the stress immediately beneath the foundation (Tomlinson, 2001). This is an important rule-of-thumb for the engineering geologist to keep in mind because it gives an indication of the minimum depth of ground to be investigated, as discussed in Chapter 4. The depth of significant stress change also depends upon the nature of the foundations required, as illustrated in Figure 2.3.

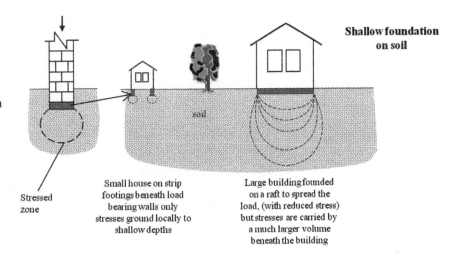

Figure 2.3 The concept of a stress bulb beneath a structure. This is based on elastic analysis of uniform materials, but is indicative and helpful. The wider the structure, the greater the volume of ground that will be stressed significantly and that must be investigated. More detail is given in Chapter 6.

Shallow foundation on soil

soil

Stressed zone

Small house on strip footings beneath load bearing walls only stresses ground locally to shallow depths

Large building founded on a raft to spread the load, (with reduced stress) but stresses are carried by a much larger volume beneath the building

For a small house, the overall load is not very great (imagine the structure collapsed as a pile of bricks and concrete which might only be a metre or two in height), so the weight of the house can usually be carried safely by narrow strip footings running beneath the load-bearing walls. For a footing of about 0.5m width, the typical load from a two-storey house might be about 50kN (5 tonnes) per metre length of wall, so the bearing pressure on the foundation would only be about 100kN/m^2 (= 10kPa), which would be safely carried by a stiff clay or dense sand (Chapter 6). For such a narrow strip footing, the appropriate depth of ground investigation for assessing potential settlement might be 1 to 2m. Such an investigation would not, however, provide adequate warning of the many other potential hazards that might affect a low-rise building, as listed in Table 2.1.

Table 2.1 Examples of hazards that might need to be considered, even for low-rise buildings.

Building on unstable ground	Investigating the adequacy of the material immediately beneath a building to carry the bearing pressure would not deal adequately with sites that are generally unstable. Such hazards can be identified by experienced people, often by air photograph interpretation. Another obvious example of sites to avoid are coastal areas where cliffs are retreating.
Old mine workings	Old mine workings can collapse causing a subsidence trough that travels gradually across the countryside and damages structures. Old shafts can open up in gardens and directly beneath buildings and other structures. Must be researched and possibly investigated by sub-surface ground investigation – possibly geophysics and drilling in areas with a mining history.
Trees taking moisture out from foundations	Trees such as willows that are too close to houses can extract water, especially in times of drought, causing clay to dry out and shrink with consequent movement in overlying foundations. This is a very common cause of damage to houses on clay-rich soils.
Services too close	Where there are back-filled trenches above services – such as pipes and cables – this can allow lateral movement of a house towards the trench.
Adverse materials	Growth of gypsum from soil containing sulphates can cause heave of foundations (e.g. Nixon, 1978; Hawkins & Pinches, 1987).
Seismicity	In seismic zones, special care is required in the design of foundations (more details in Chapter 6). In particular, saturated silt and sand underlying a structure can turn to a liquid during an earthquake (liquefaction).
Others	Other hazards to consider include flooding, impact by boulders and trees from adjacent ground, frost heave and thaw in permafrost areas and environmental hazards such as hazardous gases, especially from previous land use.

2.3.1.2 *Options for founding structures*

As discussed in detail in Chapter 6, there are two key considerations for foundation design. Firstly, there should be a check against bearing capacity failure or ultimate limit state of the underlying soil or rock. This involves analysis of the various loads and calculating the strength of the supporting ground. Generally, in traditional design methods a Factor of Safety (FoS) of between 2.5 and 3 is adopted against ultimate bearing capacity failure, i.e. the allowable bearing stress should be at least 2 or 3 times lower than the load that the ground could theoretically support without failing catastrophically.

The second check is for settlement (otherwise known as a serviceability state). Settlement is inevitable as a building is constructed and the ground loaded, but there are certain tolerances that the designer needs to be aware of. Many structures can cope with perhaps 25 mm of vertical (total) settlement and some, such as an earth embankment dam (constructed from soil and rock fill), may settle by metres without distress; the key question is usually the tolerance of a structure to differential settlement, whereby some parts of the structure settle more than others, causing shear stress between different parts. This may happen where the ground is not uniform – perhaps one corner of the building footprint is less weathered and therefore stronger and less compressible than the rest. If the design of the foundations and/or load distribution of the building does not properly account for this variability, the building will settle more towards one end. Generally, the limiting rotation for a framed structure is taken as about 1 in 500 to 1 in 300, to avoid cracking in walls and partitions (Skempton & Macdonald, 1956); for a high-rise building the tolerance may be lower. Some structures may be even more sensitive and have special requirements for restricting settlement. Generally, the structural designers will need to tell the geotechnical engineer and engineering geologist what is the tolerance for the project so that foundations can be designed accordingly.

Generally, it is cost-effective to design shallow foundations for structures. For framed structures comprising columns and beams of concrete or steel, the load is carried on the columns, which are then founded on pads of reinforced concrete. The size of pad will control the bearing pressure on the underlying soil or rock. If the ground cannot carry the applied stress from the building without unacceptable settlement, then the pad size may be increased, as illustrated in Figure 2.4. For a concrete frame structure, steel reinforcement would be placed in the columns and towards the base of the foundation where the concrete may be subject to tensile stress by bending; concrete (and rock) is relatively weak in tension – typically about one tenth of its strength in compression. Examples of design calculations are given in Tomlinson (2001). In Figure 2.4, because of the variable ground conditions below the pad, great care would be needed not to overstress any weaker zones

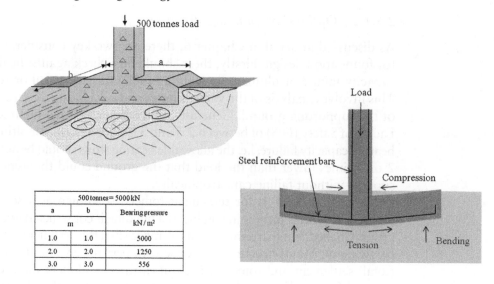

500 tonnes = 5000 kN		
a	b	Bearing pressure
m		kN/m²
1.0	1.0	5000
2.0	2.0	1250
3.0	3.0	556

Figure 2.4 Demonstration of how the bearing pressure on the ground can be reduced by increasing the dimensions of the footing whilst carrying the same building load. The weaker the ground, the larger the foundation will need to be or some other solution might be necessary, such as piling to stronger material at depth. As the size of the foundation increases, so will the cantilever effect with bending, as illustrated in the second diagram. Steel reinforcement will be necessary to resist tensile stresses throughout the structure and also to resist buckling in the column.

and it might be necessary to carry out local dentition to excavate pockets of soil or weak rock and replace with structural concrete. Excavations should be examined by a competent engineering geologist or geotechnical engineer, to check that the conditions are as good as assumed for the design. Such checks and approvals should be well documented. If the required pad sizes between individual columns are large, it may make sense to combine the foundations in a single raft over the full building footprint. It must be remembered, however, that the wider the foundation, the greater the volume of ground stressed, as illustrated in Figure 2.3, and the ground investigation must establish the nature of ground over that full depth. There are many cases where weak compressible material at depth has caused problems for foundations (e.g. Poulos, 2005).

Instead of using a raft it is often cost-effective to take the foundations deeper using piles, which might be timber, steel or concrete. The entire building load might be transferred to some stronger stratum at depth and this is called end-bearing on rockhead, a term that is discussed in Chapter 3 (Box 3-1).

Other piles rely on support from skin friction on the side of the piles, for example, by driving a pile into sand until it can be driven no further. Many piles are designed to be part end-bearing and partly relying on skin friction (Figure 2.5). The design of foundations and the many different options are discussed in more detail in Chapter 6.

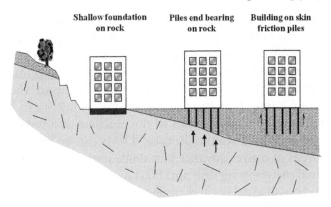

Figure 2.5 Simple foundation options. More detail is given in Chapter 6.

2.3.2   *Tunnels*

Engineering geologists are often closely involved in the investigation of tunnel routes, preparation of reference ground conditions for contracts, tunnel design and during construction. Tunnelling has been carried out since ancient times, originally probably making use of techniques developed in mining, which date back many thousands of years. Aqueduct tunnels for water supply were constructed in ancient Rome, Greece and the Middle East. Modern tunnelling really started with the development of extensive canal systems in the UK and mainland Europe in the C16 and C17 where the alternatives to tunnelling were either deep cuttings or long detours around hills. Originally, tunnels were hand-dug, using gunpowder where necessary. Many modern tunnels are constructed in a similar fashion; improvements include computer-controlled drilling for blast holes, rapid and sophisticated support methods and much better ventilation and safety systems. Generally, drill and blast tunnels involve a cycle of drilling, blasting, mucking-out and support, generally advancing a few metres per pull. Hand-excavation is also sometimes used, employing essentially mining techniques, perhaps using powerful road headers moved across the face to dislodge soil and rock but not taking out the full tunnel profile at one time. It is often the engineering geologist who, during this type of tunnelling, will examine the exposed ground and make decisions on the degree of temporary support that is required, together with any special requirements for further investigation or ground treatment before the tunnel is advanced. One advantage of drill and blast is that methods can be modified quickly to suit changing and difficult ground conditions. Another advantage is that mobilisation is fast – tunnelling can be begun quickly and carried out in remote areas of the world. The disadvantage is that it is often much slower than by using a tunnel boring machine (TBM).

TBMs were gradually introduced for excavation, in particular for the underground railway tunnels in London. In an early attempt at tunnelling beneath the Channel, a 2.13 m diameter boring machine tunnelled

1,893 m in 1881 from the UK side and a similar machine advanced 1,669 m from its portal in France. Today there is a huge range of TBMs available, ranging from ones specifically designed for hard rock, through to ones that tunnel through soft waterlogged sediments using pressurised slurry in front of the machine to support the soil. As explained in Chapter 6, precast segmental tunnel liners can be erected directly following the cutting part of the machine and bolted together with gaskets to form a watertight tube. TBMs can be highly successful with hundreds of metres advance in a single month, compared perhaps to a hundred metres using drill and blast methods; so, for long tunnels, the cost and possible delays in manufacturing a TBM for the job may be justified. Quite often, however, TBMs run into difficulties from ground conditions that can slow them down or even stop them completely, despite huge sophistication in their design. The author has experience of tunnelling through weathered rock in Singapore using a specifically designed slurry machine, where in one section of the tunnel the TBM was stopped because of the high strength of the rock and lack of natural discontinuities. Elsewhere on the same drive, the rock was weathered to a residual soil that was so clay-rich that the slurry treatment plant could not cope, again causing delays and necessitating redesign of the treatment plant. For the same machine, the machine operators had difficulty in selecting the pressure to adopt in the slurry. If the pressure was too low, the ground collapsed, if too high, slurry was ejected into the street above. Shirlaw *et al.* (2000) give examples of problems in tunnelling through weathered rock terrain, and other examples, especially in squeezing ground and zones of high stress, are given by Barla & Pelizza (2000). Further case examples are given in Chapter 7.

It is very important in tunnelling to consider all the potential hazards that might be encountered and to make sure that the TBM can cope, as addressed in Chapters 4, 6 and Appendix E. This is even more important for TBMs than drill and blast tunnels because it may be very difficult to modify the method of working and ground support. Whereas in a drill and blast tunnel the engineering geologist can examine the face and tunnel walls before and after a blast, in a tunnel excavated by TBM, all that can often be seen is the spoil being excavated (often contaminated with drilling mud), so it is rather difficult to confirm that the ground conditions are as anticipated. Engineering geologists are therefore relatively little used during TBM construction, until something goes wrong and needs investigation.

In tunnels, key aspects to consider are safety for the workers and public above the tunnel, the feasibility of different excavation techniques, limiting water ingress, stability of the face and side walls during construction and, in the longer term, the effects on surrounding structures (mainly settlement, undermining, vibration and noise) and cost.

Shallow tunnels and other excavations, such as underground railway stations, may be constructed as concrete boxes in open excavations from the ground surface (maybe 50 m deep). Immersed tube tunnels are formed from boxes of concrete constructed on land and then floated by barge to position where they are sunk to the prepared river or seabed and bolted together.

In bored tunnels in soil at relatively shallow depth, the main concerns will usually be stability of the soil in the face, inflow of water and settlement of adjacent structures. If necessary, the ground can be pre-treated with cement grout, and frozen or compressed air can be used to restrict water inflow and stabilise the ground. When using compressed air there are considerable health considerations and regulations and there is danger of a blowout occurring, particularly where running close to some pre-existing structure such as a well or borehole (Muir Wood, 2000).

In relatively shallow depths in tunnels in rock, the prime considerations will be blocks of rock falling into the opening, or encountering faults which may be full of water. Small failures are generally to be expected in drill and blast tunnels or are protected by the shield in TBM operations. Generally, in fractured rock, shotcrete is applied quickly, together with steel mesh and rock bolts as necessary to stabilise zones of potentially unstable rock. Similar temporary support systems are sometimes used in tunnels through soil, although the principles are different, as discussed in Chapter 6, and there have been many failures in soil tunnels when trying to adopt an observational approach to temporary support (essentially the New Austrian Tunnelling Method).

Tunnelling inevitably disturbs the *in situ* stress condition. Existing stresses in the ground have to flow around the created void. As discussed in Chapter 6, depending on the ratio of σ_1 (maximum principal stress) to σ_3 (minimum principal stress), tensile zones will develop. Generally, these will not cause any great problem other than some minor cracking and possibly some water ingress at shallow depths. In deep tunnels, however, the concentration of compressive stress in sidewalls to a level of the intact rock strength, combined with lack of confining stress, can lead to spalling off and rock bursts (Carter *et al.*, 2008). Such phenomena are not really a problem for most tunnels but can be significant for those constructed deep through a mountain chain, or for deep mining. Hoek (2000) reports particular problems for tunnel stability where the *in situ* stress approaches five times the rock mass strength.

2.4 Design: design codes

Building works throughout the world are generally covered by local regulations, which are mandatory, together with codes of practice and standards. Such documents cover most aspects of works, including ground engineering, and sometimes aspects of engineering geological practice. Some of the key documents that the engineering geologist in the UK needs to be aware of are listed in Table 2.2. Similar codes and standards exist for many other countries.

Table 2.2 Selected codes and standards that are useful or essential references for the engineering geologist. The bias here is towards UK practice.

---

1 CODES FOR SITE INVESTIGATION AND TESTING

1.1 INVESTIGATION

1.1.1 UK: BS 5930:1999 Code of practice for site investigations

BS 5930 (BSI, 1999) deals with the investigation of sites for civil engineering and building works in the UK; parts have been superseded by documents linked to Eurocode 7 (BSI, 2002, 2003, 2004, 2007). It encourages good practice and gives sources of information and references to original literature. In-depth guidance is given on a wide range of techniques in ground investigation, including drilling, boring, *in situ* testing and geophysical works. The code is almost a textbook in its own right and provides excellent advice on designing and managing site investigations. The engineering geologist in the UK needs to be very familiar with this code of practice. There is no equivalent advisory-style European document.

The term site investigation is used in the code in its broad sense, including desk study; the narrower subject of sub-surface exploration is termed ground investigation. Whilst common practice in the UK is covered in detail, some techniques that are used in ground investigation in other countries are dealt with only briefly or not mentioned.

BS 5930 gives guidance on standard rock and soil description for civil engineering purposes, and the terminology in the BS is used routinely in the ground investigation industry in the UK. As discussed in Chapters 3 and 4, different schemes are used in other countries, and important subjects such as rock mass classification are not covered. Amendment 1 to BS 5930 has been revised to comply with BS EN ISO 14688-1 (BSI, 2002), BS EN ISO 14688-2 (BSI, 2004) and BS EN ISO 14689-1 (BSI, 2003), which apply in Europe generally. Changes in terminology are not universally accepted as improvements (see discussions in Hencher (2008), Chapter 4 and Appendix C).

1.1.2 Other codes and standards for site investigation

There are several other codes and standards used internationally, and this applies particularly to soil and rock description, as discussed in Chapter 4 and Appendix C. Often the differences are just a matter of definition and terminology but there may also be local emphasis – for example, because of the local prevalence of weathered rock or swelling soils. It is very important that the applicable codes are used in whichever country the engineering geologist is working.

1.2 TESTING

Standard UK methods for some laboratory and *in situ* soil and rock testing are given in BS 1377 (BSI, 1990), which has been partly superseded by Eurocode 7 Part 2 (Ground Investigation and Testing) (BSI, 2007). Internationally, reference is often made to American Standards (ASTM) or to methods recommended by International Society for Rock Mechanics (ISRM) (Ulusay & Hudson, 2006) and others. The recommended methods sometimes differ in detail (such as dimensions of samples) and care must be taken to ensure that appropriate guidelines are being adopted according to the nature of the project and location. Modern, sophisticated and relatively uncommon testing practice is generally not dealt with in country standards and codes of practice, and reference must be made to the scientific literature (see also Chapter 5).

2 CODES FOR GEOTECHNICAL DESIGN

The British codes of practice discussed below are now generally withdrawn and replaced by Eurocode 7 (BSI, 2004, 2007) for geotechnical design purposes. Nevertheless, they provide general advice and guidance and therefore remain useful references on good practice based on 'well-winnowed experience' (Burland, 2007).

Table 2.2 (continued) Selected codes and standards that are useful or essential references for the engineering geologist. The bias here is towards UK practice.

2.1 FOUNDATIONS

2.1.1 BS 8004:1986 Code of practice for foundations

As with BS5930, this code of practice (BSI, 1986) gives general guidance and background information that is very useful in guiding the geotechnical engineer and engineering geologist. The code provides recommendations for the design and construction of foundations for buildings and engineering structures. It introduces general principles of design, as well as detailed consideration of the design and installation of the foundations. The code also discusses site operations and construction processes in foundation engineering, and the durability of the various materials used in foundation structures. Section 11 deals with safety issues.

In the UK, BS 8004 is superseded by BSI (2004), which adopts a limit state approach to design rather than a lumped safety factor approach, as discussed later.

2.1.2 Other codes and standards

Whereas BS 8004 has been superseded in the UK, similar codes are still used internationally. For example, CP4: 2003, the Code of Practice for Foundations in Singapore (Singapore Standards, 2003), provides general guidance on foundation design, specific to local ground conditions. In Hong Kong, comprehensive guidance is given in GEO Publication No. 1/2006 (Foundation Design and Construction).

2.2 EARTHWORKS AND RETAINING STRUCTURES

2.2.1 BS 6031:1981 Code of practice for earthworks

BSI (1981a) gives advice on formation of earthworks for civil engineering projects such as highways, railways and airfields, and on bulk excavations for foundations, pipelines and drainage works. It gives some UK-focused advice on design and construction of cuttings and embankments. Advice is also given on methods of excavating trenches, pits and temporary support to the sides, including timbering, sheet piling, diaphragm walls and contiguous bored piled walls.

2.2.2 BS 8002:1994 Code of practice for earth retaining structures

BSI (1994) is aimed at UK practitioners and provides guidance on the design and construction of retaining structures up to about 15 m high.

More detailed guidance on retaining wall design, especially where dealing with weathered rocks, is given in GEO (1993) Geoguide 1: Guide for Retaining Wall Design, which, like many other Hong Kong guides and publications, is downloadable from the Hong Kong Government Civil Engineering Design and Development website (www.cedd.gov.hk).

2.3 EUROCODE 7: GEOTECHNICAL DESIGN

The Eurocodes comprise a suite of ten standards, now adopted as British Standards, which have replaced the majority of older national codes of practice as the basis for designing buildings and civil engineering structures in the UK and in most member states of the European Community. As commented above, the superseded codes of practice still contain very useful guidance on good practice, albeit that there has been a fundamental shift in design concept from a lumped Factor of Safety (FoS) to a partial factors approach.

Fundamentally, the concepts used in the earlier codes of practice and the Eurocodes are the same: under extreme loading conditions, structures must not fail catastrophically and in day-to-day service, structures should not suffer deformations that would a) render the structure incapable of achieving the use for which it was designed or b) suffer deformations that take the structure beyond its aesthetic appearance requirements; these are different examples of limit states.

Ultimate limit state failure might include the collapse of a slope, bearing capacity failure of a building, or blocks of rock falling out of the roof of a tunnel, or might be identified as piping failure through the foundations of a dam. Serviceability limit state failure could be defined as excessive settlement, the classic example being the Leaning Tower of Pisa, which has settled dramatically but not collapsed.

Table 2.2 (continued) Selected codes and standards that are useful or essential references for the engineering geologist. The bias here is towards UK practice.

In traditional design, uncertainties are dealt with by adopting a FoS. This gives a broad protection against the inherent uncertainty in models, calculations, loads, strengths, workmanship and so on. If the site conditions, such as geological model and geotechnical parameters, are understood well or if potential consequences are minor, then a low FoS might be adopted. Where less certain or the risk is greater, then a higher FoS is adopted.

In the Eurocode approach, rather than assuming a global FoS, it has been taken as fundamental that different parts of the calculation are known with different certainties; this is certainly true in many situations and is a refinement to design philosophy. Partial factors are then applied to material properties, resistances and/or actions (loads), according to the level of uncertainty. The Eurocode clauses are written as Principles and Application Rules. Principles use the word 'shall' and are mandatory, whereas Application Rules use the words 'may', 'should' and 'can' and allow more judgement. Although this use of language suggests a more prescriptive approach than in earlier codes, in practice, the Eurocodes provide a similar level of latitude for the designer. For example, in assessing geotechnical risk, Eurocode 7 contains Application Rules that define three geotechnical categories, and alternative methods are allowed for assessing geotechnical risk. For routine design cases, the geotechnical design may be assessed by reference to past experience or qualitative assessment. For complex or high-risk situations, e.g. weak/complex ground conditions or very sensitive structures, Eurocode 7 allows the use of alternative provisions and rules to those within the Eurocode. In such situations, rational design based on site-specific testing and numerical modelling might be more appropriate. Detailed guidance is given in Bond & Harris (2008).

Limit state design approaches are used elsewhere, similar to current European practice. For example, Canadian practice has moved in that direction and AASHTO (2007) is used in the USA and internationally for the design of major projects such as the 2nd Incheon Bridge in Korea, completed in 2009 (Cho *et al.*, 2009a).

Sometimes locally mandatory codes or guidelines conflict with others prepared in other countries or by international learned societies, not least in terminology for soil and rock description and classification, with the same words used in different codes to mean different things. The engineering geologist who wishes to work in different countries needs to be aware that the standards and terms that he will need to use may change from country to country. He also needs to be aware that the advice given in codes and working party reports regarding geological matters is often generalised and sometimes difficult to adopt; for example, guidance prepared for temperate zones may not be applied readily in tropical areas and vice versa. This is addressed in more detail in Chapter 4 and Appendix C when discussing soil and rock description for engineering purposes.

2.5   Design: application of engineering geological principles

Despite codes of practice and standards, ground conditions continue to be the major source of failure in civil engineering projects – through catastrophic failure or unacceptable performance and even more commonly due to claims, delay and litigation. In hindsight, the problems can often be attributed to inadequate site investigation, incorrect interpretation of the geological conditions or inadequate design. Poor management and contractual arrangements often contribute to the problems. Ways of avoiding unexpected ground conditions are presented in Chapter 4 and case examples, mostly of projects where failures occurred, are presented in Chapter 7.

3   Geology and ground models

3.1   Concept of modelling

3.1.1   *Introduction*

The geology at a site can range from apparently simple to apparently complex scales of metres, tens or hundreds of metres (Figures 3.1 and 3.2). Geological complexity does not, however, always equate with difficulty in engineering terms. Conversely, even where the soil or rock mass is apparently relatively uniform there may be a single feature or property that will cause problems (Figure 3.3). It is the task of the engineering geologist to interpret the geology at a site and to identify those characteristics and properties that might be important to the engineering project. Much of the detail will be insignificant; the skill is in recognising what is and what is not. At some stage during the design process the geology will need to be differentiated in some way into units that can be characterised with essentially uniform mechanical properties or where the properties change in some definable way, perhaps with depth. Sometimes the way to do this is obvious – for example, a layer of fill (man-made ground) overlying alluvium, which in turn overlies bedrock, which will define the way foundations are designed – but at other locations identification of the key attributes is more difficult. Thin layers that might be overlooked in logging a borehole could turn out to be the most important features at a site.

For civil engineering, ground models need to be prepared that are simplified representations of a site and that should incorporate all the important elements relevant to design and construction. The models are generally developed from a preliminary 3D interpretation of the geology based on desk study and surface mapping and then refined by further study of environmental factors such as earthquake hazard and hydrogeology. Models will be improved by ground investigation and testing and finally presented as a design model

Figure 3.1 Massive horizontally bedded Eocene conglomerate and sandstone, unconformably overlying Triassic Lower Muschelkalk, Sierra de Montsant, north of Falset, Spain.

Figure 3.2 Folded and faulted extremely strong Devonian radiolarian chert interbedded with thin bands of extremely weak organic shale, near Cabacés, Spain.

specifically tuned to the project. The process is illustrated simply in Figure 3.4 and expanded upon later in this chapter. One of the key features of many ground models is differentiating between upper, soil-like materials, and underlying rock, with the separating boundary being called rockhead or, sometimes, 'engineering rockhead'. Care must be taken in using this term because it has various definitions and connotations and is sometimes used in an over-simplistic way for what is a complex situation. The consequences of wrong perception can be severe if, for example, soil is encountered at depth and below the water table, unexpectedly in a hard rock tunnel. Definitions of rockhead are set out in Box 3-1.

Box 3-1   Definition of rockhead

Care must be taken in using the term rockhead because it can be defined in various ways and the wrong impression may be conveyed within a geotechnical design team that things are clear-cut when they are not.

Geological definition

Rockhead is defined in BS 3618 (BSI, 1964) as 'the boundary between superficial deposits (or drift) and the underlying solid rock' and this definition is also adopted by the US Department of the Interior (Thrush *et al.*, 1968). The term solid rock is defined, in turn, in Thrush *et al.*, following Challinor (1964) as 'rock which is both consolidated and in-situ'. Solid is also generally used in a geological sense to describe formations that predate superficial deposits (Whitten & Brooks, 1972) as in solid vs. drift maps. Rockhead used in this way, essentially defines a geological boundary usually marking an unconformity. The solid rock shown on a geological map says nothing about its strength or weathering state, so rockhead does not necessarily mark a boundary between soil and rock in strength terms.

Geotechnical definition

The term rockhead or engineering rockhead is often used in geotechnical design to define a boundary between soil-like material and rock that is stronger and more resistant, whatever the geological conditions. It is also sometimes used more generally 'as the level at which the engineering parameters of the ground satisfy the design parameters for a specific project' (GEO, 2007).

Sometimes the geological profile is simple (recent soil over rock) and rockhead is readily defined, but often the situation is more complex and care must be taken not to represent a difficult and variable geological condition in over-simplified diagrams that might be misunderstood by designers. Weaker material or voids below the first occurrence of rock in a borehole might have a controlling influence on mass strength, compressibility and permeability and have severe effects on constructability, for example, collapse of pile borings or sudden inflow of soil into a tunnel.

It is particularly difficult to define a simple level for rockhead in regions of weathered rock. In the opinion of Knill (1978), in the case of karstic limestone, rockhead is the geological contact between *in situ* limestone and overlying superficial deposits (despite often great complexity due to dissolution features). Similarly, Statham & Baker (1986) define rockhead as the top of *in situ* limestone (despite the presence of sediment–infilled voids 'below rockhead'). Goodman (1993) comments that 'The unevenness of the top-of rock surface (or rock head in British usage) on karstic limestone presents obstacles for the designer.' He notes the many potential difficulties for design and construction and notes that 'Unlike most other rocks, the existence of a solid-appearing outcrop right at the location of a footing or pier does not guarantee that good rock will occur below the outcrop.' The same is true for other rock types with large corestones sitting on the ground surface, underlain by severely weathered rock, as discussed by Ruxton & Berry (1957). An example of a landslide that occurred where rockhead was misinterpreted by the slope designers on the basis of boreholes that terminated 5 m in rock, is described in Hencher & McNicholl (1995).

3.2   Relevance of geology to engineering

Attempting to form a ground model of a site based solely on descriptions from boreholes and test results, without recourse to an informed geological interpretation of the data, would be like trying to put together a complex jigsaw without having the picture on the box lid. Geologists are trained to examine rocks and soils at scales of a hand specimen or a quarry and to draw conclusions on the likely origins and

Figure 3.3
Pre-existing fault
(tight and planar)
allowing
accommodated
movement as a
counter scarp, Pos
Selim landslide,
Malaysia.

Figure 3.4 Basic
process of creating
a ground model for
a site.

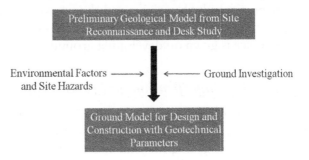

history of the sample or exposure. Then, by examining other samples
and exposures in and around a site, they can start to develop a picture
of how the various different components relate to one another.
The conceptual model for the geology at a site can be used to extrapolate and interpolate observations to make further predictions on the
basis of geological knowledge. Ground investigation can then be
designed and used to target residual unknowns.

3.3   Geological reference models

3.3.1   *A holistic approach*

Fookes *et al.* (2000) encourage a 'total geological approach' whereby any site is assessed with regard to its full geological history. That history includes original formation of the soils or rock underlying the site, tectonic events, weathering, erosion, deposition of any overlying superficial deposits, geomorphological development and anthropogenic influences. The changes that have taken place at the Earth's surface to form the landscape and the extensive time involved are almost inconceivable but must be considered when interpreting the geology at any site.

Total geological analysis should allow the distribution and nature of the various strata at the site and other features, such as hydrogeological conditions, to be explained. The assessment should be based on the extensive literature on geological and geomorphological processes, which comprises the toolbox for interpreting the conditions that are encountered at any site. Useful sources of information on geology focused on civil engineering application are Blyth & de Freitas (1984) and Goodman (1993) but often, to understand what is happening at a site, one has to refer to more fundamental geological literature. This chapter introduces aspects of geology that are relevant to engineering design and performance. It commences by considering the three basic rock types – igneous, sedimentary and metamorphic – focusing on their typical characteristics and associations that may be of particular importance to engineering. The next section introduces rock structures, particularly the origins and characteristics of discontinuities that tend to control rock mass properties. Towards the end of the chapter, guidance is given on developing ground models for a site.

3.3.2   *The need for simplification and classification*

As discussed in Chapter 4, simplified approaches are generally adopted for the description and classification of soils and rocks for engineering purposes, largely because geological detail is often irrelevant. This especially applies to logging soil and rock encountered in boreholes. Nevertheless, the engineering geologist needs to be alert to situations and cases where geological detail might be important to explain the geological situation or because particular characteristics have some special significance. In the author's experience, whilst many sites are described and characterised quite adequately using shorthand terms and classifications, occasionally one meets situations where to understand what is happening, to an adequate level for an engineering project, intensive study is necessary into geological minutiae, including chemical analysis, thin section examination and even radiometric dating.

3.3.3 Igneous rocks and their associations

Igneous rocks were once molten. As magma cools, minerals grow with an interlocking texture. As a result, most igneous rocks are strong and sometimes extremely strong in their fresh state – several times the strength of concrete. They are primarily split into intrusive rocks that solidify below the Earth's surface and extrusive rocks that form on the Earth's surface. They are then differentiated according to chemistry. Rocks with high silica content, either directly as quartz (SiO_2) or tied up in the structure of other silicate minerals, are termed acidic. Basic rocks have low silica content and ultrabasic rocks even lower. A simplified classification of igneous rocks is presented in Table 3.1.

Intrusive rocks, solidified at depth, include extensive igneous bodies now exposed at the Earth's surface, such as granite that makes up Dartmoor in the UK. This rock solidified very slowly

Table 3.1 Classification of igneous rocks.

IGNEOUS ROCKS: generally have massive structure and crystalline texture. Typically high strength in fresh state.

Volcanic rocks deposited as sediments are dealt with in Table 3.2: Sedimentary rocks. More details are given in Streckeisen (1974, 1980) and Thorpe & Brown (1985).

Grain size	ACID Much quartz		INTERMEDIATE Little quartz		BASIC[1] Little or no quartz
	Pale colour → Dark ← → Relatively light in weight → Heavy				
Coarse >2 mm	GRANITE[2] GRANODIORITE		DIORITE SYENITE		GABBRO
Medium 0.06–2 mm	MICRO-GRANITE[3] MICRO-GRANODIORITE		MICRO-DIORITE MICRO-SYENITE		DOLERITE[4]
Fine <0.06 mm	RHYOLITE[5] DACITE		ANDESITE TRACHYTE		BASALT
Glassy	OBSIDIAN		VOLCANIC GLASS		

[1] Rocks with even less silica and higher content of Fe and Mg are termed ultrabasic. Identify using standard geological terminology. Examples are peridotite, pyroxenite and norite, and these can have distinctive engineering characteristics (e.g. Dobie, 1987).

[2] Rock types are often grouped together for engineering purposes, e.g. all coarse rock with free quartz is called granite. This can be an oversimplification as distinctions can be significant. Engineering geologists should use full geological classifications, where possible.

[3] Micro-granite is sometimes termed fine-grained granite (see GeoGuide 3: GEO, 1988).

[4] Diabase in US practice.

[5] Where porphyritic, then often called quartz porphyry or feldspar porphyry, etc.

from temperatures in excess of 1,000 degrees Centigrade at depths of several kilometres in the Earth's crust and the slowness of the cooling allowed large mineral grains to grow very gradually. Since formation they have been uplifted and the overlying rocks eroded away, with huge consequential changes in stress and temperature conditions.

Granitic rocks are light-coloured and relatively light in weight (unit weight 27 kN/m^3, which is 2.7 times that of water). In terms of mineralogy, granite has a high proportion of quartz and feldspar. Quartz (SiO$_2$) is hard, has no cleavage weaknesses and is much more resistant to chemical weathering than feldspar. Feldspar (orthoclase and plagioclase) is a much more complex silicate mineral with cations of potassium, aluminium, sodium and calcium and relatively weak cleavage directions that make it prone to chemical attack. The feldspars therefore break down, primarily to form clays, which are a series of minerals of essentially the same chemical makeup as the feldspars but which are more stable at the Earth's surface temperature and chemistry. Minor minerals in many igneous rocks include biotite, hornblende and magnetite, which contain iron, which is released and then oxidised on decomposition, hence giving the rust-red of weathered rock profiles in many sub-tropical and tropical parts of the world. Iron oxide and carbonate products play an important role in cementing recently deposited sediments, as discussed later. Granitic rocks are found in continental regions and probably largely represent melted and reconstituted crust as plates are subducted beneath mountain chains or in extension zones (Davis & Reynolds, 1996).

Oceanic regions are made up of basic igneous rocks – mainly basalt – which is the fine-grained chemical equivalent of gabbro. Basalt erupts along extensional plate boundaries such as that running down the centre of the Atlantic Ocean as the European and African plates move away from the American plates at a rate of about 20 mm each year. Basic igneous rocks are darker-coloured and heavier than granite, largely because they are rich in iron and magnesium. Gabbro has a unit weight of about 30 kN/m^3. Basic rocks are sometimes extruded in continental regions where faults allow basaltic magma to rise as molten rock from great depth to the Earth's surface. Basalt rock that originated as lava fields makes up the Giant's Causeway in Ireland and much of the other Tertiary volcanics of northern Britain. The basaltic Deccan Traps in India cover an area of more than 50,000 km^2.

Lavas such as basalt are finer-grained than granite or gabbro because they cool relatively quickly; where cooled extremely quickly, say by extruding into water, they may form natural volcanic glass. Basaltic lava has relatively low viscosity and flows a long way, unlike the more viscous pale-coloured acidic lavas such as rhyolite (the fine-grained equivalent of granite). Acidic volcanoes erupt in a much more violent way than basaltic volcanoes. Other rocks associated with volcanoes are formed from the huge amounts of dust and other airborne debris in volcanic eruptions. Volcaniclastic sediments (Table 3.2) are called tuffs and often exhibit many sedimentary features, especially where deposited in water. Hot clouds of debris deposited on land become welded tuffs or ignimbrites. A welded tuff with pumice inclusions (fiamme) that were hot and plastic when deposited and then flattened by the weight of overlying material, is illustrated in Figure 3.5.

Dykes are intruded in extensional (tensile) regions and cut across other geological structures (Figure 3.6); sills follow existing geological structures such as bedding and are more concordant. Dykes can be quite local or very extensive. The Great Dyke in Zimbabwe can be traced across country for more than 450 km. Tertiary dyke swarms and associated intrusion complexes in Scotland, Ireland and northern England are shown in Figure 3.7.

Table 3.2 Simplified classification of sediments and sedimentary rock (see Tucker, 1982 for more detail).

mm	CLASTIC SOIL [1,2]	CLASTIC ROCK		VOLCANICLASTIC or PYROCLASTIC ROCK	CHEMICAL & BIOCHEMICAL ROCK
>200	BOULDERS	CONGLOMERATE (rounded clasts) BRECCIA (angular)		PYROCLASTIC BRECCIA or AGGLOMERATE	LIMESTONE (examples): Chalk
60	COBBLES				
2	GRAVEL			LAPILLI TUFF	Calcarenite (sand and gravel size)
0.06	SAND	SANDSTONE Greywacke (generally poorly sorted) Arkose (feldspathic sandstone)		COARSE ASH TUFF	Calcilutite (mud size matrix) Oolite
	SILT	SILTSTONE	MUDSTONE as general term SHALE if fissile	FINE ASH TUFF	DOLOMITE (Mg rich) EVAPORITE (salts) COAL FLINT & CHERT (Cryptocrystalline silica)
<0.002	CLAY	CLAYSTONE			

[1] Clastic means derived from fragments of other rocks. The term detrital is sometimes used, essentially synonymously. Some clay is neither of these, but newly formed, sometimes from solution.
[2] Classification and description of soil, including mixed soils, is dealt with in Chapter 4 and Appendix C.

Figure 3.5 Welded tuff with flattened fiamme that were plastic when deposited sub-aerially, Ap Li Chau, Hong Kong.

Some of the characteristics of an area intruded by igneous rocks are illustrated schematically in Figure 3.8. Hot igneous rocks affect the country rock that they intrude, giving rise to thermal metamorphism, as discussed in Section 3.3.5. Some metamorphosed zones may be much weaker than the associated igneous body, with obvious potential consequences if encountered in an engineering project. Figure 3.9 shows a zone of weak, hydrothermally altered rock encountered at a depth of about 200m in a tunnel, well below the level of anticipated weathering and therefore rather unexpected. It resulted in a minor collapse that delayed tunnelling. The extent of the zone was investigated with horizontal drillholes. Following forward probing, ground treatment and support, tunnelling was able to proceed.

3.3.4 Sediments and associations – soils and rocks

3.3.4.1 General nature and classification

Sediments and sedimentary rocks are derived from the breakdown of older rocks. Detritus is transported by gravity, water, wind and

Figure 3.6 Basalt dyke cutting granite, Hong Kong.

glaciers, before it is deposited and gradually buried and transformed into rock by diagenetic or lithification processes, as introduced in Chapter 1 and discussed further in Chapter 5. Some weathering products are carried in solution to be deposited directly on the Earth's surface, transformed into animal shells or carried by fluids to be deposited as cements in the sediment pile as it is self-compacted. During transport, detritus is sorted by size and density, largely in response to the velocity of water flow or wind. Sediment deposited close to its erosion source may have a wide range of grain sizes, as illustrated in Figure 3.10; sediments transported by water or wind tend to be winnowed to a limited range of grain sizes and are then termed well-sorted (Figure 3.11); the same material would be called poorly-graded by an earthworks engineer – he looks for soils in embankments to have a wide range of sizes so that a dense degree of compaction can be achieved. Grains are also abraded and rounded as they are blown by the wind especially and this is an important engineering characteristic. Angular sand has an internal friction angle that may be 10 degrees higher than rounded sand.

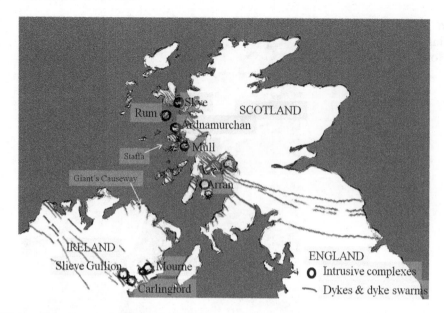

Figure 3.7 Simplified map showing Tertiary dyke swarms and igneous intrusion complexes across northern Britain. Lava flows, from the volcanoes above the igneous complexes that are now exposed (e.g. Figure 3.26) included the extensive flood basalt sheets of which the Giant's Causeway and the Isle of Staffa are part (see Figures 3.36 to 3.38). Data are from Holmes (1965) and Richey (1948).

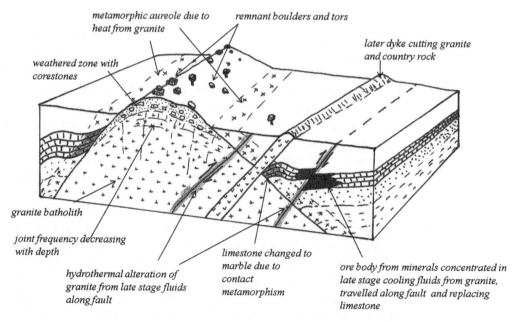

Figure 3.8 Schematic representation of igneous rock associations.

Figure 3.9
Hydrothermal zone
(darker, lower part
of tunnel face),
Black Hills
Tunnels, Hong
Kong.

Figure 3.10 Poorly
sorted cobbles and
boulders, deposited
close to source with
little transport,
Cachopo Road,
Portugal.

Figure 3.11
Uniform, well-
sorted cobbles and
gravel, Canterbury
Plain, New
Zealand.

Sediments and detrital/clastic sedimentary rocks are classified primarily on dominant grain size, as in Table 3.2. In terms of soil mechanics and geotechnical behaviour, granular soils are generally distinguished from clays in that they are non-cohesive and are comparatively inert chemically, although some authors would dispute the term cohesive being applied to clay, as discussed in Chapter 5. Granular soils include everything down to silt size and even some clay-sized soil where this is derived from mechanical breakdown (rock flour is clay-size material, generally quartz, produced by glacial abrasion). Gravel and larger grain sizes are usually made up of rock (lithic) fragments rather than mineral grains. In continental regions where granitic rocks dominate, sand and silt are often made up predominantly from quartz, the most resistant mineral from granite. In areas where chemical decomposition is inactive, feldspar might also survive (arkosic).

Many clays are distinct from granular soils, not only because of their grain size and shear behaviour, but because they comprise a new series of minerals with their own structure and chemistry; they are derived from other rocks but not only as detrital mineral grains but as precipitants from solution (Eberl, 1984). Other clays are primary, associated with igneous or hydrothermal activity or are formed by transformation of other minerals. Clays comprise three groups: phyllosilicates, weakly crystalline aluminosilicates and hydrous oxides of iron, aluminium and manganese. Selby (1993) provides a very useful review of clay mineralogy, as well as their engineering properties.

The most common clay minerals are the phyllosilicates, which, like mica, comprise sheets or layers of silica and sheets of alumina. The various clay mineral species owe their differences to the ways that sheets are arranged; substitutions of other cations are made into the structure, which produces distortion of the crystal lattice and the bonding between sheets. The type of clay that will be formed at any location depends primarily upon the source rock and climatic effects. Kaolinite and illite are relatively inactive clays and commonly produced by the weathering of granitic rocks. Montmorillonite (major member of the smectite class) is much more active and absorbs water readily, thereby swelling dramatically. When it dries it shrinks, and these characteristics have important consequences for engineering. Smectites also tend to have low shear strengths when wet (Chapter 5). They are commonly the result of weathering in basalt in tropical areas and form difficult but productive soils in many countries, including Australia, Africa, India and the USA. Soil types include vertisols and black cotton soils.

Aluminosilicate clays such as allophone develop from volcanic ash with silica content below a critical level and are important soils in New Zealand, Japan and Indonesia (Selby, 1993). Like montmorillonite, they are very responsive to changes in water content, prone to flow and liquefy when wet and to shrink and crack when dry.

Atterberg limits are measured routinely in laboratories to define the nature of soils and to interpret the likely clay mineralogy by index testing. If water is added to a clay or clay-silt mix, one can get it to flow like a liquid. The moisture content at which it changes from a plastic state to a liquid state is known as the liquid limit (LL) and can be determined using standard tests where a groove cut in the surface closes as the sample is tapped in a standard way or a cone is dropped onto the sample and penetration measured. If the sample is then dried out, the sample gets stronger and the water content at which a sausage of 3 mm diameter can be rolled without breaking is called the plastic limit (PL). The difference in moisture content between the liquid state (LL) and plastic semi-solid condition (PL) is the plasticity index (PI). The PI is found to be a good indicator of the type of material, species of clay and, hence, engineering behaviour (Figures 3.12a and b).

Clay particles are carried by rivers to estuaries where a change in salinity causes clay platelets to flocculate. Because of their shape and

Figure 3.12 (a) Chart of plasticity index vs. liquid limit used for identifying the nature of fine-grained soil using simple, standard tests (Atterberg limits). (b) Use of plasticity index to indicate clay mineral in soil sample (after Skempton, 1953).

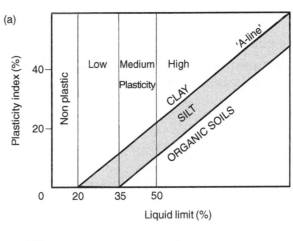

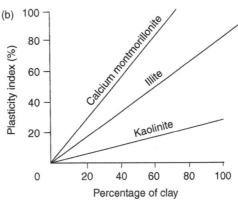

the distribution of excess electrical charges, they tend to link end-to-face and form open structures with very low density and high water content. This open structure is broken down, partly by bioturbation but also by the weight of overlying sediment. The clays develop a laminar structure and water is squeezed out. Eventually, once the clay is buried by about 2 km, theoretically there may be dry contact between clay crystals with the development of strong, covalent bonds (shared ions) (Osipov, 1975). There would also be transformation of clay, say from smectite to illite. During burial and self-weight compaction and consolidation, water may be expelled dramatically, with the formation of mud volcanoes on the sea floor. Mud volcanoes also occur on land, with the mud apparently sourced by erosion of mudstones at great depth, although the mechanism is uncertain. The LUSI mud volcano in East Java has had eruption rates of up to $180,000\,\mathrm{m}^3$ per day, continuing over several years (Davies *et al.*, 2011).

3.3.4.2   *Sedimentary environments*

The sketch in Figure 3.13 illustrates a number of sedimentary environments. The source rock, nature of weathering and erosion and especially the method by which the sediment is transported and finally deposited result in the wide range of sediments and sedimentary rocks encountered.

3.3.4.2.1   ONSHORE

Sediments deposited on land generally include colluvium (landslide deposits and slope wash) and glacial deposits, which are often poorly

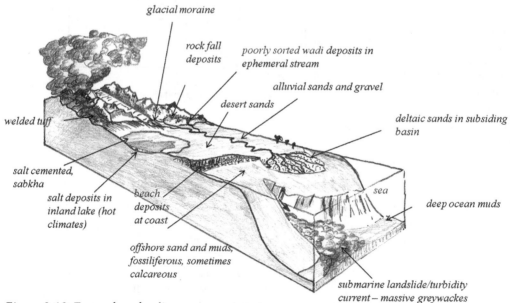

Figure 3.13 Examples of sedimentary environments.

Figure 3.14
Triassic dune-
bedded aeolian
sandstone
(Buntsandstein),
Ermita de Mare de
Déu de la Roca,
near Mointroig,
Spain.

sorted. Alluvium is deposited in rivers and if transported for any distance will become sorted and coarse clasts will be rounded. Other important soils on land include organic soils such as peat, which becomes coal when lithified.

Desert sands are distinguishable from river-transported sediments by their rounded, almost single-sized grains, which may become cemented, especially by iron oxides, giving typical red beds such as the Permian sandstone in the UK (Figures 3.14 and 3.15). They also lack mica, which is either abraded to dust or blown away. Finer wind-transported silt forms thick deposits called loess. Where cemented, loess has been exploited to construct settlements, especially in China. This can be risky as loess is prone to collapse when wetted or disturbed by earthquakes, and there are numerous instances of loss of life as a result (Derbyshire, 2001).

Glaciers have covered much of northern Europe, Canada and the northern USA several times in the last three million years. In Britain, everywhere north of a line from London to the Bristol Channel was covered in ice sheets. The rest would have been areas of permafrost

Figure 3.15
Uniform Miocene
weak sandstone
with rounded
grains, weakly
cemented with iron
oxide cement
providing some
cohesion, Algarve,
Portugal.

(periglacial conditions). Man has inhabited Britain on and off for at least 700,000 years but has had to abandon the country because of advances in ice several times, most recently about 13,000 years ago. Only for the last 11,500 years has Britain been occupied continuously (Stringer, 2006). As a consequence, much of northern Britain shows signs of glacial erosion and also materials deposited in association with the glaciers. Much of this is poorly sorted boulder clay or till but there are also great thicknesses of sorted and layered outwash sediments laid down on land or in lakes as the glaciers retreated. These materials can be very variable vertically and laterally (Fookes, 1997), so great care is needed in interpreting ground investigations. This is illustrated by one case history (piling for Drax Power Station) presented in Chapter 7. Associated with these periods of glaciations, the sea level across the world has fluctuated widely. About 20,000 years ago the sea level was almost 150 m lower than it is today (Pirazzoli, 1996). As a consequence, ancient river channels that ran across the land surface have been submerged to become infilled with marine sediments, which can prove hazardous for engineering projects such as tunnelling or foundations for bridges across estuaries. For example, during design of the recently constructed 3.24 km immersed tube tunnel as part of the Busan–Geoje fixed link crossing in South Korea, a depression was found in the sea floor, unexpectedly underlain by thick marine sediments, and this required extensive engineering works to support the tunnel, involving the use of soil-cement mixed piles. Buried channels are also found on land, often with no obvious surface expression. Krynine & Judd (1957) report several case examples, including a dam at Sitka, Alaska, where river fill was fortuitously found by drilling, extending more than 25 m below the dam foundations. If it had been missed then there would have been considerable water leakage beneath the dam, requiring remedial works. In Switzerland, during the construction of the Lötschberg rail tunnel, 25 miners were killed by inrush of glacial sediments when they blasted out of rock unexpectedly into an over-deepened glacially scoured valley about 180 m below the valley floor (Waltham, 2008). A case from Hong Kong is presented in Fletcher *et al.* (2000) and summarised in Chapter 7, where the planned construction of a tower block had to be abandoned because of the potentially huge cost of deep bored piles. A large cavernous xenolith of marble had been found unexpectedly within granite beneath the site; the caves were partially infilled with soft sediments to depths of 150 m. The formation of the cave and sediment infill certainly occurred when sea levels were much lower.

Other poorly sorted, mixed soils include landslide colluvium. One such deposit is called the Fort Canning Boulder Bed, which underlies much of the Central Business District in Singapore and has caused numerous difficulties for construction, not least because it has sometimes been misinterpreted during ground investigation as weathered rock, as discussed in Box 3-2.

Box 3-2   The Fort Canning Boulder Bed, Singapore

The Fort Canning Boulder Bed (FCBB) underlies much of the Central Business District of Singapore and is an example of a complex mixed rock and soil deposit, the geological nature and interpretation of which has had, and continues to have, great consequences for civil engineering construction. When the Fullerton Building was constructed in the 1920s, it was found that 'the foundations which consisted of clay and boulders, were of a dangerous character, and this great structure had to be placed on a concrete cellular raft, which is so designed as to give each superficial foot of soil not more than one ton to carry' (Straits Times, 27 June 1928). Since then, in several reported cases, engineers have struggled with construction on and through this material. Ground conditions have sometimes been misinterpreted following inadequate ground investigation. In 1952, the original ground investigation for the Asia Insurance Building incorrectly interpreted the site as underlain by *in situ* rock rather than FCBB. The mistake was only discovered during construction and necessitated total redesign of the foundations (Nowson, 1954). Shirlaw *et al.* (2003) provide several other examples.

The FCBB is almost 100 m thick in places, is not exposed at the ground surface, but is commonly encountered in foundation construction and tunnels (Shirlaw *et al.*, 2003). The deposit typically comprises a poorly sorted mixture of clay, silt and sand, with boulder content between 5% and 35% (Singapore Standards, 2003). The mostly angular boulder content is probably derived from the adjacent Jurong Formation of Triassic/Jurassic age, with strong sandstone predominating. Figure B3-2.1 shows boulders recovered and partially drilled in large diameter bored piles from one site near Mount Sophia. In between the boulders, the matrix sometimes comprises hard red clay with undrained shear strength up to 2 MPa (weak rock) but sometimes sandier and silty and of much lower strength. Broms & Lai (1995) also report encountering a highly permeable gravel layer within the FCBB at depth.

Figure B3-2.1 Angular boulders up to about 3 m diameter, plus cores extracted during rock coring for large diameter piles, from a site close to Mount Sophia, Singapore.

Shirlaw *et al.* (2003) suggest that the FCBB is a colluvial deposit originating from Fort Canning Hill (Figure B3-2.2) but the low height of the hill relative to the thickness of the deposit makes this somewhat doubtful and the red clay is often much stronger than would be expected, even for clay that has undergone quite deep burial. Furthermore, the clay is clearly

Figure B3-2.2 View towards Fort Canning Hill from the Central Business District, and a few locations where FCBB have been encountered (according to Shirlaw *et al.*, 2003).

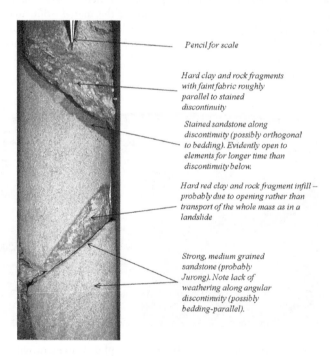

Figure B3-2.3 Section of core (about 60 mm diameter) showing slightly stained sandstone with apparent soil infill.

not a weathered product of the sandstone and, anyway, the contained boulders in the FCBB generally show little weathering. The red clay sometimes appears as an infill to apparent relic joint structure in sandstone blocks (Figure B3-2.3).

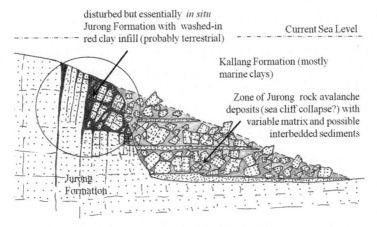

disturbed but essentially *in situ*
Jurong Formation with washed-in
red clay infill (probably terrestrial)

Current Sea Level

Kallang Formation (mostly
marine clays)

Zone of Jurong rock avalanche
deposits (sea cliff collapse?) with
variable matrix and possible
interbedded sediments

Jurong
Formation

Figure B3-2.4 Schematic geological model that can explain the features of the Fort Canning Boulder Bed. The hard red clay is probably infill, washed in to open joints and fissures in deteriorating rock cliffs, and probably derived from weathered mudstone units in the Jurong. The clay became hardened and cemented in a terrestrial environment in the same way as terra rossa in weathered limestone environments. The colluvial facies of the FCBB was probably deposited as rock and soil avalanches, at the coast. It remains unsorted, although may be intercalated with alluvial or beach deposits. The FCBB was later buried locally by Kallang marine clay and fluvial horizons and by reclamation.

Broms & Lai (1995) present two sketches of caisson excavations for the Republic Plaza, one of which appears to have a vertical fabric. If this was a transported colluvial deposit, boulders would be expected to be essentially random or preferentially flat lying. The same sketch shows vertical zones containing few boulders, with a form similar to gulls found in areas of cambering (Parks, 1991).

An alternative model for the origin of the FCBB, therefore, is rock cliff deterioration, regression and local collapse as sea levels rose from -150 m at the end of the Holocene glacial periods. The red clay is probably partly washed-in infill to the open fractures within the deteriorating rock mass, in a terrestrial environment, as illustrated schematically in Figure B3-2.4. If the model is correct, the collapsed colluvial facies of the FCBB would grade laterally into essentially *in situ* deteriorated Jurong Formation rocks.

Whatever the truth regarding origin, the FCBB is a most difficult founding stratum and one that might be called unforgiving in the context proposed in Chapter 4, but situated in an area of one of the most valuable real estates in the world. Clearly, if dealing with a site which comprised weathered but essentially undisturbed Jurong, one might use rather higher design parameters than for a colluvial soil that might contain weaker horizons. Similarly, tunnelling through the weathered Jurong, the potential hazards might be different from tunnelling through a colluvial and possibly interbedded section of the FCBB with weaker horizons that could flow or ravel. The literature has many examples of problems that arise through getting the ground model wrong and as Shirlaw *et al.* (2003) note, identification of the FCBB can be difficult if adopting a routine approach to ground investigation.

3.3.4.2.2   OFFSHORE

Sediments are deposited offshore in tectonically controlled, continually subsiding basins and can be thousands of metres thick (Leeder, 1999). Skempton (1970) compiled data on rates of deposition for argillaceous sediments (clay rich) and reported these as ranging from 0.03 m/1,000 years for deep sea conditions, about 1–2.5 m/1,000 years for estuaries and shallow marine conditions and up to 120 m/1,000 years for deltaic situations. Sand deposited close to the shoreline tends to be quite well sorted, bedded and often fossiliferous (Figure 3.16). Offshore, below the continental shelf, thick, often relatively poorly sorted sandstones are sometimes deposited very rapidly from dense turbidity currents arising from submarine landslides. Rocks formed in this way (once lithified) include greywacke. They are often bedded and cyclic with predominantly finer-grained horizons followed by predominantly coarser beds (Figure 3.17). Submarine landslides are a major hazard to offshore engineering works, including pipelines, cables and oil rigs, and are often triggered by earthquakes, although they may simply occur as the sediment accumulation on the continental shelf reaches a critically unstable geometry.

Figure 3.16
Bedded Miocene sandstone, Capela da Sra da Rocha, Nr Porches, Algarve, Portugal.

Figure 3.17
Carboniferous schistose turbidites, Cachopo Road, Portugal.

Other sediments are deposited as a result of chemical processes. Many limestones, including chalk, are essentially biochemical, involving microplankton and algae in their formation. Other types include bioclastic and reef limestone. Limestone can be fine-grained rock, uniform and extremely strong, as per the Carboniferous Great Scar Limestone of northern England. Other limestones such as oolite can be weak and porous. Limestone is readily dissolved by acidic water, leading to karstic conditions (irregular rockhead and cavernous conditions) with obvious implications for site formation and design. Rocks susceptible to dissolution, including limestone, dolomite and salts (discussed below), can constitute natural hazards due to sudden collapse of sinkholes and general subsidence. Relatively minor geological differences can give rise to very different engineering properties. Within the chalk of southern Britain there is a gradational change from the characteristic White Chalk, which is relatively strong and brittle, down into grey, clay-rich rocks. The latter, Chalk Marl, intermediate between the overlying fractured, brittle White Chalk and underlying Gault Clay, was recognised as the ideal tunnelling medium for constructing the Channel Tunnel (Varley & Warren, 1996). The presence of hard flint and chert bands in chalk can give rise to considerable difficulties during construction, because of abrasion and wear on cutting tools, as illustrated by a case example in Chapter 7.

Limestone can become dolomitised, which involves partial replacement of calcium carbonate by magnesium carbonate and leads to greater porosity. Dolomitisation often occurs in the presence of evaporates (salt deposits) and dissolution of underlying salts such as gypsum can lead to the development of sinkholes and brecciation of the overlying rock, followed by re-cementation. This process is described for the Magnesian Limestone in the UK by Dearman & Coffey (1981). Similarly, the Miocene limestone and dolomite sequences in Qatar and Saudi Arabia are complex and extremely variable due to their post-formation dolomitisation, collapse and re-cementation (Sadiq & Nasir, 2002). The rock is sometimes strong, elsewhere very weak, and can be cavernous. This is important for founding engineering structures and for other activities, including dredging (Vervoort & De Wit, 1997). Limestone is an important rock economically, particularly as a source of cement aggregate and, where massive and strong, is commonly used as armourstone for breakwaters.

Deposits of salt are formed by evaporation of lakes and even seas (the Mediterranean completely dried up about five million years ago), and are also very important economically as source rocks for chemical industries such as fertilisers. They are significant for the oil and gas industry because they have low density and low permeability and gradually rise through the overlying denser country rock as diapiric structures. Traps for oil and gas at the boundaries of these diapiric structures are searched for using geophysical methods and then targeted by drilling. Salt deposits are also considered to have great potential as nuclear waste repositories because of their low permeability, although there are also reservations because of potential dissolution, mobility and influence of heat (e.g. Krauskopf, 1988). Salt also provides a cementing medium for sediments in some environments (sabkha) and can be very important for founding structures, as in the Gatch underlying Kuwait (Al-Sanad *et al.*, 1990).

Engineering properties of sediments and sedimentary rocks and their investigation are addressed in Chapters 4 and 5.

3.3.5   *Metamorphic rocks and their associations*

Metamorphic rocks are rocks that have been changed by heat, pressure or both. By definition, they do not include low-temperature and low-pressure diagenetic processes affecting soils, as discussed in Chapter 5.

A general classification of metamorphic rocks is presented in Table 3.3. Contact metamorphism occurs at the host boundaries of igneous bodies, as illustrated in Figure 3.8. The metamorphic aureole surrounding a major granite or gabbro pluton can extend for hundreds of metres. The greatest effect is on sedimentary rocks; recrystallised, generally very strong rock found close to large plutonic igneous rock bodies is called hornfels. At greater distances the effect of the intrusion is less – often the only indication of change being the growth of new minerals such as kyanite or cordierite in the otherwise largely unaltered

Table 3.3 Simplified classification of metamorphic rocks. Refer to Fry (1984) for more detail.

	FOLIATED	NON-FOLIATED
Coarse to medium	GNEISS Often widely spaced and irregular foliation MIGMATITE Mixed schist and gneiss SCHIST Strong foliation	MARBLE Derived from limestone/dolomite QUARTZITE Recrystallised sandstone
Fine <0.06 mm	PHYLLITE Undulose foliation. Often micaceous, shiny SLATE Planar cleavage MYLONITE Fault gouge	HORNFELS Generally recrystallised contact rock SERPENTINITE Metamorphosed peridotite/norite

country rock. A good field example is the Skiddaw Granite in the Lake District, UK, where the crystalline hornfels zone extends more than 3 km from the exposed granite outcrop (the granite shallowly underlies the ground surface). The limit of metamorphism can be traced up to about 5 km away from the exposed granite; the metamorphic aureole, measured at right angles to the granite, is about 1 km thick (Institute of Geological Sciences, 1971).

Marble forms from the metamorphism of limestone through heat and pressure, often during mountain-building processes. Metamorphism of sandstone can form quartzite, in which the sand grains are welded together. This rock is often extremely strong and abrasive to drills and tunnel boring machines.

Minor intrusive rocks such as dykes and sills, though very hot when emplaced, often cause little metamorphism, because of their relatively small volumes, as illustrated in Figure 3.18 for a basalt dyke cutting limestone.

Regional pressure during mountain building can impose a marked cleavage or schistosity perpendicular to the maximum compressive stress. Rock types formed in this way range from relatively

Figure 3.18
Tertiary dyke through Jurassic limestone, Island of Muck, Scotland.

low-temperature slate, through phyllite, to high-temperature schist. In all cases, the rocks are recrystallised and the new fabric and structure dominates their mechanical properties, although original bedding may be evident. Phyllite is intermediate between slate and schist and generally has shiny, low-friction foliation because of the presence of minerals such as mica and chlorite. In schist, the original bedding may still be broadly recognised by chemical layering throughout the rock mass – some zones could be richer in silica (originally sandstone), others might be graphitic (the original rock perhaps having been organic mudstone with coal). Sometimes there have been several phases of metamorphism with several different cleavage or schistosity foliations imposed on the same rock mass, leading to blocky rock, which may cause difficulties for underground excavations, as occurred for the power house at Kariba Dam, Zambia (Blyth & de Freitas, 1984).

Mineralogical, grain size and shear strength variability along schistose foliation can result in joint styles changing very rapidly from layer to layer, as illustrated in Figure 3.19, which causes obvious difficulties for characterisation of the fracture network. Schist is sometimes associated with thin (say 100 mm) shear zones of low-frictional strength (15–25 degrees) often running roughly parallel to foliation and sometimes extending laterally for more than one kilometre (Deere, 1971). Not surprisingly, these often cause problems for engineering structures, including tunnels and slopes. Deere gives several examples, particularly of tunnels running parallel to the strike of steeply dipping schistosity. There are various possible origins for these shear zones but many are probably the result of slippage along foliation during folding – similar in origin to intra-formational shear zones in folded sedimentary rock sequences, as discussed below.

Figure 3.19 Joint system geometry varying with each stratum. About half way along Sector 9A of the Via Algarviana – São Bartolomeu de Messines to Barragem do Funcho, Algarve, Portugal.

3.4   Geological structures

3.4.1   *Introduction*

Where plates collide, large compressional stresses are generated, as along the western coast of North and South America. The consequence is uplift of the Rocky and Andes mountains and earthquakes on active faults such as the San Andreas in California and on the subducting plate beneath Chile and Peru. The rocks are squeezed and are either deformed plastically (see Figure 3.2), where temperatures and confining pressures are high, or fractured or both (Figure 3.20). Folding may control the disposition of the various rocks at a site and a specific geotechnical hazard involved with folding is intra-formational slip. As the rocks are folded, different layers slip relative to one another, possibly resulting in highly polished planes of low shear strength (Salehy *et al.*, 1977; Kovacevic *et al.*, 2007). Such highly polished intra-formational shear surfaces are common in the Coal Measures in the UK and have been responsible for large landslides.

For geotechnical engineering, the geological structures that are of prime importance are called discontinuities. These are fundamentally important to the mechanical properties of rock and some soil masses and how they perform in engineering projects.

Figure 3.20
Severely folded and thrusted sandstone of the Table Mountain Group, Cogmanskloof, South Africa.

3.4.2   *Types of discontinuity*

For geotechnical purposes, a discontinuity may be defined as a boundary or break within the soil or rock mass which marks a change in engineering characteristics or which itself results in a marked change in the mass properties. At a macroscopic scale, the most important discontinuities that engineering geologists need to consider are:

1. Geological interfaces such as bedding, pluton boundaries, dykes and sills, and unconformities
2. Faults
3. Joints and other fractures

3.4.3   *Geological interfaces*

Geological interfaces are the main boundaries mapped by geologists and therefore fundamental to unravelling the geological history at a site, interpolation between boreholes or field observations and extrapolation to some other location. The boundaries such as unconformities and dyke boundaries often represent a major gap in time, sometimes of many millions of years. There may be sudden contrasts in rock and soil type and in degree of fracturing across the boundary (Figure 3.21). Such contrast is often mapped in the field by lines of seepage and marked changes in slope. Often, however, rock strata boundaries, especially where unweathered, are of little engineering consequence, as illustrated in Figure 3.22.

3.4.4   *Faults*

Faults are geological fractures on which there has been demonstrable shear displacement. They range from minor breaks, with only a few

Figure 3.21 Signal Hill, Cape Town, South Africa. Joints orthogonal to horizontal bedding in Table Mountain Sandstone, unconformably overlying Cape Granite.

Figure 3.22 Fused boundary, between volcanic rock and granitic intrusion, Anderson Road Quarry, Hong Kong.

mm movements, to breaks in the Earth's crust extending many kilometres laterally and vertically and with cumulative displacements over many years, also of many kilometres. Faults can form in compressive regions (reverse faults) and in extensional zones (normal faults). The term 'normal' originates from coal mining in the UK, because most fault blocks dropped away, down the dip of the fault, and the miners knew in which direction the productive coal seam was likely to be found. Faults in the upper 10 km or so of the Earth's crust break in a brittle manner, producing fractured, brecciated rock. At greater depth, where the temperature and stress is much higher, faults occur more plastically. Typical features of brittle and plastic fault zones, now exposed at the Earth's surface, are illustrated in Figure 3.23. Fault zones can be very extensive, with metres of broken rock and gouge between the walls of the fault (Figure 3.24) but can otherwise be represented by a single surface with very little gouge (Figure 3.25). One example presented in Chapter 7 (TBM tunnel collapse) serves as an illustration of how important it is to know the nature of any fault zone. Sometimes the fault movement results in a

Figure 3.23 Schematic representation of ductile and brittle fault zones now exposed at the Earth's surface (after Fletcher, 2004).

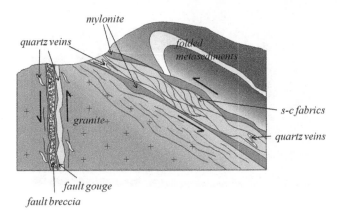

Figure 3.24 Thick fault zone, about 3m across – note discontinuous pale strata, Pos Selim landslide, Malaysia.

Figure 3.25 Thin fault zone with clay gouge, hammer for scale, Pos Selim landslide, Malaysia.

highly fractured shear zone, rather than a discrete plane with gouge, and such zones are often highly permeable.

Faults are a particular concern in geotechnical engineering in that they can be associated in sudden and often rather unexpected changes in rock quality. They may act as barriers to flow (termed fault seals in oil reservoirs) or, conversely, they can be highly permeable zones, full of water, and lead to a sudden inrush of water into tunnels. Faults are also, of course, the main source of earthquakes. By definition, faults disrupt the rock mass and may throw rocks of very different engineering characteristics together. As a result, a tunnel may pass from hard and good rock to extremely poor rock conditions over a very short distance and without warning. Such situations can be very difficult to deal with, necessitating a change in excavation methods, support requirements and sometimes a complete rethink of a project (e.g. Ping Lin Tunnel – a case study in Chapter 7). For foundations, there may be a sudden change over a few metres from simple pad foundations resting on rock to the need for deep piles to carry the load of a structure.

Not all faults cause problems for projects, so there is a danger of being over-cautious, leading to over-expensive investigation and unnecessary allowance in design for potential poor ground. Furthermore, faults shown on geological maps are sometimes conjectural, inferred by the mapping geologist on some topographic feature such as a valley or other lineament. However, lineaments and river systems can reflect geological features other than faults and the drainage system may owe its geometry to ancient geological history.

3.4.5   *Periglacial shears*

Another type of fault that can cause considerable problems because of low shear strength is that formed close to the Earth's surface due to periglacial processes. Such shear surfaces can be formed by a number of different mechanisms and can be extensive laterally (Spink, 1991). Numerous failures of slopes and embankments have been attributed to their presence (Early & Skempton, 1972), including Carsington Dam, during construction as described in Chapter 7. Skempton *et al.* (1991) note that solifluction surfaces are often difficult to find, even when you are fairly certain that they are there. Patient and detailed logging will be required, possibly with trial pits left open for several weeks to allow the shear surfaces to become apparent as the ground dries out and stress relief occurs.

3.4.6   *Joints*

Joints are fractures in rock that, by definition, show no discernible displacement relating to their time of origin, which distinguishes them from faults. All fractures are the consequence of overstressing the rock or soil material. The nature of joints – their orientations, roughness and persistence – is controlled by the local stress conditions that caused them to develop, together with the strength of the rock, other conditions, including temperature and water pressure, and subsequent history. Many joints occur as sets of fractures, pervasive through large volumes of rock, and owe their origins to processes such as cooling, burial or orogenic events (e.g. Hancock, 1985; Mandl, 2005). A set comprises a roughly parallel series of joints. Sets that are apparently related in terms of origin are called systematic. Joints can also be non-systematic or random. Joints can be regarded as essentially:

- primary – associated with the geological formation of rock
- secondary – caused by tectonic and gravitational stress including the result of uplift and bending or
- tertiary – due to local geomorphological or weathering influences.

Many of these begin as proto-joints that develop with time – they begin as general planes of weakness, which only become visible traces and, later, mechanical discontinuities, on uplift and exposure. Rock masses that have few or no joints (as visible traces or mechanical fractures) include deep-seated, unweathered igneous plutons (Martin, 1994), as illustrated in Figure 3.26. The water-lain sandstone in Figure 3.27 also has very few visible joints – presumably the rock was not sufficiently over-stressed during burial for the formation of hydraulic joints, as discussed later. Furthermore, the later uplift that must have occurred did not involve tension, bending or relaxation to cause differential stresses sufficient to induce fracturing.

Much effort is made to try to characterise joint networks in rock masses in geotechnical engineering – orientation, spacing, persistence and aperture, especially. Guidance is given in BS5930 (BSI, 1999),

Figure 3.26
Massive layered gabbro with no visible joints, Loch Scavaig, Isle of Skye, Scotland.

Figure 3.27
Miocene sandstone
with very few
visible joint traces,
Capela da Sra da
Rocha, Nr Porches,
Algarve, Portugal.

ISRM (1978) and by Priest (1993). The standard guidelines for practitioners, however, treat joints largely as statistical entities rather than geological features, and this is an area where geology has the potential to offer great insight and time-saving in geotechnical engineering; an opportunity that has been rather disregarded to date. This is partly because joint origin is still a difficult, rather poorly understood and highly debated subject (Pollard & Aydin, 1988).

The following discussion refers to the stress conditions that initiate fracturing, relative to the strength of the soil or rock at the time of joint formation. This is explained, through reference to Mohr's stress circles, in some detail in Chapter 5 in introducing triaxial testing, and the reader is recommended to go through that section in order to understand the following discussion. Mohr's circles are also well explained in most soil mechanics textbooks (Craig, 1992), rock mechanics textbooks (Hudson & Harrison, 1997) and structural geology textbooks (Davis & Reynolds, 1996), which demonstrates the importance of these concepts to different scientific disciplines.

Most joints are thought to develop as extensional fractures (in tension), parallel to a compressive major principle stress, σ_1. In a cooling igneous body, the extensional stresses might be due to contraction. Alternatively, the tensile stress σ_3 might be tectonic due to pulling apart of plates, as at the Mid-Atlantic Ridge or along the East African Rift Valley, or the result of bending and relaxation during uplift or exhumation (Price, 1959; Price & Cosgrove, 1990; Rives *et al.*, 1994). It can also be due to excess pore water pressures in thick sediment piles (Engelder & Peacock, 2001). The condition for formation of extensional joints is illustrated by the left-most Mohr's circle in the lower Figure 3.28, where the minimum principle stress is tensile and equal to the tensile strength of the soil or rock mass and the maximum principle compressive stress (assumed vertical) is less than three times the tensile strength of the intact rock. Under those stress conditions, joints would form in the plane of σ_1 and σ_3 with the dihedral angle, 2θ, equal to 0 (Joint E in the upper diagram). At higher levels of differential stress, shear fractures (S in the upper diagram) would form as illustrated by the right-most Mohr's circle in Figure 3.28. The stress circle is tangential to the rock failure envelope with a dihedral angle at 60 degrees (assuming a friction angle of 30 degrees). In between these extremes

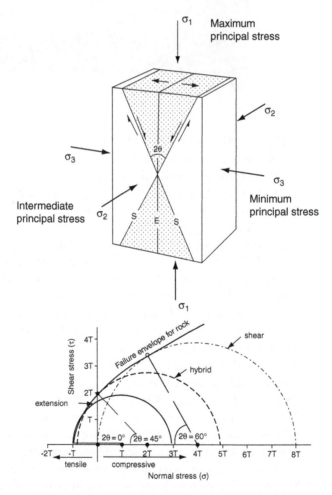

Figure 3.28 Stress conditions expressed as Mohr's circles for the formation of extensional (tension), hybrid and shear joints. The tangential line to the various circles is the overall failure envelope for the rock under different stress conditions (after Hancock, 1985).

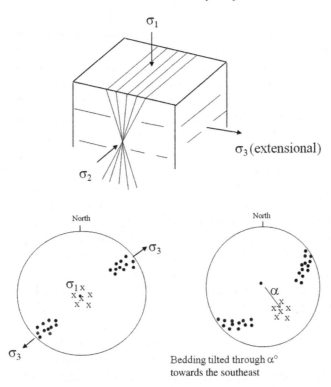

Joint spectrum of
hybrid joints

Figure 3.29 Stress conditions for the formation of hybrid joints and how these would appear on a stereographic representation. The centre crosses represent a horizontal set of discontinuities (say bedding in a sediment or flow banding in an igneous rock). The hybrid system plots from the circumference (vertical, extensional) inwards, up to dips of about 60 degrees to horizontal. The lower right figure shows how this pattern might appear if the whole rock mass was tilted through α degrees (after Hancock, 1991).

there is a possibility for hybrid fractures to form with dihedral angles between 0 and 60 degrees (sub vertical), and a potential joint spectrum, as illustrated in Figure 3.29. Details are given in Hancock (1985), who notes that the regular arrangement of structures such as joints within large areas (>1,000 km²) of weakly deformed rocks, gives confidence that they are indeed linked to tectonic processes, as per theory. Engelder (1999), however, questions the predictive validity of the Mohr-Coulomb approach in detail.

The concept of fracture formation in a large mass of rock governed by a Mohr-Coulomb strength law and under uniform stress conditions, is helpful in explaining joint formation at a site but generally geological history and local stress conditions and constraints means that the situation is more complex. The soil or rock mass is unlikely to be uniform and will include intrinsic flaws, pre-existing discontinuities and variable hydraulic conductivity (controlling effective stress). Pre-existing bedding, schistosity and faults, will have a controlling influence on the way joints develop (e.g. Rawnsley *et al.*, 1992). Where the rock mass has a long and complex geological history, there may be several generations of fracturing, each influenced by the former condition (see Rawnsley *et al.*, 1990). Deciphering that history is made more difficult once it is appreciated that all fractures that we now see as obvious visible mechanical discontinuities at the Earth's surface, may have only been incipient or integral proto-joints at the time of later joint formation and therefore might have had little influence on the formation of the later joints (Hencher & Knipe, 2007).

The proto-joint network provides relatively easy directions for breaking otherwise massive rock, such as the rift and grain directions in granite quarrying (Fujii *et al.*, 2007) or as preferential directions for breakage in laboratory testing (Douglas & Voight, 1969). Proto-joints develop as persistent mechanical fractures later, following the pre-imposed geological blueprint (location, orientation and spacing), through weathering processes and/or stress changes. The development of each joint is progressive as microfractures merge and extend over geological time (Hencher, 1987; Selby, 1993; Rogers & Engelder, 2004; Hencher, 2006; Hencher & Knipe, 2007). At any particular moment, a joint may be made up of open sections, sections where a trace is visible but where there is still considerable tensile strength and sections where the rock is apparently intact (rock bridges). Only microfractures mark the line of the future development of a mechanical fracture. That this is so, is evident from the obvious tensile strength of many rock joints, even though they are clearly visible as traces (Figures 3.30 and 3.31).

Figure 3.30 Partly developed joints through granite, north of Seoul, South Korea.

Figure 3.31
Columnar joints
through rhyolite,
High Island
Reservoir, Hong
Kong. Evidently,
the traces are not
fully developed as
persistent
mechanical
fractures, otherwise
the undercut
columns would not
remain in place
(gravity being what
it is).

3.4.7   *Differentiation into sets*

Joint sets are generally differentiated for rock mechanics analysis according to their geometries. The basics of stereographic projection representation of rock joints is dealt with in Chapter 4 and explained in detail in Wyllie & Mah (2004). If the poles representing discontinuities plot closely on a stereographic projection, then they are considered to comprise the same set. For example, when one defines a set in the rock mass modelling programme, UDEC (Chapter 6), this is done by inputting a mean dip and dip direction (perhaps plus or minus 5 degrees). Programs such as Dips from Rocscience (Chapter 6) can be used to identify sets statistically, according to various methods such as the Fisher distribution. This is a useful tool but a number of things must be borne in mind:

1. The original data might be biased or partial. Some joint sets may not be fully developed at the point of observation or might be misrepresented in terms of population, because of the geometry of the exposure.
2. Joints of similar geometry might include different sets geologically (in terms of time of formation) which have characteristics that are quite different despite their parallel orientation.
3. Important but rare geological features, such as a fault, might be overshadowed by the rest of the data and even removed from consideration by statistical manipulation, as discussed by Hencher (1985).
4. A better approach is first to try to interpret the distribution and nature of joints in terms of a model for the geological history at the site (Rawnsley *et al.*, 1990; Hencher & Knipe, 2007). Data such as surface textures and mineral coating can also be very helpful for differentiating between joint sets, especially for high-level interpretations such as for nuclear waste studies (e.g. Bridges, 1990).

Some aids for understanding joint origin on the basis of their geometrical expression, as seen in stereographic projections, are set out below. These are often very helpful for interpreting geological history, but as Price & Cosgrove (1990) put it: 'be warned – many fractures resist all attempts at interpretation'.

3.4.8   *Orthogonal systematic*

Many joint sets are orthogonal; two sets occur perpendicular to one another and perpendicular to some planar fabric such as bedding, schistosity, or flow banding in an igneous pluton. Examples of such joint sets in sandstone and granite are presented in Figures 3.32 and 3.33. The formation history can be quite complex, with one set being formed initially, the second following stress relief due to the development of the first set or perhaps a general stress reversal, as discussed by Rives *et al.* (1994). Interpretation may need detailed study of cross-cutting relationships. For the practising engineering geologist, the important thing is that this joint arrangement is very common in a variety of rock types and this can aid in interpretation of sets from field

Figure 3.32 Orthogonal fractures in sandstone, Table Mountain, Cape Town, South Africa.

Figure 3.33
Orthogonal
jointing in granite,
Mount Butler
Quarry, Hong
Kong.

Figure 3.34
Schematic
representation of
orthogonal
fractures and their
representation on a
stereographic
projection
assuming one set
(e.g. bedding or
cooling surface) is
horizontal. Lower
right diagram
shows typical
pattern of poles to
orthogonal fracture
sets if the model
were tilted through
α degrees by
folding.

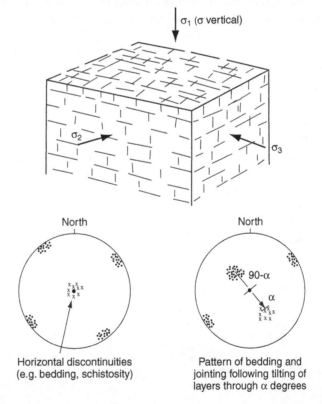

data. Figure 3.34 shows the typical distribution of poles that might be
seen in a stereographic projection of horizontally bedded strata with
two orthogonal joint sets and the geometrical expression if the strata
have been tilted. The interpreter should always be looking for angular

relationships between sets and linking these back to possible modes of origin. Orthogonal sets may be essentially primary, formed orthogonal to the intermediate and minor principal stress directions, during burial and diagenesis of sedimentary rock (hydraulic joints of Engelder, 1985) or during cooling in igneous rock. It has also been demonstrated experimentally that orthogonal sets can be formed as a secondary phenomenon by flexing layers (Rives *et al.*, 1994), and this is probably a common origin during erosion, uplift and general doming (Price & Cosgrove, 1990).

3.4.9   Non-orthogonal, systematic

These joints are formed where the stresses perpendicular to the maximum principal stress, σ_1, are uniform. The most common joints in this category are those that develop in extensive sheet lavas, as illustrated in Figures 3.35–3.37. They form as tensile, planar zones of microfractures arranged around centres of cooling, as primary structures, and may develop incrementally during cooling (DeGraff & Aydin, 1987). Figure 3.38 shows deformed columnar jointed basalt lava, and one explanation is that a vertical columnar framework was established at an early stage of cooling but then part of the lava sheet collapsed whilst the rock was still plastic and the full joints developed as fractures later. Figure 3.39 shows how columnar joints in a general lava flow would appear on a stereographic projection originally and when tilted.

Non-orthogonal, systematic columnar fractures can occur in rocks other than lavas. Young (2008) describes their occurrence in sandstone. In Chapter 7, the case example is presented of the Pos Selim landslide in Malaysia. The pervasive jointing, which played a

Figure 3.35
Vertical jointing through andesite lava, Hallasan, Cheju Island, South Korea.

Figure 3.36
Columnar joints in basalt lava flow, Isle of Staffa, Scotland.

Figure 3.37
Columnar joints, as seen at right angles to cooling surface (looking down, vertically). There is no directional control on joint formation in the horizontal plane ($\sigma_2 = \sigma_3$). Isle of Staffa, Scotland.

Figure 3.38 Collapsed columnar joints. This has apparently happened after the pattern of columns had been defined in the cooling lava sheet but whilst the lava was still plastic. *Am Buachaille* (the Herdsman), Isle of Staffa, Scotland.

major role in allowing the failure to develop, had no preferred orientation, other than it was at right angles to the planar shistocity (Figure 3.40). These are secondary joints, probably formed during a relatively late stage of the regional tectonics responsible for the schistosity.

3.4.10   Shear joints

Pollard & Aydin (1988) dismiss the concept of shear joints as 'sheer nonsense' but this seems to be a bit tongue-in-cheek (the paradox being that once shear takes place a joint becomes a fault by definition). Fractures certainly do develop in shear directions as they do in triaxial testing (Chapter 5). In the example shown in Figure 3.41, some sections of the shear joints show no visible displacement but over other lengths of the same discontinuity, there is displacement. The argument about shear joints is largely academic in that whereas the joints propagate in the shearing direction predicted from Mohr's circle representation, in detail, the joint is probably made up of coalesced sections, which are strictly tensile, originating from minor flaws in the rock (Kulander & Dean, 1995). Engelder (1999) extends the discussion to hybrid joints. The appearance of shear joints on a stereonet, before and after tilting, is shown in Figure 3.42.

3.4.11   Complex geometries

As discussed above, many joints follow some systematic geometrical pattern relating to the principal stress directions and magnitudes at the time of their formation. In some field exposures, however, the fracture network can be very complex and difficult to unravel, especially when a rock mass has been through several structural events, with each event

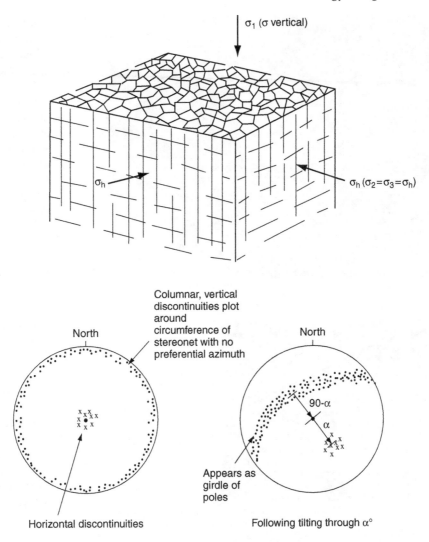

Figure 3.39 Schematic representation of columnar fractures and their representation on a stereographic projection, assuming one set (e.g. schistosity or cooling surface) is horizontal and formed in an isotropic stress field in the horizontal direction. Second diagram shows typical pattern of poles if the model were tilted, say through folding. The original pattern of vertical fractures, plotting around the circumference, are expressed as a girdle of poles following a great circle, centred on the originally horizontal set.

resulting in a new episode of fracturing, which will be influenced by any pre-existing fractures (Rawnsley *et al.*, 1990). Examples of major structures influencing the geometrical development of joints are given by Rawnsley *et al.* (1992) and one of these is illustrated in Figure 3.43. The joint geometries clearly follow stress trajectories that were strongly influenced by the pre-existing major fault.

Figure 3.40 Joint pattern in schist. All joints are approximately at right angles to the schistosity but otherwise random in orientation. Pos Selim landslide, Malaysia. See Chapter 7 for more details of this case history.

No visible displacement

Local normal faulting

Figure 3.41 Joints with conjugate shear arrangement. At some locations there are measurable displacements so the shear joints grade into small displacement faults. Near Austin, Texas, USA.

3.4.12   Sheeting joints

Sheeting joints, which are sometimes also referred to as exfoliation fractures, are unlike other joints in that their geometry is not pre-defined by ancient geological history but instead they develop in response to near-surface stress conditions reflecting locally prevailing topography (Hencher *et al.*, 2011). These are tertiary joints, as defined earlier. Sheeting joints are a striking feature of many landscapes and they have been studied for more than two centuries (Twidale, 1973). They run roughly parallel to the ground surface in flat-lying and

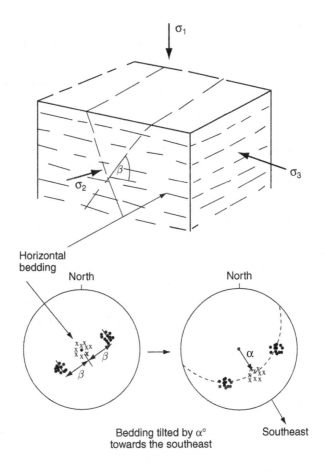

Figure 3.42
Schematic representation of shear joints and their representation on a stereographic projection, assuming one set (e.g. bedding or cooling surface) is horizontal. Second diagram shows typical pattern of poles if the model were tilted, say through folding.

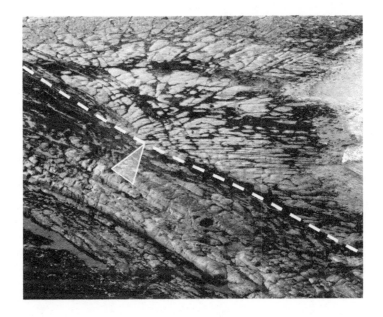

Figure 3.43
Systematic joints, whose geometries of formation were clearly influenced by a pre-existing mechanical discontinuity (the Peak Fault). Ravenscar, near Robin Hood's Bay, North Yorkshire, England.

steeply inclined terrain and generally occur close to the surface, typically at less than 30m depth. They can often be traced laterally for hundreds of metres. Most sheeting joints are young, geologically, and some have been observed to develop explosively and rapidly as tensile fractures in response to unloading (Nichols, 1980). Others are propagated to assist in quarrying, using heat or hydraulic pressure (Holzhausen, 1989). Their recent origins and long persistence without rock bridges differentiates them from most other joints and from most bedding, cleavage or schistosity-parallel discontinuities.

Sheeting joints are common in granite and other massive igneous rocks but also develop more rarely in other rock types, including sandstone and conglomerate (Figures 3.44 and 3.45). Some sheeting joints develop at shallow dip angles, for instance, during quarrying, where high horizontal compressive stresses are locked in at shallow depths. In Southern Ontario, Canada, for example, high horizontal stresses locked in following glacial unloading, often give rise to quarry floor heave and pop-up structures accompanied by opening up of pre-existing incipient discontinuities such as bedding planes and schistose cleavage (Roorda *et al.*, 1982). Where there are no pre-existing weakness directions, new sub-horizontal fractures may develop in otherwise

Figure 3.44
Sheeting joint through arkosic sandstone, Uluru (Ayers Rock), Northern Territory, Australia.

Figure 3.45
Sheeting joints
through
conglomerate. Kata
Tjuta (the Olgas),
Northern Territory,
Australia.

Figure 3.45
Sheeting joints
through
conglomerate. Kata
Tjuta (the Olgas),
Northern Territory,
Australia.

unfractured rock. Holzhausen (1989) describes propagation of new sheeting joints under a horizontal stress of about 17 MPa at a depth of only 4 m, where the vertical confining stress due to self-weight of the rock is only about 100 kPa. The mechanism is similar to a uniaxial compressive strength test where tensile fracture propagates parallel to the maximum principal stress (σ_1). Such exfoliation and tensile development of sheeting joints is analogous to the sometimes explosive spalling and slabbing often seen in deep mines (Hoek, 1968; Diederichs, 2003).

From a worldwide perspective, however, the joints most commonly recognised as sheeting structures are observed in steep natural slopes. These joints are also thought to develop as tensile fractures where the maximum compressive stress due to gravity is reoriented to run parallel to the slope, as demonstrated by numerical models (Yu & Coates, 1970; Selby, 1993) and discussed in detail by Bahat *et al.* (1999). Sheeting joints also develop parallel to the stress trajectories that curve under valleys where there has been rapid glacial unloading or valley downcutting. Failure and erosion is a continuing process, with the formation of new sheeting joints following the failure of sheet-bounded slabs. Wakasa *et al.* (2006) calculated an average erosion rate of 56 m in one million years from measurements of exposed sheeting joints in granite in Korea (Figure 3.46), which is significantly higher than erosion rates on other slopes without sheeting joints. Whilst many exposed sheeting joints are evidently very recent, others are much older. Jahns (1943) and Martel (2006) note the apparent dissection of landscapes post-dating sheet joint formation. Antiquity is also sometimes indicated by degree of weathering. Additional evidence for the great age of some sheeting joints is the fact that they can sometimes be observed cutting through otherwise highly fractured rock. Most sheeting joints occur in massive strong rock and it is argued that if the rock mass had been already highly fractured or weathered then the topographic stresses would be accommodated by

Figure 3.46
Sheeting joints in
granite, Mount
Bukansan, near
Seoul, South Korea.
Climbers show the
scale.

movements within the weak mass rather than by initiating a new
tensile fracture (Vidal Romani & Twidale, 1999). Therefore, where
sheeting joints are found in highly fractured rock masses, it is likely
that they predate the gradual development of the other joints, as
mechanical fractures during unloading and weathering (Hencher,
2006; Hencher & Knipe, 2007). Figure 3.47 shows the stereographic
representation of sheeting joints at a site in Hong Kong, together with
cross joints, perpendicular to the sheeting joints and at right angles to
the azimuth of dip, which indicates their likely tensile origin.

3.4.13   *Morphology of discontinuity surfaces*

The shape of discontinuity surfaces is important to rock engineering,
not least because of its influence on shear strength, and this is dealt

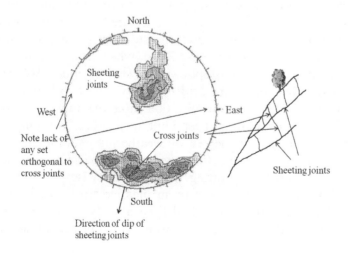

Figure 3.47
Stereonet showing
sheeting joints and
cross joints at 90
degrees. Tuen Mun
Highway, Hong
Kong (after
Hencher *et al.*,
2011).

with in Chapters 5 and 6. As for overall joint set orientation and spacing, discontinuity morphology is related to geological history and mode of propagation, and surface features should be observed and described. The description, analysis and interpretation of discontinuity surface morphologies and the ways these can be linked to interpretation of causative stresses, mechanisms and subsequent evolution, is a relatively undeveloped area of scientific study called fractography (Ameen, 1995). It is difficult to be specific, that one type of joint can be expected to have certain morphological characteristics compared to others, but it is generally recognised that tensile fractures are particularly rough and variable. Generally, roughness is characterised by measurement. Waviness is the deviation from mean direction at the scale of metres. Smaller scale roughness (sometimes termed second-order) is either measured objectively or estimated with reference to a scale called Joint Roughness Coefficient (JRC). Details are given in ISRM (1978) and Barton (1973, 1990). Roughness is described in the field using terms such as stepped or slickensided (British Standards Institution (1999)). At the smallest scale, surface texture, together with mineralogy, will control basic friction.

3.4.13.1   Sedimentary rocks

Sedimentary bedding planes may show a variety of distinct morphologies, as illustrated by Stow (2005). Many are flat and planar but others are rough. Ripple marks, as seen on beaches, are commonly preserved in sandstone; in turbidite surfaces, scour marks can make them very rough, interlocking with the overlying layer (e.g. Figure 3.48). Bedding surfaces can also be roughened by biological activity such as burrowing. Sometimes, bedding surfaces are exposed, weathered and eroded before the overlying rocks are

Figure 3.48
Complex load/
scour markings at
base of sandstone
layer, Texas, USA.

emplaced and, again, such features need to be interpreted correctly, both in terms of understanding the geological situation and because they will affect geotechnical properties.

3.4.13.2   *Tension fractures*

The first major work on fracture morphology was by Woodworth (1896). He observed and illustrated what are now considered classic expressions of fractures propagating from a flaw, with typical features including feather markings and arrest or hackle marks (Figure 3.49). Most researchers consider that most such markings involve extension (tension), although other features probably involve some shear. They are recognised in all rocks and even drying sediments. They are also commonly observed in drill core, and distinguishing natural from induced fractures is an important task for those logging the recovered rock (Kulander *et al.*, 1990). Such features are significant in their own right geotechnically but are also important for the interpretation of the geological history of a site.

Figure 3.49
Shallow bedding plane with ripple marks. Near-vertical surface with typical arrest and hackle marks associated with tensile fracture propagation. Tsau-Ling landslide, Taiwan.

3.5   Weathering

3.5.1   *Weathering processes*

Weathering is the process by which rock deteriorates until it eventually breaks down to a soil. It occurs close to the Earth's surface and depends very much on climatic influences: rainfall and temperature. Ollier (1975) and Selby (1993) provide good overviews.

In hot, humid climates the following are the most important processes

- *Decomposition*: the result of chemical changes on exposure to the atmosphere (H_2O, CO_2 and O_2). The original rock minerals, stable at the temperatures and pressures operative at the time of formation, break down at the Earth's surface to sand, clay and silt.
- *Disintegration*: inter- and intra-grain crack growth and coalescence of cracks to form fissures and propagation of large-scale joints.
- *Eluviation*: the soft, disintegrated (or dissolved) material is washed out from the parent rock fabric through open joints or from the porous skeletal structure and deposited elsewhere (illuviation).

Weathering affects not only strong rocks but weak masses, including materials that might be regarded as engineering soils, even in their fresh state. Processes include softening and chemical change (e.g. Moore & Brunsden, 1996; Picarelli & Di Maio, 2010).

The rock mass in tropical areas is commonly severely weathered to depths of tens of metres and occasionally over 100 m. Weathering is manifested by changes from the original rock state (fresh), including mineralogy, colour, degree of fracturing, porosity and, thereby, density, strength, compressibility and permeability.

In colder climates, chemical decomposition is less active and rocks tend to deteriorate due to frost and ice action (Figure 3.50). Mechanical deterioration is also the dominant process in desert

Figure 3.50 Disintegrated granite with corestones, above Lake Tahoe, Nevada, USA.

environments, sometimes associated with expansive formation of salt in pores. The depth of weathering in cold climates is far less than in tropical environments.

In temperate climates such as the UK, weathering is rarely very significant and certainly far less so than in countries such as Brazil, Malaysia and Singapore, where weathering has extremely important consequences for investigation, design and construction (e.g. Shirlaw *et al.*, 2000). In the UK and much of northern Europe, most weathered rocks were stripped from the landscape by recent glaciations.

3.5.2   *Weathering profiles*

Weathering reduces the strength of rock material, as illustrated in Figure 1.5 and discussed in Chapter 5 in terms of geotechnical parameters. Weathering processes generally operate at upper levels in the saturated zone and in the vadose zone above the water table, although it should be noted that water tables change periodically and have done so over the millions of years that it will have taken for the development of some thick weathered profiles. Therefore, current water levels may not be related to depth of weathering at a site.

As a general rule, weathering works in from free surfaces where chemicals in water (including the water itself) can attack the parent rock (Figure 3.51). Eventually, it may leave a framework of corestones of less weathered rock separated by severely weathered zones marking out the loci of the original joints (Figure 3.52). The process is illustrated in Figure 3.53. The wide varieties of conditions that can be encountered in weathered terrain are discussed by Ruxton & Berry (1957), and Figure 3.54 is based on their interpretation of one type of weathered profile from Hong Kong. Despite the conceptual usefulness of Ruxton & Berry's profiles, exposures are often encountered that do not conform and such exposures provide challenges to mass weathering classifications such as the current European standard, as discussed in Chapter 4 and Appendix C. Furthermore, other rock types often weather without the development of corestone-type profiles. Mudstone sequences tend to develop a gradational weathering profile, as characterised for the Keuper Marl by Chandler (1969). Limestone is often karstic, with large caves and open joints, as illustrated in Figure 3.55 and classified by Fookes & Hawkins (1988).

At any particular location, the weathering profile is a function of parent geology, groundwater conditions and the geological and geomorphologic history of the site. The profiles may be ancient and bear little relationship to current geomorphologic setting. Given these and other factors, weathering profiles can be rather unpredictable from examination of the current topography. Valleys might be associated with deep weathering along faults (e.g. Shaw & Owen 2000), but not always.

Figure 3.51
Stained joints,
volcanics, Hong
Kong.

Figure 3.52
Corestone
development in
granite, Stubbs
Road, Hong Kong.

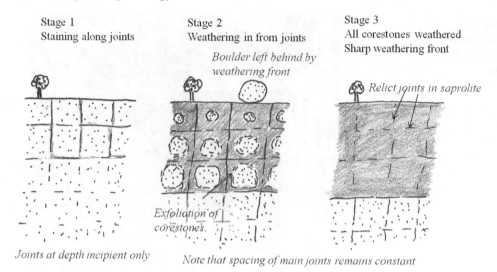

Stage 1
Staining along joints

Stage 2
Weathering in from joints

*Boulder left behind by
weathering front*

Stage 3
All corestones weathered
Sharp weathering front

Relict joints in saprolite

*Exfoliation of
corestones.*

Joints at depth incipient only

Note that spacing of main joints remains constant

Figure 3.53 Corestone development. Note that boulders can be left behind at the surface as the weathering front penetrates in to the rock mass.

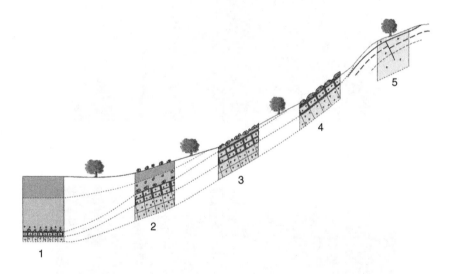

Figure 3.54 Schematic weathering profile with corestone development, after Ruxton & Berry (1957), modified by Hencher *et al.* (2011).

Chemical weathering rarely occurs to depths of more than 100 m, even in tropical and sub-tropical areas, but this should only be used as a general guide to current situations; sometimes weathered profiles are encountered at much greater depths, having their origins in some past landscape and time. For example, Younger & Manning (2010) describe tests on a highly permeable zone in granite at a depth of 410 m and attribute the high permeability to weathering during the Devonian when the granite was exposed at the surface. Ollier (2010) provides other examples of ancient weathering profiles buried by later sediments.

Figure 3.55
Limestone
pavement with
protojoints etched
out by dissolution
weathering. Above
Malham, West
Yorkshire.

3.6 Water

3.6.1 Introduction

Water is critically important to many geotechnical projects. Surface water causes erosion and flooding; groundwater controls effective stress and, therefore, frictional strength, and can be a major problem for tunnelling. Water problems for most civil engineering design and construction is dealt with by measurement and monitoring using piezometers, backed up by numerical modelling and analysis (Chapters 4 and 6). Groundwater levels do not generally fluctuate too much in response to individual rainstorms, other than close to the surface, which is significant for shallow landslides, as discussed below, but not for most other engineering works. More significant are groundwater changes brought about by engineering works, either deliberately (e.g. dewatering to carry out excavations in the dry) or as an unintended consequence, for example, where tunnelling below a site. Lowering of the water table inevitably causes water migration and potentially internal erosion, a loss of buoyancy, increase in effective stress and self-weight compaction of soil and rock. This may result in settlement and damage to adjacent structures; piles can become overloaded by negative skin friction (Chapter 6). Similarly, rising water levels can cause difficulties due to buoyancy, which could require holding down anchors or piles. Rising water might also weaken the ground supporting a structure, sometimes due to dissolution of cementing agents.

For tunnels and other underground structures, inflows can be very difficult to predict accurately because it depends so much on the geological situation, which is rarely well understood before the works commence. Given knowledge of water pressure and hydraulic conductivity, then inflows can be predicted following standard equations or by numerical simulation, but the controlling parameters are difficult to measure or estimate and predictions can be wildly out, as found for trials for nuclear waste investigations (e.g. Olsson & Gale, 1995). This is especially so for tunnels passing through variable geology, as discussed by Masset & Loew (2010). The pragmatic solution is to probe ahead of the tunnel face periodically, and if inflows from the probe holes are high, then to improve the ground in front of the tunnel, usually by injecting cement or silica grout to reduce the hydraulic conductivity. Alternatives are dewatering the ground or freezing the ground temporarily. If predictions of groundwater conditions are badly incorrect, this can have major consequences for the suitability of tunnelling method or machinery (e.g. the degree of waterproofing of equipment). Where there is a risk of high water inflow, it is normal practice to drive the tunnel uphill to reduce the risk of inundation, danger to workers and damage to machines. Where tunnelling under the sea, lakes or rivers, there may be a risk of disastrous inflows, as occurred during the construction of the Seikan Tunnel in Japan (Matsuo, 1986; Tsuji *et al.*, 1996). Other examples where the severity of groundwater conditions was underestimated with severe consequences include the SSDS tunnels in Hong Kong and Ping Lin Tunnel in Taiwan, which are discussed in Chapter 7.

3.6.2   *Groundwater response to rainfall*

Most landslides are caused by rainfall and in Hong Kong, for example, rises in water level during a storm of more than 10 m have been recorded (Sweeney & Robertson, 1979). Therefore, there is great interest in trying to predict changes that might occur, as these will greatly affect any numerical calculations of slope stability, as well as other engineering projects. Geological profiles are generally depicted for groundwater modelling, as made up of discrete, homogeneous and often isotropic units of given hydraulic conductivity (Todd, 1980), and most commercially available, hydrogeological software only deal with homogeneous units. To be more realistic, models may need to incorporate local barriers such as fault seals, fracture flow or more variable geological conditions such as local lithofacies (e.g. Fogg *et al.*, 1998).

The wetting band theory, first proposed by Lumb (1962), is still used for estimating the likely depth of ground that might be affected by a rainstorm. Given a rainstorm of a certain duration, and knowledge of the original saturation of the ground and porosity, the thickness of the surface zone of saturation can be estimated. It is assumed that the saturated layer will then descend until it meets the groundwater table, resulting in a rise in groundwater table, equal to the thickness of the wetting band. This provides a tool for assessing the design groundwater condition, albeit rather crude. Some of the geological conditions that will conspire to make such simple approaches unrealistic are illustrated in Figure 3.56.

More sophisticated attempts have been made to model infiltration and pressure diffusion processes in pressure head response and the triggering of landslides, mathematically (see Iverson, 2000). Such methods are useful in visualising mechanisms but again rely on

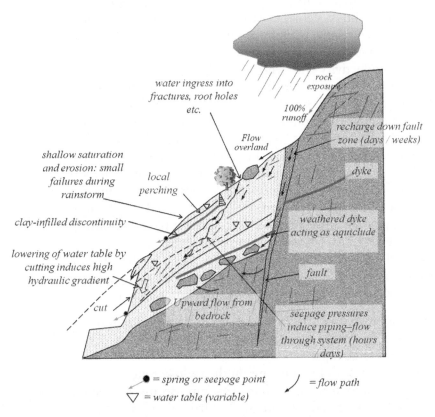

Figure 3.56 Schematic model of water runoff, inflow and throughflow in weathered profile. Note the importance of minor geological features such as dykes, clay-infilled joints or simple permeability contrasts in the profile and the development of natural pipes. In rock, water flow will be controlled by joints and specific channels along those joints. There is often a zone of more highly fractured rock just below rockhead with preferential flow and sometimes upward flow into overlying soil profile (Hencher, 2010).

generalised parameters such as hydraulic diffusivity, which are often difficult to define and only apply to simple geological conditions. More recent modelling methods are described by Blake *et al.* (2003).

It is often argued that it is more realistic to instrument a site and then to extrapolate rainfall response to a design rainstorm (GCO, 1984a). However, instrumenting slopes to measure critical water pressures is not easy because instruments must be installed (and monitored remotely) at the precise locations where water pressures will develop. Quite often rock and soil profiles are compartmentalised and water flow carried in narrow channels, pipes or through particularly permeable zones, as discussed below, so the success of any instrumentation programme will depend very much on the geological and hydrogeological insights and skills of the investigating team, as well as a degree of good fortune.

3.6.3   *Preferential flow paths through soil*

Soil formed by sedimentation might be expected to suit isotropic modelling but intrinsic and systematic variability is to be anticipated. Sediment piles of even relatively uniform sand can be expected to contain intercalations of finer material that will result in heterogeneous flow characteristics. Conductivity will generally be higher in the horizontal direction than vertical. For soils derived from the weathering of rock, the mass can be incredibly complex and likewise the hydraulic conductivity. Degree of weathering varies from place to place and the soil mass will contain remnant fabric and relict joints from the parent rock that will probably control water throughflow. In practice and empirically, the mass might be represented by a simple set of parameters characterised for a representative elemental volume (REV), but those parameters probably do not actually represent the physical and mechanical processes taking place in anything other than very simple situations.

Richards & Reddy (2007) provide a comprehensive review of piping, particularly as related to earth dam construction, which is where piping was first recognised as an important phenomenon. They define several types of piping, which are also relevant to natural ground, namely:

1. Suffosion (or eluviation): the washing out or dissolution of material en-masse, leaving a loose framework of granular material, which is prone to collapse.
2. Dispersion of clay soils by rainwater in the vadose zone.
3. Backwards erosion from a spring. The pipe forms (for some reason) and then material is gradually lost from that opening.
4. Erosion along some pre-existing opening such as a master joint.

The majority of pipes investigated by geomorphologists are confined to the upper few metres in the ground profile (e.g. Jones, 1971; Uchida *et al.*, 2001). They are particularly common in forested areas within shallow soil profiles and are associated with shallow landslides

(Pierson, 1983). Anderson *et al.* (2008) used dyes (as have others) to trace pipe networks exposed by excavation. Not surprisingly, the size and connectivity of these shallow features was related to surface catchments. The same is probably not true of deeper features.

Channelised fracture networks from the parent rock often persist through the various stages of weathering. This type of preferential flow needs to be considered in investigation, hydrogeological modelling and design. Such natural pipes probably follow original structural paths (especially master joint or fault intersections), but may also be formed by seepage pressure in weak saprolite or in superficial soils such as colluvium. They also develop at permeability contrasts (e.g. colluvium overlying saprolite). More details are given in Hencher (2010).

Pipes are commonly seen associated with many deep-seated failures in weathered terrain (Hencher, 2006). It is implied that the development of pipes at depth may be linked to early stages of progressive failure, as the rock mass dilates and ground water exploits the dilating and deteriorating rock mass. Such deep pipes are probably distinct in origin from pipes found in upper soil horizons. Fletcher (2004) reports that infilled pipes are sometimes encountered up to depths of 80 m below present sea level, and these must be associated with ancient lowstand levels.

3.6.4   *Preferential flow paths through rock*

Fracture flow in rock is poorly understood and difficult to investigate, characterise or model in any real sense (Black, 2010). Interpretation of even sophisticated test data is not straightforward, depending upon whether one assumes that measured flow volume into or away from a test location is three-dimensional, planar (along a planar feature such as might be assumed for a fault or major joint) or essentially linear along a preferential channel. In simple terms, transmissivity depends upon the aperture (degree of closure of rock walls) and roughness of the discontinuity walls and lengths of fractures, together with their intersections. The flow paths are tortuous and extremely difficult to identify. Dershowitz & LaPointe (1994) report how new oil wells caused large drops in production to existing wells at distances of several kilometres, within two days, whist other wells between them were unaffected. Such behaviour could not be predicted without extremely good knowledge of the fracture network and understanding of potential connectivity, which is unlikely ever to be the case. To do so with any hope of reasonable success would require a good understanding of the fracture origins and this is quite unlikely given the current state of geological knowledge. Most attempts are simplistic, extrapolating from superficial, statistical and poorly constrained observations in exposures or boreholes, to draw implications for the rock mass at some distance. Thomas & La Pointe (1995) attempted to discriminate between dry and flowing fractures in the drift at Kiamichi

Mine in Japan, on the basis of descriptive parameters such as roughness, aperture and orientation, as observed in the drift (at the exposed
traces of discontinuities in the mine). They used a neural network
approach to train the analysis, but with limited success. Indeed, it
seems clear that connectivity is the most important factor that controls
flow through rock, rather than locally measurable characteristics such
as fracture intensity, aperture or spacing. The fact that the most
transmissive and well-connected features might not be adequately
sampled during investigation, is a major problem for all geotechnical
projects, not least for potential nuclear waste repositories, where it is
recognised that large-scale conductive features need to be identified
and dealt with in a deterministic manner (Black *et al.*, 2007; Nirex,
2007).

In weathered rock profiles, water flow is sometimes concentrated in
fractured rock, underlying weathered saprolite or colluvium, and this
can result in transmission of water (and high water pressures) from one
part of a hillside to another, or even between catchments, where the
water then feeds into the overlying weathered mantle and may trigger
landslides. During the Mid Levels study in Hong Kong, it was recognised that confined conditions could occur where material of lower
conductivity overlies rock of higher conductivity. Strong upward
hydraulic gradients from bedrock to the decomposed rock aquifer
were identified in some areas (GCO, 1984b), and these observations
were used in setting up a numerical model of the hydrogeology (Leach
& Herbert, 1982). Jiao & Malone (2000) and Jiao *et al.* (2005, 2006)
have extended this concept of a highly transmissive zone at depth to
explain several deep-seated landslides and evidence of artesian pressure in Hong Kong. Montgomery *et al.* (2002) report an intensely
instrumented site (more than 100 shallow piezometers) in Oregon,
USA, and similarly noted artesian flow from the underlying bedrock,
which they found surprising for such a steep hillside. They also commented that, whilst the seepage from bedrock might effectively determine the specific locations where debris flows might initiate, the
distribution and connectivity of the near-surface bedrock fracture
system are almost impossible to predict. Similar upward flows from
bedrock into the overlying soil mantle are reported for steep granitic
terrain by Katsura *et al.* (2008).

3.7   Geological hazards

3.7.1   *Introduction*

Numerous hazards can be regarded as essentially geological, including
the potential for subsidence, swelling, clay shrinkage, natural noxious
gases and mining, as generally addressed in Chapter 4, when considering elements to be targeted during site investigation. The most

important natural hazards in terms of loss of life, however, are landslides, earthquakes and volcanoes.

3.7.2   *Landslides in natural terrain*

There are more than 300 fatal landslides each year on average (Petley, 2011). Some of these are in man-made slopes and therefore a matter for engineering design (Chapter 6). Many, however, are in natural slopes and single incidents can cause many deaths. The debris flow that struck Gansu, China, in August 2010, killed more than 1,200 people and destroyed at least 300 buildings. A total of 45,000 people had to be evacuated. The risk to sites from natural terrain landslides is therefore an important consideration at many sites.

3.7.2.1   *Modes of failure*

Natural terrain landslides can be split into those where the detached debris directly impacts a site through gravity and those where the debris becomes channelised and flows down a valley (Figure 3.57). Channelised landslides are relatively easy to deal with conceptually, in that the pathway for the debris, which often becomes saturated and flows, is easily predicted, even if the size of the event is not. The best thing to do is to avoid the outlet of any valley, but if this is not possible,

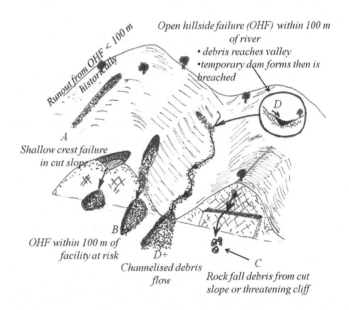

Figure 3.57 Run-out models for shallow landslides. Most landslides are only hazardous where they directly threaten a facility. Remote landslides can, however, feed into a stream channel where they can be channelised and flow great distances. Things may be made worse where the initial failure produces a landslide dam, behind which a lake forms. When the temporary dam is breached, a sudden discharge of fluidised debris is released.

for example, where building a road or railway, then the size of the design event may be predicted by historical studies or from first principles. Hungr *et al.* (2005a) present a useful review of landslide characteristics that might be considered for design. In assessing existing facilities and structures, sometimes these can be protected by barriers and other engineering devices, but occasionally the risks are so high and cost of mitigation too expensive, so that relocation is the only real solution.

Open hillside landslides and rockfalls have a much more limited distance of travel. For most landslides in Hong Kong, debris travel distance is less than 100 m, so the area of concern is quite obvious, both in terms of source of landslides and structures at risk. It does not make them less dangerous, just that the nature of the hazard and focus for analysis is clear. Hazard assessment can follow standard methods of investigation, analysis and design, as outlined in Chapter 6. Risk review can be used to justify cost of mitigation works or taking no action – despite there being a clear and obvious danger as, for example, along many roads through mountainous areas. The practical application of risk concepts are reviewed by many authors in Hungr *et al.* (2005b). Fell *et al.*, (2005) and Wong (2005) are particularly useful.

3.7.2.2   *Slope deterioration and progressive failure*

The concept of ripening of slopes prior to failure has been a useful idea for many years, but recently evidence for progressive deterioration of slopes prior to detachment has become better documented. This applies to both natural slopes and cut slopes (Malone, 1998; Hencher, 2000; Parry *et al.*, 2000). Factors involved in slope deterioration are illustrated schematically in Figure 3.58, and some of the factors triggering natural terrain failures are illustrated in Figure 3.59.

The gradual deterioration can be represented by a curve in which the Factor of Safety reduces over a period of time, which may be hundreds of years (Figure 3.60). The vertical lines represent temporary reductions in Factor of Safety caused by relatively short-term, transient events (days). In the course of time, the slope will deteriorate to the point where it is vulnerable to a transient event – causing a reduction in the Factor of Safety below 1.0. Whether that event results in catastrophic failure, or only minor movement and internal deformation, depends on many factors, including the severity of the triggering event and how long it lasts. The concept of ripening and progressive failure is discussed in more detail in Hencher (2006). Similar concepts are discussed for claystone slopes by Picarelli & Di Maio (2010).

Signs of gradual deterioration or, more likely, the cumulative effect of intermittent triggering events, can be seen in many exposures and these can be used during ground investigation to help judge whether a failure is imminent, although this may still not be straightforward, as discussed in Chapter 6, Box 6-4.

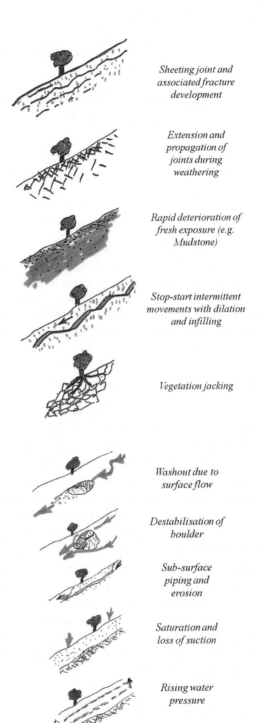

Figure 3.58
Deterioration
factors in slopes.

Sheeting joint and
associated fracture
development

Extension and
propagation of
joints during
weathering

Rapid deterioration of
fresh exposure (e.g.
Mudstone)

Stop-start intermittent
movements with dilation
and infilling

Vegetation jacking

Figure 3.59 Water
triggering
mechanisms of
shallow landslides.

Washout due to
surface flow

Destabilisation of
boulder

Sub-surface
piping and
erosion

Saturation and
loss of suction

Rising water
pressure

Cleft water
pressure in
discontinuities

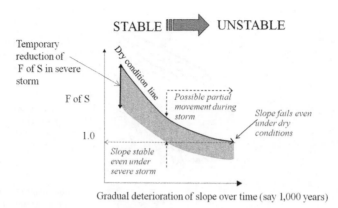

Figure 3.60 The concept of gradual ripening of slopes prior to the development of a full landslide. At some stage, the slope will reach the point where it may be moved a little by some transient process such as intense rainfall or an earthquake. Deterioration will then continue and probably accelerate until eventually full detachment occurs. As Hencher (2006) notes, hillsides can be regarded as having an inventory of different parcels of ground, all at different stages of deterioration and, therefore, susceptibility to a particular triggering event. Depending on the severity of the event, one, two or many landslides will occur.

3.7.3   Earthquakes and volcanoes

Volcanic risk is a clear problem, but sources of the hazards are generally well known, although surprises do occur, as in the case of the mud volcano that erupted disastrously in 2006 (Davies *et al.*, 2011). Clearly, if a volcano is active then construction should avoid the potential zone of travel and deposition of very hazardous materials such as lava and ignimbrite. Landslides associated with volcanoes are called lahars, which can travel great distances and be hugely damaging. Noxious gases are produced by volcanic activity. A tragic case at Lake Nyos in Cameroon, Africa, in 1986, involved the eruption of a bubble of carbon dioxide that suffocated more than 1,700 people and 3,500 livestock in nearby villages. Avoidance is again the only real option.

Earthquakes are rather more of a general hazard in that they can occur anywhere in the world, though seismic activity is concentrated along active plate boundaries. The process of assessing earthquake hazard for a site and then design to withstand the potential shaking are dealt with in Chapters 4 and 6.

3.8   Ground models for engineering projects

3.8.1   Introduction

Ground modelling is an essential part of engineering design. The ground model for a project will mainly comprise a simplified representation of the site geology that should include all aspects that are likely to affect the project or to be affected by the project. A useful review is given in GEO (2007) and some principles are set out in Box 3-3. Ground models in this context are essentially 3D models of the geological conditions at a site, together with environmental influences and hazards. Numerical and physical models can be designed based on the conceptual ground model and might be used in the development of ground models, for example, in simulating the development of *in situ* stresses.

Box 3-3   Principles of ground modelling

1. Ground models for a site should be based on *adequate* interpretations of geology and hydrogeological conditions; adequate with respect to the engineering circumstances and project requirements.
2. Models should be extensive enough to include all the ground that will be affected by the works. For example, a building will stress the ground significantly to a depth of up to twice the breadth of the foundation footprint. A new dam and reservoir may influence the terrain and environment for a large area, many kilometres from the actual dam, through changes in ground water levels and perhaps induced seismicity.
3. Engineering geological models should make sense in terms of the geological history at a site. This test is sometimes failed in ground models produced by non-geologists.
4. Make sure that the models incorporate all the features of the ground important to physical performance as conditions change (e.g. increased or decreased loading by the engineering works or the application of fluid pressures).
5. The creation and testing of several simple models exploring the sensitivity of the site to various assumptions will often be more revealing than a single complex model. This is certainly true when it comes to using numerical and physical models.
6. Geotechnical engineers and engineering geologists should act as detectives in characterising a site, studying the evidence, hypothesising and testing hypotheses through the collection of additional data, including the output from numerical analyses. Several iterations may be necessary before the models are adequate (Starfield & Cundall, 1988).

A preliminary geological model based on desk study, together with geomorphological interpretation, should be used for planning ground investigation, which will allow the model to be checked and refined. Fookes (1997) suggests that a model can simply be a written description or presented as cross sections or block diagrams and plans. It might be focused on some aspect such as groundwater, geomorphology or rock structure but should be targeted at the engineering needs of the project. At a later stage, the geological model may be split into units which can be characterised in terms of engineering properties and anticipated performance. In some locations and for particular projects, rocks that are quite different in origin and age might be lumped together because they can be expected to behave in a similar manner. Complex geology does not always equate with difficult geotechnical behaviour. Conversely, an apparently simple geological profile may have subtle variations that affect the success, or otherwise, of a project.

Once the model includes the range of engineering parameters and ground conditions that need to be considered, it becomes a design model (Knill, 2002). It might then be used, for example, to decide what foundation system might be required to carry the load from a building. It is used for making decisions on how to deal with the ground conditions. The full model will include not only geological features but also other site factors, including environmental conditions and influences such as groundwater, rainfall, wind and earthquake loading, as well as anthropogenic influences such as blasting and traffic vibration.

3.8.2   *General procedures for creating a model*

The starting point of a model should usually be a three-dimensional representation of the geology of the area and to the depth that will be affected by the project or which the project may affect. The first attempt at a geological model for a site will usually be an interpretation of published maps and the interpretation of aerial photographs and satellite imagery, depending upon the location of the project. Unfortunately, for some projects, that is as far as the geological interpretation goes, sometimes with disastrous results because the maps are either incorrect or at such small scale that they cannot represent the site-scale features that will affect the project. Those working in civil engineering need to appreciate that all published geological maps are professional interpretations of relatively small pieces of reliable data that are then interpolated and extrapolated. Faults may have been interpreted from lineaments and might not exist in reality. Conversely, published maps will certainly not show all the major geological discontinuities that may be significant for an engineering project. Most features on geological maps are generally marked as uncertain or inferred but that does not stop the unwary assuming that they are accurate. In all cases, maps and plans that are not site- or project-specific, should be taken as indicative only and a starting point for detailed investigation, as discussed in more detail in Chapter 4. Despite the inevitable limitations of published information, a broad understanding of the geological and geomorphological setting can be used to make predictions of what might be encountered at the site through experience and training. For example, if the site includes a granitic intrusion, then one might expect certain joint styles in the granite, a metamorphic aureole around the intrusion where the granite has cooked the country rock, and associated minor dykes and hydrothermal alteration, as illustrated in earlier sections of this chapter. In limestone country, one should expect caves and open fissures, perhaps infilled with secondary sediments, even where they have not actually been sampled at the site. The use of earth science skills to interpret the available data as a site history is clearly important, yet sometimes lacking in civil engineering practice (Brunsden, 2002). The best source of information on what might be anticipated is the geological literature – textbooks on physical and structural geology and sedimentology in particular, as background, together with local geological reports and memoirs, and there are no real short cuts.

3.8.3   Fracture networks

A particular problem with modelling rock masses is defining the fracture network. As addressed earlier in this chapter, consideration of geological origin and an appreciation of the history of development of fractures can be important for creating a realistic model. In reality, most fracture models, be they for assessing rock strength or permeability, are generated statistically based on orientation data. Persistence is extremely difficult to judge and most such models start off essentially as geological guesswork that can be adjusted and modified as field test data are collected, say in the petroleum industry or from large-scale pump tests associated with water supply or nuclear waste investigations. Particular techniques are used for discrete fracture network (DFN) modelling as in software packages such as FracMan.

3.8.4   Examples of models

A simple model for a cut slope alongside a road is shown in Figure 3.61. The mass of rock and soil has been split into five units, largely on the basis of strength factors, as discussed in later chapters, but discrete and possibly important elements, such as major adverse discontinuities along which a landslide could occur, are identified for special consideration. A simple model of ground conditions for the design of foundations of a building is illustrated in Figure 3.62. The various units will each give some support to the building and the ground-structure interactions need to be assessed if the foundations are to be designed cost-effectively. Models should not be overly complex but must account for all the important features at the site, including apparently 'minor geological details' that have 'major geotechnical importance', such as individual weak discontinuities along which slippage could occur (Terzhagi, 1929; Baecher & Christian, 2003). Many such features can be searched for specifically during investigation, provided there is a proper appreciation of the geological and anthropogenic history of the site, as discussed in Chapter 4.

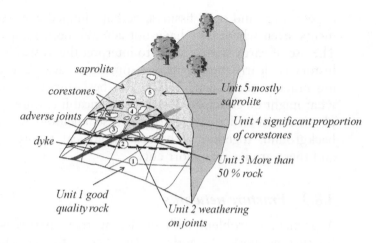

Figure 3.61 A schematic ground model for a weathered rock slope with units defined by degree of weathering, percentage of corestones and degree of fracturing. Particular features, such as dykes, faults and adverse master joints, need to be included as individual entities.

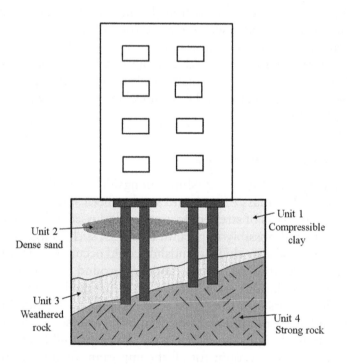

Figure 3.62 Preliminary ground model for foundation design, with key geological elements identified for geotechnical characterisation re physical parameters and behaviour.

The model with assigned geotechnical parameters becomes a design model, which is then used to predict the interaction between the structure and the ground, for example, from a building load, to ensure that failure will not occur (ultimate limit state) and that deformation will be within the tolerance of the structure (serviceability limit state). It must allow failure and deformation mechanisms to be correctly identified – a mechanism model – whereby performance is predicted in response to changes brought about by the engineering works or in the long term from environmental impacts, including rainfall, rises in groundwater pressures and vibration shocks. There may be a long sequence of events contributing to the outcome, and the geotechnical team need to predict this sequence or to recognise the sequence when carrying out forensic studies of failures. The need for a dynamic approach rather than just static is illustrated by the concept of bore pile design. A pile can be designed to carry load through skin friction, as well as through end bearing, but the designer needs to take account of how the load is taken up sequentially. In reality, depending on the geological conditions and geometry of piles, the skin friction in the upper part of a long pile will come into play in carrying the load from the structure, long before any load will reach the toe of the pile; appreciation of this process will help the designer produce a cost-effective solution.

The design model might be used as the basis for predictive numerical modelling, but quite often a conceptual model can be used in its own right to identify the hazards and the best way to proceed at a site. When failures occur, it is often the conceptual engineering geological model that helps to explain what has happened. The parameters are very much secondary – it is rare to be able to be very certain regarding parameters such as strength and water pressures and often a range of possible solutions will fit the facts equally well (Lerouiel & Tavernas, 1981). An example of an engineering geological model used to explain a major landslide is given in Chapter 7 (the Pos Selim landslide, Malaysia) and discussed in Malone *et al.* (2008). Experience tells us when a model makes sense in terms of likely strengths and mechanisms. A complex 3D modelling exercise would often only serve to confirm what we can already tell by judgement. Examples of conceptual models used to explain the observed facts from two landslides are given in Boxes 3-4 and 3-5.

Box 3-4   Mechanism model for a landslide on Tsing Yi Island, Hong Kong

Facts to be taken into account

- The failure occurred in August 1982, several days after heavy rain (Choot, 1983). It constituted general distress in an elliptically shaped area in the centre of a cut slope, with kicking out at the toe by about 750 mm (Figures B3-4.1 and B3-4.2).
- No movement had been detected previously, despite even more severe rainfall in May 1982. However, one piezometer in the centre of the distressed area had begun to show positive pore pressures since the May rainstorm.
- The failure was in severely weathered granite with daylighting (partly clay-infilled discontinuities).

Dammed, boulder filled river valley, issuing water

Area of distress

Figure B3-4.1 View of distressed area with dammed river valley in background (as discussed below).

Figure B3-4.2 View of landslide from below. The toe has kicked out about 750 mm and water was seen issuing from this surface for several days.

- The rear scarp was defined by a near vertical set of joints with thick slickensided infill of kaolin and manganese dioxide, indicative of intermittent opening up and movement (Figure B3-4.3).
- Water was observed issuing from the toe of the distressed area soon after the movement but this dried up after several days.
- The permanent water table was several metres below the level of distress, as indicated by constant seepage from the lower part of the cut slope.

- To the left of the rear scarp, facing the slope, soil was exposed containing 70% angular boulders and cobbles.
- The boulder-rich area extended to the left of the distressed area, though this was mostly covered with concrete. The boulder zone was discrete, V-shaped and wet, with vegetation growing from cracks (Figure B3-4.4).

Figure B3-4.3 Relict joints in the rear of the landslide with thick, slickensided kaolin and manganese dioxide infill.

Figure B3-4.4 View of dammed river valley to left of failure.

Developing a mechanism model

What is the slip surface?	Apparently sliding on relict joints.
What triggered the movement?	Timing suggests it was associated with heavy rainfall, but delayed, and this needs to be explained. Also, there is suspicion that some opening up might have been caused by the more intense May storm, because of the water measured in the failure zone, although site staff had not reported any signs of distress.
If it was triggered by groundwater, then why did movement stop?	Probably because the rock mass dilated, reducing water pressure, together with a reduction in water supply to the distressed area. This is supported by the observed cessation of water flow from the toe of the distressed area, some days after the movement.
So where was the water from?	The permanent water table is below the level of failure so the most likely source is throughflow from the boulder-filled feature to the left of failure.
So what is that feature?	Aerial photo interpretation indicates that the boulder zone is the cross section of an old river valley adjacent to the landslide.

So the explanation is probably as illustrated in Figure B3-4.5.

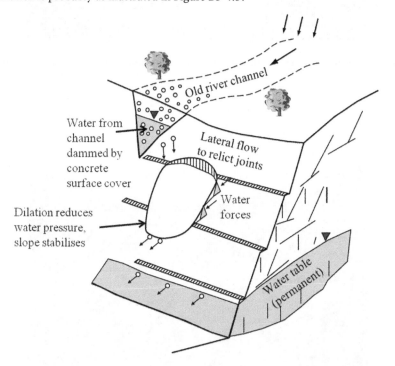

Figure B3-4.5 Proposed model for the failure.

Predisposing factors

1.   Weathered granite with adverse daylighting joints, some clay-infilled (possibly associated with long-term deterioration).
2.   River valley with large catchment, adjacent to slope.

Likely mechanism

1. After cutting of slope for road construction, the slope was coated with concrete. The concrete dammed the truncated river valley seen in Figure B3-4.4.
2. Probably minor movements had been occurring intermittently for many years, as evidenced by infilled and slickensided joints.
3. The May 1982 rainstorm probably resulted in high dammed water levels in the old river valley, feeding water laterally into the eventual failure area. This led to dilation of the rock mass and increased permeability, so that a piezometer that had been dry became responsive after May 1982.
4. The August 1982 storm led to a rise in water in the dammed river valley. It took several days for the critical pressures to develop.
5. Water moved laterally into the adjacent rock mass. Water pressures reduced effective stress and initiated failure.
6. The slope reached a metastable condition as the rock mass dilated and water pressures were dissipated.

Coda

The distressed area was cleaned off and the slope further investigated. Despite a lack of any 'useful piezometer information', the engineers responsible for the slope decided to install more than 100 deep and expensive caissons and interconnecting drains in three lines, right across the slope, rather than concentrating on the most likely sources of shallow infiltration and throughflow, such as the old river valley. In August 1984, during re-profiling, major wedge failures occurred by sliding on clay-filled joints (Figure B3-4.6). The remedial measure adopted (apart from the deep drainage system already instigated) was to cut back the slope as far as feasible and to install more than 100 permanent ground anchors, each stressed to more than 1,000 kN through a grillage of ground beams (Buttling, 1986).

Figure B3-4.6 View of slope during remedial works and new, joint-controlled failures.

Box 3-5   Mechanism model for a landslide on Tuen Mun Highway, Hong Kong

Facts to be taken into account

−   The failure occurred in May 1982, following about 400 mm of rain in a few hours.
−   Three failures occurred in the upper level of a cut slope (Figure B3-5.1).
−   A piezometer had been installed through the site of the main failure prior to construction of the road. Measurements indicated that the main water table was below the failure levels and this was confirmed by seepage points in the lower slopes.
−   The failures were in predominantly highly and completely decomposed granite (mostly sand-sized). Two dolerite dykes in the vicinity of the failure were decomposed to fairly uniform silt-sized material (Figures B3-5.2 and 5.3).

Figure B3-5.1 View of section of Tuen Mun Highway at Ch. 6750 with two major and one minor landslide.

Figure B3-5.2 View of main landslide with dolerite dyke shallowly dipping out of slope through weathered granite.

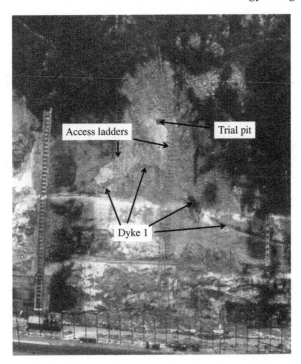

Figure B3-5.3 Close up of left flank of landslide with two dykes, the main one dipping out of the slope and a second dipping back into the slope. Piezometer tube was installed at time of road construction (about 1976).

– A series of triaxial and direct shear tests were carried out. Back-analysis demonstrated that the strength of intact material was too high to have permitted failure in the absence of adverse groundwater pressures.
– Adverse discontinuities were recorded in the second largest failure but were not persistent.

Develop the mechanism model

Mechanism?	Apparently through the intact weathered rock for the main failure; relict discontinuities may have reduced the mass strength for the second failure.
What triggered the movement?	The failures were associated with intense rainfall, although timing was not known precisely. Vegetation above the slope was flattened towards the slope, suggesting considerable surface flow.
But measured and observed water table was below failure surface?	Nevertheless, back analysis suggests that failure was impossible without adverse positive water pressure.

So the explanation is probably the following.

Predisposing factors

Weathered granite with cross-cutting dolerite dykes. There would be a permeability contrast, the dykes acting as aquitards.

Mechanism

1. The geological model of the main failure is shown in Figure B3-5.4.
2. The main landslide was probably caused by direct infiltration, during or shortly after the intense rainfall.

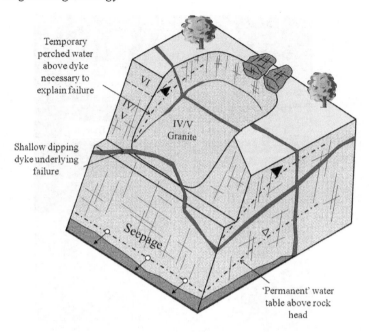

Figure B3-5.4 Geological model of landslide with normal and temporary perched water tables invoked to explain failure.

3. Water infiltrated as a wetting band, until it reached the shallow dipping dyke underlying the main landslide scar, leading to increased water pressure.

4. Figure B3-5.5 shows a cross section through the failure with three postulated piezometric surfaces, and Figure B3-5.6 shows the results from numerical analysis; the main curves show solutions for Factor of Safety = 1.0 for different strength conditions and for each trial piezometric surface individually. Although a range of strengths were measured for the decomposed granite, the most likely strength was $\varphi' = 36$ degrees, $c' = 5$ kPa. If that was correct, then the failure would have been triggered by a piezometric pressure somewhere between surfaces 2 and 3.

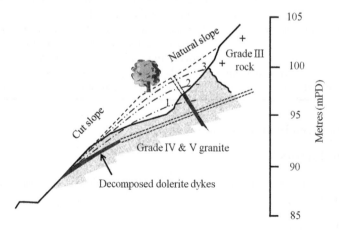

Figure B3-5.5 Cross section through failure with various levels of water used in back analysis to try to explain the likely conditions when the landslide occurred.

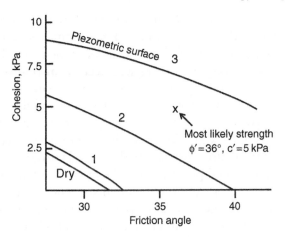

Figure B3-5.6 Results from numerical analysis. Each point along the lines for the various trial piezometric surfaces gives a FoS of 1.0. From testing and empirical data, the most likely field strength is φ = 36 degrees and c´ = 5 kPa, which would indicate that the likely perched water pressure was somewhere between levels 2 and 3 at the time of failure (Hencher (1983b), Hencher & Martin (1985)).

An example of a simple ground model prepared for the design of a real project is given in Figure 3.63. When designing a tunnel one needs to predict the ground conditions along the route so that one can decide what tunnelling method needs to be adopted, as discussed in detail in

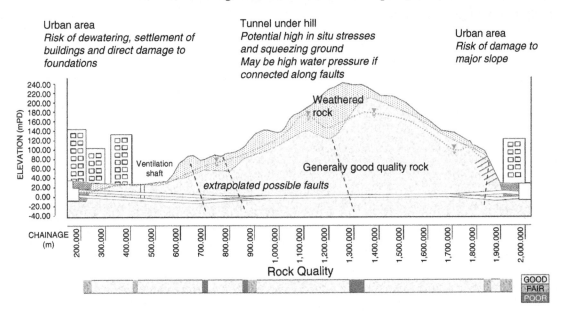

Figure 3.63 Ground model for tunnel. Predicted rock mass characteristics can be used to estimate the amounts of reinforcement, such as rock bolts and shotcrete, that will be required. It will also be used during construction, as part of the risk control, probing ahead, as necessary, to establish zones of hazardous ground. As illustrated in Chapter 7, tunnelling remains a risky endeavour because very rarely are ground investigations adequate for characterising the ground along their length.

Chapter 6. Another requirement of the model is to allow the support requirements to be predicted along the tunnel. For example, in a long drill and blast/hand-excavated tunnel, some sections of the ground, in good rock, may need little support, others will need local reinforcement to prevent rock blocks falling. In other sections, through weak ground or where there are high water pressures, the tunnel might need a thick reinforced concrete liner. A ground model needs to be prepared that includes predictions of rock quality along the route. These predictions are needed so that the team constructing the tunnel know what to expect and where to take special precautions. The predictions are also necessary so that the contractor can prepare a realistic tender price for the construction and as a reference so that all parties can judge whether conditions were more difficult than might have been anticipated, so that additional payment can be made to the contractor, if appropriate, as discussed in Chapter 2. The model also needs to allow the influence of the works on other structures to be assessed. In Figure 3.63, each end of the tunnel terminates in urban areas. The problems of noise, vibrations, dewatering and physical interaction with existing foundations and slopes need to be considered and these are all aspects where a comprehensive ground model is essential. Where the tunnel passes underneath the hill, there may be unusually high lateral stresses causing squeezing on the tunnel. The possible faults identified along the route may be associated with particularly poor ground and possibly high inflow of groundwater.

4   Site investigation

'... if you do not know what you should be looking for in a site investigation you are not likely to find much of value.'

(Glossop, 1968)

This much-quoted quote is worth repeating because it sums up the philosophy of site investigation very well. Critical features need to be anticipated and looked for. Without care, the important details might be hidden within a pile of essentially irrelevant information. The difficulty and skill, of course, is in recognising what is critical.

4.1   Nature of site investigation

At any site, the ground conditions need to be assessed to enable safe and cost-effective design, construction and operation of civil engineering projects. This will generally include sub-surface ground investigation (GI), which needs to be focused on the particular project needs and unknowns. The requirements for GI will be very different for a tunnel compared to the design of foundations for a high-rise building or for stability assessment of a cut slope. There needs to be a preliminary review of the nature of the project, the constraints for construction and the uncertainties about the engineering geological conditions at the site. The British Code of Practice for Site Investigation, BS 5930 (BSI, 1999), sets out the objectives broadly as follows:

1. *Suitability*: to assess the general suitability of a site and its environs for the proposed works.
2. *Design*: to enable an adequate and economic design, including for temporary works.
3. *Construction*: To plan the best method of construction and, for some projects, to identify sources of suitable materials such as concrete aggregate and fill and to locate sites for disposal of waste.
4. *Effect of changes*: to consider ground and environmental changes on the works (e.g. intense rainfall and earthquakes) and to assess the impact of the works on adjacent properties and on the environment.
5. *Choice of site*: where appropriate, to identify alternative sites or to allow optimal planning of the works.

4.2   Scope and extent of ground investigation

4.2.1   *Scope and programme of investigation*

The scope of site investigation is set out in Box 4-1. This should include everything relevant to use of the site, including site history and long-term environmental hazards and not just geology. All authorities (e.g. AGS, 2006) agree that site investigation should, ideally, be carried out in stages, each building on the information gained at the previous stage, as outlined in Box 4-2. A preliminary engineering geological model should be developed for the site from desk study and field reconnaissance, as outlined in Chapter 3. That model should then be used to consider the project constraints and optimisation (e.g. the likely need for deep foundations or the best location for a dam) and for designing the first phase of GI. For a large project, this first phase is usually carried out during the conceptual phase. Further GI campaigns might be carried out for basic design, for detailed design and often additional works during construction. Engineering geologists should readily appreciate that all sites do not require the same level of ground investigation. Some have simple ground conditions, others more complex. At some locations, existing exposures will allow the broad geology to be assessed and reduce the need for GI. Projects may be situated in areas where the geology and ground conditions are already well understood. For example, if designing piles in London Clay, because of the wealth of published data and industry experience, GI requirements should be fairly routine[1] – little should be needed in the way of testing to determine parameters for design.

Taking this further, experience shows that the majority of sites world-wide do not have any particularly inherently hazardous conditions and might be categorised as *forgiving*. Even with no, or no competent investigation, the project is often completed without geotechnical difficulty. Such sites need little investigation – enough to establish that there are no particularly adverse hazards. In a review of the scope of ground investigations for foundation projects in the UK, Egan (2008) found that GI was either not conducted or was lacking borehole plans for 30% out of 221 projects, but he reported no adverse consequences. In other words, the engineers took a risk, perhaps on the basis of previous experience in an area, and apparently got away with it, although, as Egan points out, a ground investigation might have allowed more cost-effective solutions. Unfortunately, the world also has relatively rare *unforgiving* sites with inherently difficult geotechnical conditions that need careful and insightful investigation if problems are to be

---

[1] It does not follow that London Clay is without hazards for construction projects, for example, the Heathrow Express Tunnel collapsed during construction, as discussed in Chapter 7. De Freitas (2009) also provides a warning over geological variation through the London Clay stratum and argues that data banks of geotechnical properties need to be used with care from one area to another.

avoided. The big problem is identifying whether any particular site is unforgiving and in what way. It is the task of the engineering geologist, through his knowledge of geological processes, to anticipate hazardous geological conditions and to make sure that a GI is properly focused. A checklist approach to hazard prediction is advocated below.

Box 4-1   Overall scope of site investigation

1. **Hazards and constraints during construction and in the longer term**
   - Previous site use – obstructions, contamination
   - Any history of mining or other underlying or adjacent projects (e.g. tunnels or pipelines)
   - Sensitive receivers – such as neighbours that might be affected by noise, dust, vibration and changes in water levels
   - Regulatory restrictions
   - Natural hazards, including flooding, wind, earthquakes, subsidence and landslides

2. **Assess and record site characteristics**
   - Access constraints for investigation and construction
   - Need for traffic control, access for plant and waste disposal
   - Access to services
   - Site condition survey (partly as a record for any future dispute)

3. **Geological profile at site**
   - Distribution and nature of soil and rock underlying the site, to an adequate degree, to allow safe and cost-effective design
   - Usually this will require a sub-surface ground investigation

4. **Physical properties of soil and rock units and design parameters**

   Key parameters:
   - mass strength (to avoid failure)
   - deformability (to ensure movements are tolerable)
   - permeability (flow to and from site, response to rainfall and loading/unloading)

   Other factors:
   - chemical stability (e.g. reactivity in concrete, potential for dissolution)
   - potential for piping and collapse
   - abrasivity (sometimes a major consideration for construction)

5. **Changes with time**
   - install instruments to check physical nature of the site – e.g. groundwater response to rainfall
   - install instruments to monitor settlement and effect on adjacent structures during construction
   - consider the potential for deterioration and need for maintenance

Box 4-2   Stages in a site investigation

Stage 1: Desk study at project conception stage

- Identification of key geological and environmental hazards at optional sites based on broad desk study and possibly site visits.
- Consider site constraints, engineering considerations and economic factors.

Stage 2: Detailed desk study and reconnaissance survey

- Collect and review all documents relevant to the preferred site, including topographic and geological maps, aerial and terrestrial photographs and any previous investigation reports. Review site history including previous building works and mining. Look for hazards such as landslides.
- Site mapping, possibly with advance contract allowing safe access, vegetation clearance and trial pits or trenches.

The Preliminary Ground Model

Develop a preliminary geological and geotechnical working ground model that can be used as a reference for the rest of the ground investigation.

This preliminary model should be used as a reference by all the team, including those logging boreholes and trial pits. The loggers need to know what to expect and to be able to identify anything that necessitates revisions to the ground model.

Site-specific ground investigation should be aimed at verifying the model, answering any unknowns and allowing design parameters to be derived.

Stage 3: Preliminary ground investigation linked to basic engineering design

- Consider use of geophysical techniques to investigate large areas and volumes.
- Preliminary boreholes designed to prove geological model (rather than design parameters).
- Instrumentation as appropriate (e.g. to establish groundwater conditions and seismicity).

Stage 4: Detailed ground investigation

- Further investigation to prepare detailed ground model and allow detailed design.
- *In situ* and laboratory testing to establish parameters.
- Detailed instrumentation and monitoring.

Stage 5: Construction

- Review of ground models during construction (including logging of excavations).
- Testing to confirm design parameters.
- Instrumentation to monitor behaviour and check performance against predictions.
- Revision to design as necessary.

Stage 6: Maintenance

- Ongoing review – e.g. of settlement, slope distortion, groundwater changes and other environmental impacts, possibly linked to a risk management system.

Typically, the cost of a site investigation is only a small part of the overall project cost (less than a few percent), yet clients often require some persuasion that the money will be well spent and might be especially reluctant to allow a staged approach because of the impact on programme. He might be reluctant to allow thinking and planning time as the GI data are received and especially unwilling to pay for a revised design as the ground models are developed and refined. Sometimes the engineer might adopt a fast-track approach whereby GI, design and construction are carried out concurrently, although this approach carries the risk that information gained later might impact on earlier parts of the design and even on constructed parts of the works. The programming can sometimes go awry, as on a site in Algeria where the author was trying to set out locations for drilling rigs in the same area as a contractor was preparing to construct foundations which obviously did not make sense. It turned out that design engineers had made assumptions about the ground conditions without waiting for GI, thinking that surface footings would be adequate. This proved incorrect and the design needed complete revision. In a similar manner to fast tracking, an observational approach is sometimes adopted, especially for tunnelling, whereby ground conditions are predicted, often on rather sparse data, and provisions made for change if and when ground conditions turn out to be different from those anticipated (Powderham, 1994). The observational method often relies on instrumentation of ground movements, measured loads in structural members, or water levels, whereby performance is checked against predictions. This can go seriously wrong where the ground behaves outside predictions – perhaps because the geological model is fundamentally incorrect or because instrument systems fail or are not reacted to quickly enough. Examples where instruments were not reacted to early enough include the Heathrow Express Tunnel (Muir Wood, 2000) and the Nicholl Highway collapse in Singapore (Hight, 2009); these are described in some detail in Chapter 7. An observational approach should also generally be adopted for rock slope construction, although it is seldom referred to as such. Basically, it is very difficult to characterise the complete rock fracture network from a few boreholes and therefore it is very important to check any design assumptions during construction and to be prepared to come up with different solutions for stabilisation as the rock is exposed and structures identified and mapped (see Box 1-1).

4.2.2   *Extent of ground investigation*

A large part of any site investigation budget will generally be taken up in sub-surface investigation and characterisation of the ground conditions (Items 3 and 4 in Box 4-1). Important questions are, how much ground investigation is required and how should it be done? There are no hard and fast rules, even though some authors try to provide guidance on the basis of site area or volume for particular types of operation (e.g. Figure 4.1 for dredging) or on hypothetical considerations (e.g. Jaksa *et al.*, 2005). In reality, it depends upon the complexity of the geology at the site, how much is already known about the area, the nature of the project and cost. For sites with simple geology, the plan might be for boreholes at 10m to 30m spacing, for discrete structures like a building (BS 5939: 1999). For a linear structure like a road or railway project, the spacing might be anywhere between 30 and 300m spacing, depending on perceived variability (Clayton *et al.*, 1995). West *et al.* (1981) consider the particular difficulties in planning investigations for tunnels. So much depends upon the depth of tunnel, the topography and variability of geology. Often, considerable reliance is made on aerial photography interpretation, geological mapping, a few widely spaced preliminary boreholes and other boreholes targeted at particular perceived hazards such as faults that might be associated with poor quality rock and high water inflows. For example, Figure 4.2 shows the route of a planned tunnel in Hong Kong, with potential hazards identified, together with a rationale for their mitigation and additional GI. Where steeply dipping geological structures such as faults are anticipated, inclined boreholes may be required. Figure 4.3 shows an

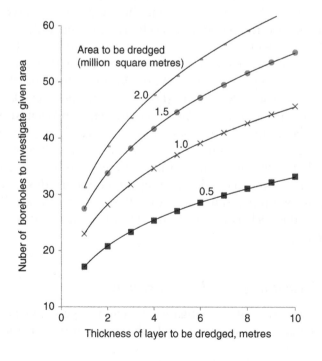

Figure 4.1 Number of boreholes for dredging area (in millions of square metres) vs. average thickness of material to be removed, based on equation of Bates (1981), as presented in PIANC (2000). Other factors that should be taken into account are variability of ground conditions and existing knowledge about the area.

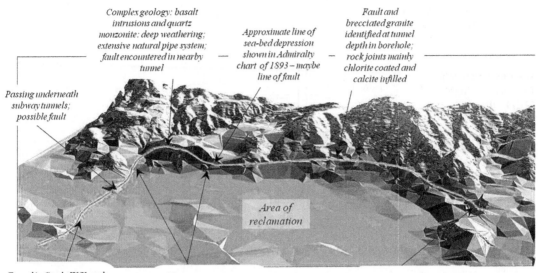

Complex geology: basalt intrusions and quartz monzonite: deep weathering; extensive natural pipe system; fault encountered in nearby tunnel

Approximate line of sea-bed depression shown in Admiralty chart of 1893 – maybe line of fault

Fault and brecciated granite identified at tunnel depth in borehole; rock joints mainly chlorite coated and calcite infilled

Passing underneath subway tunnels; possible fault

Area of reclamation

Tunnel in Grade IV/V rock below newly reclaimed areas with high water pressure
- Delay, water inflow and settlement; difficult formation of shaft and launching TBM
- Pretreatment and local pressure relief
- Slurry mode excavation

Valleys with shallow rock cover and comparatively thick soil mantle
- Instability, water inflow, delay and may affect existing subway tunnels
- Further GI required
- Ground pre-treatment
- Instrumentation and close monitoring

Complex geology: 3 inferred faults, rock joints locally infilled with kaolin and calcite and chlorite coated
- Instability, water inflow & delay; difficult formation of shaft and retrieving TBM
- Further GI required

Figure 4.2 Preliminary assessment of ground investigation requirements for a new tunnel, Hong Kong.

assessment of possible conditions under the Eastern Tower of Stonecutters Bridge in Hong Kong at tender stage, based on desk study together with a proposed borehole investigation targeted at likely faults and zones of deep weathering. Some broad details of what was actually found are given in Fletcher (2004) and consequences by Tapley *et al.* (2006).

Requirements and practice for GI vary around the world. In Hong Kong, for example, it is normal practice to put down a borehole at the location of every bored pile (called a pre-drill). Elsewhere, a pattern of perhaps three, four or five boreholes might be adopted below each pile cap for a major structure. For example, for the 2nd Incheon Bridge in South Korea, opened in 2009, for each of the main cable stay bridge towers there were four boreholes per pile cap, each of which was about 70 by 25 m in plan and supported by 24 large-diameter bored piles. For the Busan-Geoje fixed link crossing, completed in 2010, also in South Korea, there were two cable-stay bridge sections, one with two towers and main span of 475 m, the other with three towers. The towers were founded on gravity caissons sitting

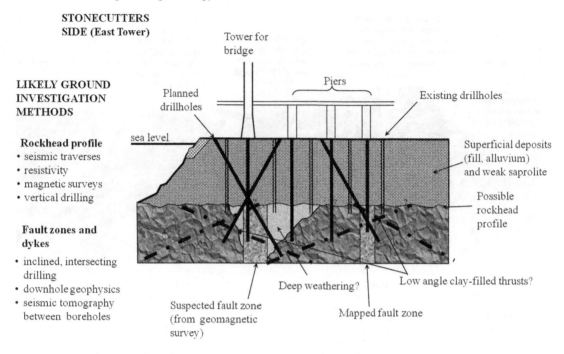

STONECUTTERS
SIDE (East Tower)

Tower for
bridge

Piers

Planned
drillholes

Existing drillholes

**LIKELY GROUND
INVESTIGATION
METHODS**

sea level

Superficial deposits
(fill, alluvium)
and weak saprolite

Rockhead profile
• seismic traverses
• resistivity
• magnetic surveys
• vertical drilling

Possible
rockhead
profile

**Fault zones and
dykes**

• inclined, intersecting
  drilling
• downhole geophysics
• seismic tomography
  between boreholes

Deep weathering?

Low angle clay-filled thrusts?

Suspected fault zone
(from geomagnetic
survey)

Mapped fault zone

Figure 4.3 Preliminary assessment of ground conditions by Halcrow for East Tower of Stonecutters Bridge, Hong Kong, and need for inclined boreholes to investigate major fault structures.

on excavated rock (Chapter 6) and with plan dimensions of up to 40× 20m. For each of these foundations, there were usually about six boreholes, typically one put down at the conceptual stage, three for the basic design and two for detailed design. For most of the other viaduct piers with plan caisson dimensions of 17 × 17m, there were from one to five boreholes – less where the geology was better known, close to shore.

Obviously, where the site reconnaissance, together with desk study or findings from preliminary boreholes, indicate potentially complex and hazardous conditions, it may prove necessary to put down far more boreholes. For the design of the new South West Transport Corridor near Brisbane, Australia, the preliminary investigation over a critical section of more than 500 m comprised five boreholes and a few trial pits, mostly along the centre line of the road. As the earth works were approaching completion, minor landslides occurred at road level, together with some indications of deeper-seated movements. Over the next few months, an additional 70+ deep boreholes were put down, 56 trial pits and 54 inclinometers installed, despite almost 100% rock exposure in the cuttings (which was carefully examined and mapped). This intensive investigation allowed the landslide mechanisms to be identified in this very complex site and

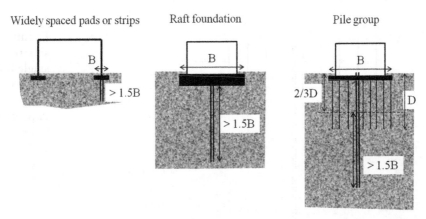

Figure 4.4 Criteria usually adopted for investigating the ground for foundations. Where geology is or may be complex, ground conditions might need to be proved to greater depth and several boreholes might be required. Similarly, these criteria do not apply or limit the need to consider particular site hazards, such as slope stability above or below the site.

remedial works to be implemented, which permitted the project to be completed on time (Starr *et al.*, 2010). In hindsight, the preliminary boreholes, which would have been more than adequate for a normal stretch of road, gave no indication of the degree of difficulty and complexity at this unforgiving site, which only became clear following intensive work involving a wide range of experts. In a similar manner, the landslide at Pos Selim, Malaysia, described in Chapter 7, could not have been anticipated from a few boreholes. The mechanism was at a very large scale and involved too many components to have been understood before the major displacements occurred.

As a general rule, at any site, at least one borehole should be put down to prove ground conditions to a depth far greater than the depth of ground to be stressed significantly by the works. Generally, for foundations, at least one borehole should be taken to at least 1.5 times the breadth (B) of the foundation (Figure 4.4). For pile groups, it is generally assumed that there is an equivalent raft at a depth of 2/3D where D is the length of piles and the ground should be proved to at least 1.5B below that level. This is only a general guideline – if there is any reason to suspect more variable conditions and, where the geology is non-uniform, one borehole will probably not be enough (Figure 4.5). Poulos (2005) discusses the consequences of 'geological imperfections' on pile design and performance. Boreholes are often terminated once rock has been proved to at least 5 m, but this may be inadequate to prove bedrock in weathered terrain (Hencher & McNicholl, 1995). Whether or not one has reached *in situ* bedrock might be established by geological interpretation of consistent rock fabric or structure across a site, but elsewhere it may be more difficult, in which case it

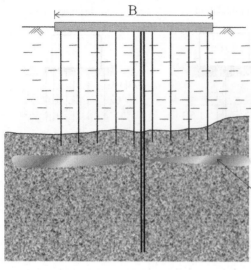

Borehole meets usual criteria in terms of depth

soft stratum
missed by
borehole
(perhaps as
lost core)

Figure 4.5
Example of
situations where a
single borehole (or
few boreholes)
might miss
important
information that
will affect the
integrity of the
structure.

is best to take one or more boreholes even deeper if important to the design.

4.3   Procedures for site investigation

4.3.1   General

Guidance on procedures and methodologies for site investigation is given for the UK by Clayton *et al.* (1995) and for the USA and more broadly by Hunt (2005). The British Code of Practice for Site Investigations, BS 5930 (BSI, 1999), provides comprehensive advice on procedures and techniques and for soil and rock description for the UK. Other codes exist for different countries (e.g. Australia, China and New Zealand). Generally, there is consistent advice over the overall approach to site investigation, although terminology and recommended techniques differ. All agree, however, that the first step should be a comprehensive review of all available maps and documents pertaining to a site – this is called a desk study.

4.3.2   Desk study

4.3.2.1   Sources of information

For any site, it is important to conduct a thorough document search. This should include topographic and geological maps. Hazard maps are sometimes available. These include broad seismic zoning maps for countries linked to seismic design codes. In some countries, there are also local seismic micro-zoning maps showing locations of active faults and hazards such as liquefaction susceptibility. Sources of information for the UK are given in BS 5930 (BSI, 1999) and Clayton *et al.*, 1995. The Association of Geotechnical and Geoenvironmental Specialists (AGS), whose contact details are given in Appendix A, also give useful advice and sources of reference. Records of historical mining activity and previous land use are especially important. In the UK, the British Geological Survey (BGS) has made available a digital atlas of hazards, including mining (but not coal), collapsible materials, swelling and compressible soils, landslides and noxious gas. Landslide hazard maps are published in the USA for southwest California and in Hong Kong, as discussed below.

4.3.2.2 Air photograph interpretation

Air photographs can be extremely useful for examining sites. Pairs of overlapping photographs can be examined in 3D using stereographic viewers, and skilled operators can provide many insights into the geology and geomorphological conditions (Allum, 1966; Dumbleton & West, 1970). Historical sets of photographs help to reveal the site development and to assess the risk from natural hazards such as landslides. In Hong Kong, it is normal practice to set out the site history for any new project through air photo interpretation (API) of sets of photos dating back to the 1920s. The role of API in helping to assess the ground conditions at a site is illustrated in Box 4-3.

Box 4-3 Role of air photo interpretation (API)

Overlapping air photos allows a skilled earth scientist to examine the site topography in three dimensions. According to Styles (personal communication), in order to do it well you must put yourself on the ground mentally and walk across the terrain looking around in oblique perspective. Topographic expression and other features such as the presence of boulders, hummocky ground, arcuate steps and vegetation, can be interpreted in terms of terrain components and geomorphological development: landslide morphology, degree of weathering, and distribution of superficial deposits such as colluvium and alluvium. Broad geological structure such as major joint systems, faults and folds, may be observed, interpreted and measured in a way that would be more difficult working only by mapping exposures on the ground (Figure B4-3.2).

Where landslides are identified on photographs, debris run-out can be measured, which may help in assessing the degree of risk for existing and future developments. River channels can be traced and catchments measured. Where a series of historical photographs is available, an inventory of landslide events can be compiled and related to historical rainfall records. Anthropogenic development and use of sites can be documented.

It is important that API is checked by examination in the field and this is known as ground truthing, which is an integral part of site reconnaissance and field mapping. Similarly, interpreted site history should be checked and correlated against other documentary evidence such as old maps and photographs. The preliminary ground model developed from API and field studies can then be investigated further by trial pits and boreholes, as necessary. Conversely, a ground investigation in an area of variable topo-graphy, without prior API, reconnaissance and desk study, may be ineffective and poorly focused. An introduction to the use of air photographs, with particular consideration of landslide investigations, is given by Ho *et al.* (2006).

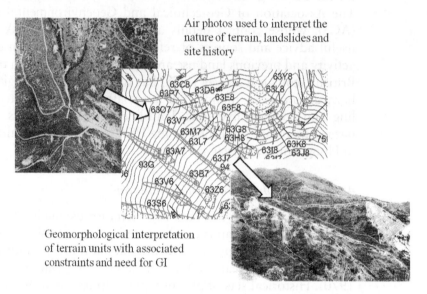

Air photos used to interpret the nature of terrain, landslides and site history

Geomorphological interpretation of terrain units with associated constraints and need for GI

Figure B4-3.1 Process of API. Pairs of overlapping photographs can be examined stereographically to give a 3D image. Major terrain features can be identified and if historical series of photographs are available, then land development and site history can be ascertained, in this example, in terms of landslide history. In the second image above, interpreted landslides have been mapped (with date of the photo in which the landslide is first seen). These interpretations can then be checked in the field (Devonald *et al.*, 2009). In addition, terrain can be split into units on the basis of surface expression, underlying geology, activity and vegetation, as described by Burnett *et al.* (1985). Third photo and overlay provided by K. Styles.

Victoria Falls

Figure B4-3.2 Major structural lineations visible in aerial photograph and controlling river development of Zambezi River above Victoria Falls between Zimbabwe and Zambia.

Even with little training, the importance of air photographs can
be immediately clear, as in Figure 4.6, which is an air photograph
from 1949 on to which has been marked the route of the Ching
Cheung Road in Hong Kong, constructed in 1963. Various ground
hazards are evident in the photo (landslides and deep gulleying) and
it is no surprise that these led to later problems with the road,
as addressed in Chapter 7 and discussed by Hudson & Hencher
(1984).

Systematic interpretation of air photographs for determining
geotechnical hazards has been carried out in several countries.
For example, the whole of Hong Kong was mapped, in terms of
perceived geotechnical hazard, from air photographs in the 1980s
at a 1:20,000 scale and locally at 1:2,500 and, whilst never
intended for site-specific interpretation, these were very useful for
urban planning (Burnett *et al.*, 1985; Styles & Hansen, 1989). Air
photos can be used for detailed measurement by those trained to do
so. Figure 4.7 shows displacement vectors for the slow-moving rock
landslide at Pos Selim, Malaysia. The 3D image was prepared from
as-built drawings and oblique air photographs taken from a heli-
copter and linked to surveyed control points. The vectors produced
(up to 15 m drop in the rear scarp) are considered accurate to about
0.2 m. Topographic surveys can also be carried out using terrestrial
or airborne LIDAR surveys and these can be repeated to monitor
ongoing movements in landslides or in volcanic eruptions
(e.g. Jones, 2006). In some situations, especially for remote sites
lacking good air photo coverage, satellite images may be helpful,
although often the scale is not large enough to provide the detailed
interpretation required and stereo imagery is impossible – unlike for
purpose-flown aerial photograph sequences. Use of false spectral

Figure 4.7
Visualisation of Pos
Selim landslide,
Malaysia, showing
displacement
vectors over a
two-year period
(after Malone *et al.*,
2008).

images such as infra-red can help interpretation, for example, of vegetation and seepage.

4.3.3 *Planning a ground investigation*

BS 5930 and most textbooks on site investigation provide good information on techniques and procedures but little advice on how to plan a ground investigation or on how to separate and characterise geotechnical units within a geological model. They also say little about how to anticipate hazards, which is a key task for the engineering geologist. It is important to take a holistic view of the geological and hydrogeological setting – the 'total geological model' approach of Fookes *et al.* (2000), as discussed in Chapter 3 – but the geological data need to be prioritised to identify what is really important to the project and to obtain the relevant parameters for safe design.

The problem is that there are so many things that might potentially go wrong at sites and with alternatives for cost-effective design that it is sometimes difficult to know where to start in collecting information. One might hope that simply by following a code of practice, that would be enough, but, in practice, the critical detail may be overshadowed by relatively irrelevant information collected following routine drilling and logging methodology.

One approach that can be useful for planning and reviewing data from a ground investigation, and focusing on critical information, is to consider the different aspects of the site and how they might affect the project in a checklist manner (Knill, 1976, 2002; Hencher & Daughton, 2000; Hencher, 2007). The various components and aspects of the project and how different site conditions might affect its success are considered one by one and in an integrated way. This is similar to the rock engineering systems methodology of Hudson (1992), in which the various parameters of a project are set out and their influence judged and measured in a relative way (Hudson & Harrison, 1992). This is also akin to the concept of a risk register for a civil engineering project at the design and construction stages, whereby each potential hazard and its consequence is identified and plans made for how those risks might be mitigated and managed. This is addressed in Chapter 6.

The three verbal equations of Knill (1976) are set out in Table 4.1. The first part is to consider geological factors: material and mass strengths and other properties. The second is to assess the influence of environmental factors such as *in situ* stress, water and earthquakes. The final consideration is how these factors affect, and are affected by the construction works. A very similar process has been proposed for addressing risk by Pöschl & Kleberger (2004), particularly for tunnels.

4.3.3.1 *Equation 1: geological factors*

The first equation encourages the investigator to consider the ground profile (geology) and its properties at both the material and mass scales.

Table 4.1 Engineering geology expressed as three verbal equations (after Knill, 1976).

Equation 1 GEOLOGY
MATERIAL PROPERTIES + MASS FABRIC ⇒ MASS PROPERTIES
The first equation includes the geology of the site and concerns the physical, chemical and engineering properties of the ground at small and large scales. It essentially constitutes the soil and rock ground conditions.
Equation 2 + ENVIRONMENT
MASS PROPERTIES + ENVIRONMENT ⇒ ENGINEERING GEOLOGICAL SITUATION
The second equation relates to the geological setting within the environment. Environmental factors include climatic influences, groundwater, stress, time and natural hazards.
Equation 3 + CONSTRUCTION
ENGINEERING GEOLOGICAL SITUATION + INFLUENCE OF ENGINEERING WORKS ⇒ ENGINEERING BEHAVIOUR OF GROUND.
The third equation relates to changes caused by the engineering works such as loading, unloading and changes to the groundwater levels. It is the job of the engineer to ensure that the changes are within acceptable limits.

MATERIAL SCALE

The material scale is that of the intact soil and rock making up the site. It is also the scale of laboratory testing, which is usually the source of engineering parameters for design. Typical factors to review are given in Table 4.2. They include the chemistry, density and strength of the various geological materials and contained fluids making up the geological profile. Hazards might include adverse chemical attack on foundations or ground anchors, liquefaction during an earthquake, swelling or low shear strength due to the presence of smectite clays, abrasivity or potential for piping failure. Inherent site hazards associated with geology include harmful minerals such as asbestos and erionite. Granitic areas, phosphates, shale and old mine tailings are sometimes linked to relatively high levels of radon gas, which is estimated to cause between 1,000 and 2,000 deaths each year in the UK (Health Protection Agency). Talbot *et al.* (1997) describe investigations for radon during tunnelling. Gas hazards are especially important considerations for tunnelling and mining but are also an issue for completed structures, as illustrated by the Abbeystead disaster of 1984 when methane that migrated from coal-bearing strata accumulated in a valve house and exploded killing 16 people (Health and Safety Executive, 1985). These are all material-scale factors linked to the geological nature of the rocks at a site.

Locating sources for aggregate, armourstone and other building materials is often a task for an engineering geologist. Other than the obvious considerations of ensuring adequate reserves and cost, one must consider durability and reactivity, and this will involve geological characterisation and probably testing. Two examples in Chapter 7 relate how adverse material properties of sourced fill and aggregate material led to severe consequences. At Carsington Dam, UK, a

Table 4.2 Examples of material-scale factors that should be considered for a project

FACTOR	CONSIDERATIONS	EXAMPLES OF ROCK TYPES/SITUATIONS
mineral hardness	abrasivity, damage to drilling equipment	silica-rich rocks and soils (e.g. quartzite, flint, chert)
mineral chemistry	reaction in concrete	olivine, high temperature quartz, etc.
	oxidation – acids	pyrites
	swelling, squeezing dissolution	mudrocks, salts, limestone
	low friction	clay-infilled discontinuities, chlorite coating
loose, open texture	collapse on disturbance or overloading, liquefaction, piping, low shear strength	poorly cemented sandstone, completely weathered rocks (V); loess; quick clays

chemical reaction was set up between the various rocks used to construct the dam, which resulted in acid pollution of river courses and the production of hazardous gas, with the death of two workers; at Pracana Dam, Portugal, the use of reactive aggregate led to rapid deterioration of the concrete. The latter phenomenon has been reported from many locations around the world and is associated with a variety of minerals, including cryptocrystalline silica (some types of flint), high-temperature quartz, opal and rock types ranging from greywacke to andesite. Details of how to investigate whether aggregate may be reactive and actions to take are given in RILEM (2003).

MASS SCALE

Mass-scale factors include the distribution of different materials in different weathering zones or structural regimes, as successive strata or as intrusions. It includes structural geological features such as folds, faults, unconformities and joints (Table 4.3). Discontinuities very commonly control the mechanical behaviour of rock masses and some soils. They strongly influence strength, deformability and hydraulic conductivity.

Table 4.3 Examples of mass-scale factors that should be considered for a project

FACTOR	CONSIDERATIONS	EXAMPLES OF ROCK TYPES SITUATIONS
lithological heterogeneity	difficulty in establishing engineering properties, construction problems (plant and methodology)	colluvium, un-engineered fill, interbedded strong and weak strata, soft ground with hard corestones
joints/natural fractures	sliding or toppling of blocks, deformation, water inflows, leakage/migration of radioactive fluids	slopes, foundations, tunnels and reservoirs, nuclear repository
faults	as joints, sudden changes in conditions, displacement, dynamic loads	tunnels, foundations, seismically active areas
structural boundaries, folds, intrusions	heterogeneity, local stress concentrations, changes in permeability – water inflows	all rocks/soils
weathering (mass scale)	mass weakening; heterogeneity (hard in soft matrix), local water inflow, unloading fractures	all rocks and soils close to Earth's surface, especially in tropical zones; ravelling in disintegrated rock masses
hydrothermal alteration	as weathering, low strength and prone to collapse especially below water table	generally for igneous rocks especially near contacts

One of the main geological hazards to engineering projects at the mass scale is faults. Faults can be associated with zones of fractured and weathered material, high permeability and earthquakes. Alternatively, faults can be tight, cemented and actually act as barriers to flow, as natural dams rather than zones of high permeability. Faults should always be looked for and their influence considered. There are many cases of unwary constructers building on or across faults, with severe consequences, sometimes leading to delays to projects or a need for redesign. Consequence is sometimes difficult to predict but should be considered and investigated. Other examples of mass factors that would significantly affect projects include boulders in otherwise weak soil, which might preclude the use of driven piles or would comprise a hazard on a steep slope.

An example of where a formal review of the potential for large-scale structural control might have helped is provided by the investigation for a potential nuclear waste repository at Sellafield in the UK, as explained in Box 4-4. It appears that early boreholes and tests did not sample relatively widely spaced master joints within the stratum and, therefore, an incomplete picture was formed of the factors controlling mass permeability. In hindsight, the true nature of the rock might have been anticipated by desk study and field reconnaissance of exposures.

Box 4-4   Anticipating mass characteristics: the Brockram and the Sellafield Investigations

The UK Government specification for acceptable risk from any nuclear waste repository was set to be extremely onerous and necessitated intensive investigation combined with intensive modelling. Ground investigation has been conducted at Sellafield, Cumbria, since 1989, aimed at determining whether or not the site is suitable as a repository for radioactive waste. The target host rock is the Borrowdale Volcanics at a depth of more than 500 m. Part of the modelling has involved trying to predict groundwater flow and the movement of radio-nucleides. For this, a good ground model was necessary with estimates of permeability for the full rock sequence. Several high-quality boreholes have been put down at the site and logged very carefully. A general model has been developed, as illustrated in Figure B4-4.1 (ENE is to the right). The geological model has the Borrowdale Volcanics, which contain saline water, separated from the overlying sandstones, containing fresh water, by a bed called the Brockram, which is typically 25–100 m in thickness and cut by faults. The Brockram and associated evaporites and shale further west evidently play a very important potential role as a barrier to flow of groundwater (flow into the repository) and, hence, radio-nucleides migrating away from the repository.

Early modelling

For most early numerical simulations, the Brockram was modelled with very low conductivity (2×10^{-10} to 1×10^{-9} m/s), based largely on borehole tests and 'expert elucidation' (Heathcote *et al.*,

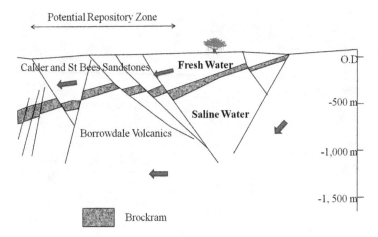

Figure B4-4.1 Cross section across the potential repository zone, showing basic geology and directions of flow (modified from Chaplow, 1996).

1996). These values are similar to those measured for the Borrowdale Volcanics – 50% measured over 50 m lengths, with conductivity $< 1 \times 10^{-10}$ m/s, according to Chaplow (1996).

Later tests

At a later stage, field tests were carried out that yielded 'significant flows' in the Brockram, and the earlier modelling had to be revised. Michie (1996) reports hydraulic conductivity measurements within the Brockram with a maximum of 1×10^{-5} m/s, i.e. four orders of magnitude higher than adopted for the early models.

A surprise?

The changed perception for this important stratum might be considered just part of what is to be expected in any progressive ground investigation. However, the potential for locally high permeability associated with extremely widely spaced and persistent joints, at spacing such that they will be rarely sampled in boreholes, could have been anticipated, partly because such joints can be observed directly at exposures in the Lake District. At Hoff's Quarry to the east of the Lake District, the rock can be examined, and at a material scale a low permeability would be anticipated (Figure B4-4.2).

However, at a larger scale, the rock at Hoff can be seen cut by near-vertical master joints which would affect the mass permeability in a dramatic way (as evidenced from the Sellafield test). There were also indications from the literature that the Brockram might be permeable at a scale of hundreds of metres. For example Trotter *et al.* (1937) commented on the possibility of pathways through the Brockram, with reference to the distribution of haematite mines within the Carboniferous Limestone underlying the Brockram.

Lessons: it is very important not simply to rely on site-specific data when elucidating parameters for design. There is a need to consider the geological setting, origins and history – with all that entails – as per the 'total geological approach' advocated by Fookes *et al.* (2000). Furthermore, when looking at data from boreholes, especially ones with a strong directional bias, one should consider all the field evidence that might offer some clues as to the validity of the expert elucidation process.

Figure B4-4.2 Close up of Brockram rock, at Hoff's Quarry, Vale of Eden, UK. The rock is a cross-bedded, limestone-rich, well-cemented breccia. It contains fossiliferous blocks of Carboniferous Limestone as well as more rare rocks such as Whin Sill dolerite. It has the appearance of a wadi-type deposit – poorly sorted, probably rapidly deposited by flash floods. From field assessment, it has a low permeability at the material scale. Lens cap (58 mm) for scale.

Figure B4-4.3 More distant view of Brockram at Hoff's Quarry. Note the fully persistent, near-vertical master joints about 40 m apart, which will control mass permeability. Evidently joints from this set would only be intersected using inclined rather than vertical boreholes.

4.3.3.2   *Equation 2: environmental factors*

Environmental factors, some of which are listed in Table 4.4, including hydrogeological conditions, should be considered part of the ground model for a site, but are best reviewed separately from the basic geology, although the two are closely interrelated. The environmental factors to be accounted for depend largely on the nature, sensitivity and design life of structures and the consequence of failure. It is usual practice to design structures to some return period criterion such as a 1 in 100 year storm or 1 in 1,000 year earthquake, the parameters for which are determined statistically through historical review. In some cases, engineers will also want to know the largest magnitude event that might occur, given the location of the site and the geological situation. Then some thought can be given as to whether or not it is possible to make some provision for that maximum credible event. For earthquakes, for example, a structure might be designed to behave

Table 4.4 Examples of environmental factors that should be considered for a project

FACTOR	CONSIDERATIONS	EXAMPLES OF ROCK TYPES SITUATIONS
in situ stresses	high stress: squeezing, overstressing, rockbursts	mountain slopes and at depth, shield areas, seismically active areas
	low stress: open fractures, high inflows, roof collapse in tunnels	extensional tectonic zones, unloaded zones, hillside ridges
natural gases	methane, radon	coal measures, granite, black shales
seismicity	design loading, liquefaction, landslides	seismically active zones, high consequence situation in low seismic zones
influenced by man	unexpectedly weak rocks, collapse structures	undermined areas
	gases and leachate	landfills, industrial areas
groundwater chemistry	chemical attack on anchors/nails foundations/materials	acidic groundwater, salt water
groundwater pressure	effective stress, head driving inflow, settlement if drawn down	all soils and rocks
ice	ground heave, special problems in permafrost/tundra areas, freeze-thaw jacking and disintegration	anywhere out of tropics
biogenic factors	physical weathering by vegetation, rotted roots leading to piping, insect attack	near-surface slopes weathered rocks causing tree collapse

elastically (without permanent damage) for a 1 in 1,000 year event but for a, very unlikely, maximum credible event, some degree of damage would be accepted.

The factors to review at this stage include natural hazards such as earthquake loading, strong winds, heavy rain and high groundwater pressures or flooding. Anthropogenic factors to consider include industrial contamination and proximity of other structures and any constraints that they may impose.

4.3.3.3　*Equation 3: construction-related factors*

The third verbal equation of Knill & Price (Knill, 2002) addresses the interaction between the geological and environmental conditions at a site and the construction and operation constraints (Hencher & Daughton, 2000). Excavation will always give rise to changes in stresses, and the ground may need to be supported. Excavations may also result in changes in groundwater, and the consequences need to be addressed and mitigated if potentially harmful. Similarly, loading from structures has to be thought through, not only because of deformations but also because of potentially raising water pressures, albeit temporarily.

There will also be hazards associated specifically with the way the project is to be carried out. For example, a drill and blast tunnel is very different to one excavated by a tunnel boring machine and will have specific ground hazards associated with its construction (Chapter 6). Similarly, the construction constraints are very different for bored piling compared to driven piles (Table 4.5). The systematic review and investigation of site geology and environmental factors, discussed earlier, needs to be conducted with specific reference to the project at hand. This will hopefully allow the key hazards to be identified and design to be robust yet cost-effective. Nevertheless, models are always simplifications, and the engineer must adopt a cautious and robust approach when designing, especially where the geological conditions are potentially variable and where that variability might cause difficulties, as illustrated by the case of a tunnel failure reported in Chapter 7 (Grose & Benton, 2005).

Table 4.5 Examples of the influence of engineering works

FACTOR	CONSIDERATIONS
loading/unloading – static/dynamic	settlement, failure, opening of joints, increased permeability in cut slopes, blast vibrations
change in water table	increased or decreased pressure head, change in effective stress, drawdown leading to settlement, induced seismicity from reservoir loading
denudation or land clearance	increased infiltration, erosion, landsliding

4.3.3.4   *Discussion*

It is evident that site investigation cannot provide a fully detailed picture of the ground conditions to be faced. This is particularly true for tunnelling, because of the length of ground to be traversed, the volume of rock to be excavated and often the nature of the terrain, which prevents boreholes being put down to tunnel level or makes their cost unjustifiable. Instead, reliance must be placed on engineering geological interpretation of available information, prediction on the basis of known geological relationships and careful interpolation and extrapolation of data by experienced practitioners. Factors crucial to the success of the operation, need to be judged and consideration given to the question: *what if?* It is generally too late to introduce major changes to the methods of working, support measures, etc. at the construction stage, without serious cost implications.

Site investigation must be targeted at establishing those factors that are important to the project and not to waste money and time investigating and testing aspects that can be readily estimated to an acceptable level or aspects that are simply irrelevant. This requires a careful review of geotechnical hazards, as advocated above. Even then, one must remain wary of the unknowns and consider ways in which residual risks can be investigated further and mitigated, perhaps during construction, as addressed in Chapter 6.

There is a somewhat unhealthy belief that standardisation (for example, using British Standards, Eurocodes, Geoguides and ISRM Standard Methods) will provide protection against ground condition hazards. Whilst most standards certainly encompass and encourage good practice, they often do so in a generic way that may not always be appropriate to the project at hand and they may not provide specific advice for coping with a particular situation. Ground investigations are often designed on the basis of some kind of norm – a one-size-fits-all approach to ground investigation. It is imagined that a certain number of boreholes and tests will suffice for a particular project, essentially irrespective of the actual ground conditions at the site. This ignores the fact that ground investigations of average scope are probably unnecessary for many sites but will fail to identify the actual ground condition hazards at rare, but less forgiving sites. Similarly, an averaging-type approach will mean that many irrelevant and unnecessary samples are taken and tested whilst the most important aspects of a site are perhaps missed or poorly appreciated. This is, unfortunately, commonplace.

If the hazards are considered in a systematic way, as discussed earlier, then the risks can be thought through fully and this will help the ground investigation to be better focused. The process is illustrated in Box 4-5 for a hydroelectric scheme involving the construction of a dam, reservoir, power station and associated infrastructure.

Box 4-5   Planning a site investigation for a new hydroelectric scheme

Project concept: high arch dam with high-pressure penstock tunnels (120m hydraulic head) leading to underground power house, tailrace tunnels and surge chamber. Structures to be considered include reservoir, ancillary buildings, roads, power lines and diversion tunnel. Sources of concrete aggregate need to be identified, as well as locations for disposing construction waste.

General setting: valley with narrowing point suitable for arch dam (high stresses). Topography and hydrology adequate for reservoir capacity. Steep slopes above reservoir.

Geology from preliminary desk study: major fault along valley, maybe more. Right abutment (looking downstream) in granitic rock, sometimes deeply weathered. Left side, ancient schist, greywacke, mudstone and some limestone. Folded and faulted with many joints. Alluvial sediments along valley.

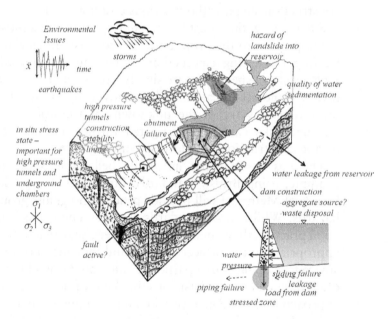

Figure B4-5.1 Schematic model of site for new hydroelectric scheme with some of the most important hazards that need to be quantified during the site investigation.

Key issues for investigation:

Dam: stability of foundations and abutments, settlement, leakage, overtopping from landslide into reservoir, silting up.

Tunnels and powerhouse: rock quality, *in situ* stress state, construction method, stability, lining and support requirements.

Reservoir: leakage, siltation, water quality.

Construction: source of aggregate, waste disposal, access, river diversion.

Main geotechnical considerations when conducting site investigation

Issues	Equation 1		Equation 2	Equation 3
	Geology		Environment	Construction
	Material	Mass		
Arch dam stability and construction	Strength, deformability and durability of foundation materials, including highly stressed abutments	Geological profile, depth to bedrock Presence of discontinuities, allowing failure in abutments or sliding failure below dam Fault reactivation	Seismicity Water pressure in foundations and abutments Check history of mining	Adequate source of non-reactive aggregate Waste disposal locations (fill embankments)
Leakage below dam and from reservoir	Permeability (need for grouting/cut offs) Potential for piping	Leakage on main fault and other faults/weathered zones Limestone might be karstic	Groundwater profile in surrounding terrain Existing throughflow paths	Options for grouting and/or cut-off structures
Landslides into reservoir	Material strength	Adverse discontinuities, aquitards causing perched water pressure to develop Landslide history	Response of groundwater to storms and to lowering of water in reservoir Seismic loading	Need for stabilisation such as drainage or option to remove hazardous ground
Powerhouse and high-pressure tunnels	Rock strength Abrasivity for tunnel equipment	Fracturing (rock mass classification allows judgement of stabilisation required) Weathered zones	*In situ* stress state (potential squeezing or leakage and need for steel liners) Groundwater pressure and permeability (inflows or water loss for operating tunnels)	Method of excavation Ground movement due to excavation Blasting vibration Groundwater changes

4.4 Field reconnaissance and mapping

4.4.1 General

At many sites, geologists can get a great deal from examining the landscape, mapping and interpolating information from exposures, and this is one of the most important aspects of geological education and training. This, together with desk study information, should allow preliminary ground models to be developed, which can then be used to form the basis for planning any necessary ground investigation. The preliminary model should allow an initial layout of the components of the project and, for buildings, some insight into the types of foundation that might be required. For tunnels, decisions can be made on locations for portals and access shafts. The degree to which walk-over studies and field mapping can be cost-effective is often overlooked, as illustrated by a case example in Box 4-6.

Box 4-6   Case example: cost-effectiveness of site reconnaissance – bridge abutment, Lake District, UK

The first ground investigation that the author was involved with was for a bridge abutment in the Lake District, UK. Figure B4-6.1 is a view of the rock cliff that was to form the abutment, and halfway down the cliff is a platform. Figure B4-6.2 is a side view of the platform. The man in the middle of the photograph is logging a borehole, using a periscope that has been inserted into a hole, inclined at about 45 degrees, drilled into the rock from the same platform. In the foreground, rock can be seen with a fabric dipping roughly parallel to the cliff. For reasons that are unimportant now, a question arose regarding the geological structure being logged by the periscope.

   The site engineer was asked for his geological map of the rock along the river (including the 100% exposed cliff). He replied, 'what map?'

Figure B4-6.1 Drilling platform on cliff.

Figure B4-6.2 Borehole periscope in use.

There we were, perched on a precarious and extremely expensive platform. A drilling rig had been brought in and lowered down the cliff to drill an inclined borehole of perhaps 73 mm diameter, at great cost, and we had been brought to the site from London to log the hole using a periscope. Meanwhile, the full rock exposure was available to be mapped and interpreted at very little cost, which would have allowed a much better and more reliable interpretation of the geological structure than was possible from a single borehole.

Lesson: Use the freely available information first (desk study and walk-over/mapping) before deciding on what ground investigation is necessary at a site.

Mapping can be done in the traditional geological manner, using base maps and plans, or on air photographs, which may need to be rectified for scale. Observations such as spring lines (Figure 4.8) are not only important in delineating probable geological boundaries but also in their own right for hydrogeological modelling. Observation points can be marked in the field, to be picked up accurately later by surveyors. Alternatively, locations can be recorded by GPS and input directly into a computer, as illustrated in Figure 4.9. The success of preliminary mapping can be enhanced by letting an early contract to clear vegetation, allow safe access and to put down trial pits and trenches on the instruction of the mapping geologist (Figure 4.10).

Soils and rock can be examined, described and characterised in natural exposures and in trial pits and trenches, and full descriptions should be provided, as discussed later. Samples can be cut by hand for transfer to the laboratory, with relatively slight disturbance (Figures 4.11 and 4.12).

Access can be facilitated by using hydraulic platforms or by temporary scaffolding (Figure 4.13). Trial pits and trenches should not be entered unless properly supported, and care must be taken in examining any steep exposure; as a general rule, for safety reasons, field work should be conducted by teams of at least two people.

Figure 4.8 Spring line revealed following heavy rain at base of Carboniferous Limestone, north of Kilnsey Crag, West Yorkshire, UK.

Figure 4.9 Hand-held computer with ortho-corrected air photographs and terrain maps, used to locate and map natural terrain landslide. GPS used to get accurate locations of identified features.

Apart from the general benefits to be gained from mapping freely available or cheaply created surface exposures to determine local geology, they are particularly important for characterising aspects of rock structure such as roughness and persistence of discontinuities, which cannot be determined in boreholes. As for all measurements, however, extrapolation should only be made with caution and with awareness that structure and rock quality may change rapidly from location to location (Piteau, 1973). Exposed soil may be desiccated and stronger than soil at depth; exposed rock will often be more weathered with closer and more persistent fractures than rock only a few metres in from the exposed surface.

Figure 4.10 Local labourers employed to dig some trial pits during preliminary field mapping. Tlemcen University and Hospital site, Algeria.

Information gained from desk study and site reconnaissance can be analysed and draped over 3D digital models using GIS, as illustrated in Figure 4.14, which greatly assists visualisation, interpretation and planning of GI, including access.

4.4.2 Describing field exposures

The task of describing a large field exposure, say in a cut slope, can be daunting, and the following procedure is recommended. The exposure (natural or man-made) should be split initially into zones, layers or units, by eye. The primary division will often be geological, i.e. rock and soil units of different age, but then differentiated by rock or soil mass quality such as degree of weathering or closeness of fracturing. Differentiation on strength can be made quickly by simple index tests such as hitting or pushing in a hammer. The split might be on structural regime, i.e. style and orientation of discontinuities. The process is

Figure 4.11 Hand trimming a sample to size in the field, for transportation to laboratory and triaxial testing.

Figure 4.12 (a) Block sample cut into grade IV weathered sedimentary rock and transported to the laboratory. (b) The sample trimmed by hand to fit into a Leeds direct shear box.

Figure 4.13 Cherry picker platform used to examine recently failed rock slope to allow remedial action to be determined, Hong Kong.

Figure 4.14 Surface geology draped onto topographic representation, for assessment of new road.

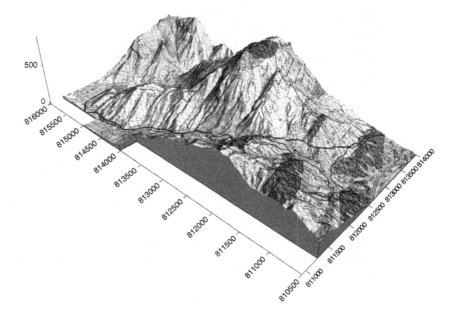

illustrated in Figure 4.15. Once the broad units or zone boundaries have been identified, then each needs to be characterised by systematic description and measurement, as shown schematically in Figure 4.16. Evidence of seepage should be noted; lush vegetation can be indicative of groundwater. The distinction between engineering geological mapping and normal geological practice is the emphasis on characterising units in terms of strength, deformability and permeability, rather than just age (Dearman & Fookes, 1974).

Some of the equipment that might be used in field characterisation of exposures includes safety harness, tape measures, hammer, knife, hand penetrometer, Schmidt hammers (type N and L), compass/clinometer

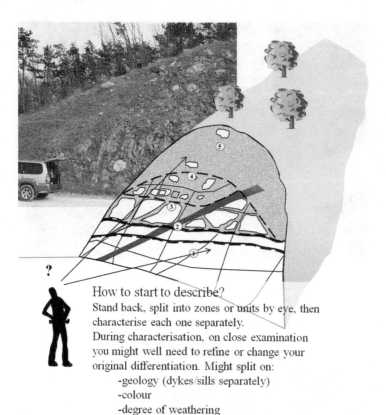

Figure 4.15 Approach to characterise rock mass. First stage is to split into units by eye. Units/zones will be used in later analysis and design.

?

How to start to describe?
Stand back, split into zones or units by eye, then characterise each one separately.
During characterisation, on close examination you might well need to refine or change your original differentiation. Might split on:
  -geology (dykes/sills separately)
  -colour
  -degree of weathering
  -percentage of included boulders
  -jointing style
  -perceived hazards

---

1. FIRST SPLIT EXPOSURE INTO UNITS
- *Geology*
- *Percentage corestones / boulders*
- *Jointing style and intensity; structure including faults, folds*
- *Weathering grades*
- *Other characteristics such as seepage, vegetation, hazards*

2. WITHIN EACH MAPPABLE UNIT
- *Geological origin*
- *Where heterogeneous, the proportion of fine fraction to coarse fraction*
- *Shape, size and distribution of coarse fraction*
- *Jointing pattern, structures (characterise these)*

3. DESCRIBE EACH MATERIAL IN EACH UNIT (as appropriate)

Colour	*Colour codes as appropriate*
Grain size	*Textures/fabric* *Particle size distribution*
Strength	*Field tests (hammer, knife)* *Schmidt hammer* *Penetrometer, vane*
Cohesion	*Slake test*
Permeability	*Infiltration test*
Mechanical Decomposition	*Degree of microfracturing*
Chemical Decomposition	*Scratcheability* *Decomposition grade*

Figure 4.16 Once the broad units/layers have been identified, each needs to be characterised.

and hand lens. Water and a container are useful for conducting index tests such as slake tests and for making estimates of soil plasticity and grading. Where appropriate, strength can be measured using such tools as a hand vane, and point load testing, which can be carried out on irregular lumps of rock. Whatever measurements are taken at exposures, the end user needs to be aware that it may be inappropriate to extrapolate properties because of the effects of drying out or softening from seepage and possibly the effects of weathering.

Guidance on geological mapping and description is given in a five-volume, well-illustrated handbook series by the Geological Society of London, which deals with Basic Mapping, the Field Description of Igneous, Sedimentary and Metamorphic Rocks (referenced in Chapter 3) and Mapping of Geological Structure, each with more than 100 pages (www.geolsoc.org.uk). Much of the detail that could be recorded by a geologist, however, might prove irrelevant to an engineering project, but what is or is not important might not be immediately obvious. It is worth bearing in mind the observation of Burland (2007):

> 'It is vital to understand the geological processes and man-made activities that formed the ground profile; i.e. its genesis. I am convinced that nine times out of ten, the major design decisions can be made on the basis of a good ground profile. Similarly, nine failures out of ten result from a lack of knowledge about the ground profile.'

Despite this observation, current standards codes and textbooks dealing with ground investigation tend to take a very simplified, prescriptive, formulaic approach in their recommendations for the description of geological materials and structure. The reason dates back to the 1960s when Deere (1968) noted that:

> 'Workers in rock mechanics have often found such a classification system [geological] to be inadequate or at least disappointing, in that rocks of the same lithology may exhibit an extremely large range in mechanical properties. The suggestion has even been made that such geologic names be abandoned and that a new classification system be adopted in which only mechanical properties are used.'

Deere went on to introduce classifications based on compressive strength and elastic modulus and the Rock Quality Designation (RQD), and these or similar classifications are now used almost exclusively for logging rock core, with geological detail rarely recorded.

Deere at the same time noted, however, 'the importance to consider the distribution of the different geologic elements which occur at the site'. This sentiment would have been supported by Terzhagi (1929), some of whose insightful observations on the importance of geological detail are revisited by Goodman (2002, 2003). Restricting geological description to a few coded classifications, as in industry standards, is over-simplistic but it is a fine balance between providing too much geological information and too little.

Generally, GI loggers tend to provide minimal summary descriptions, as per the examples given in BS5930 and other standards, and avoid commenting on unusual features, although it varies from company to company and, of course, the knowledge and insight of the logger. Some guidance on standard logging is given in Appendix C and examples of borehole logs are provided in Appendix D and discussed later. Fletcher (2004) provides many examples of the kind of geological information that can be obtained from logging of cores for engineering projects, most of which would be missed if following standard guidelines for engineering description and classification.

There is much to be said for the engineer informing the GI contractor of his preliminary ideas regarding the ground model, based on desk study and reconnaissance, so that the contractor knows what to look for and can update the model as information is gained.

Rock exposures are particularly important for characterising fracture networks. Orientations are usually measured using a compass clinometer, as illustrated in Figure 4.17, with different diameter plates used to help characterise the variable roughness at different scales (Fecker & Rengers, 1971). Electronic compass/clinometers are under

Figure 4.17 Joint survey underway using Clar compass clinometer attached to aluminium plates. Investigation for Glensanda Super Quarry, Scotland.

development, which will avoid the need to level the instrument, which can be difficult, especially in the underground mapping of tunnels.

Data are usually collected by systematic scan-line or window surveys but these are tedious to carry out, seem to be routine to the unknowledgeable, and therefore sometimes delegated to junior staff who may be unable or reticent to exercise independent judgement on what is or is not significant. Such surveys can give a false impression of rigorous characterisation, whilst the important element of geological interpretation, best done in the field, is lacking. Experienced engineering geologists with training in structural geology should be able to assess the rock conditions by eye, both with respect to the geological conditions and potential for instability in a slope, and therefore can carry out a subjective survey (Figure 4.18). The recommended approach for collection and interpretation of discontinuity data from rock exposures is set out in Box 4-7.

Figure 4.18 Distinction between objective surveys (line/window) and subjective surveys (from Hencher, 1987).

SYSTEMATIC

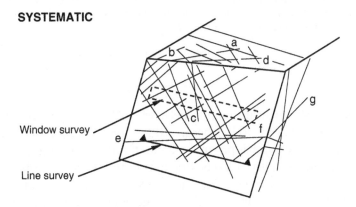

Data collected systematically for all discontinuities intersected along line of survey or within window

SUBJECTIVE

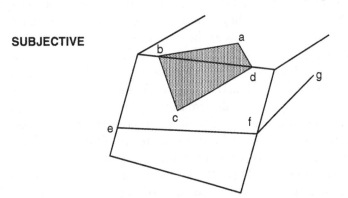

Potential for wedge and planar failures identified in field and data collected specifically for those adverse joints

Box 4-7   Collection of discontinuity data in exposures (modified from Hencher & Knipe, 2007)

1. First take an overview of the exposure. Examine it from different directions.
2. Develop a preliminary geological model and split it into structural and weathering zones units. Sketch the model.
3. Broadly identify those joint sets that are present, where they occur, how they relate to geological variation and what their main characteristics are, including spacing, openness as mechanical fractures (or otherwise), roughness, infill and cross cutting or terminations in intact rock or against other discontinuities. Surface roughness characteristics such as hackle marks should be noted as these are indicative of origin and help differentiate between sets.
4. Measure sufficient data to characterise each set geologically and geotechnically. Record locations on plans and on photographs. This might be done using line and window surveys but quite often these are time consuming and not very productive. It is generally best to decide what to measure and then measure it, rather than hope that the answer will be revealed from a statistical sample.
5. Plot data and look at geometrical relationships. Consider how the various sets relate to one another and to geological history as evidenced from faults, folds and intrusions (Chapter 3).
6. Search for missing sets that might have been expected given the geological setting.
7. Analyse and reassess whether additional data are required to characterise those joints that are most significant to the engineering problem.

Where the data collection point is distant from the project location, consider whether the collected data might be unrepresentative.

Remote measurement of fracture networks is becoming more reliable using photogrammetry (Haneberg, 2008) or ground-based radar (Figure 4.19) and research is progressing into the automatic interpretation of laser-scanned data into rock sets (orientation and spacing) (Slob, 2010). Currently, this approach, however, lacks any link to an interpretation of origin of the discontinuities and their geological inter-relationships (Chapter 3), which would make it much more valuable. In the author's opinion, probably the best use for laser scanning at the moment is as an aid to the field team, in particular for measuring data in areas of an exposure with difficult access, but they cannot replace mapping and characterisation by experienced persons at the current stage of development.

Rock joint data are generally represented on stereographic projections, as illustrated in Figure 4.20. The technique allows sophisticated analysis of geological discontinuity data (Phillips, 1973), but its most common use in engineering geology is for determining the potential for specific rock discontinuities to cause a failure in a cut slope or in an underground opening (Hoek & Bray, 1974 and Chapter 6). Plotting of data, statistical grouping and comparison to slope geometry is now easily done using software such as Dips (Rocscience), but care should be taken in interpretation and especially against masking important but relatively rare data (Hencher, 1985). Bridges (1990) demonstrates the importance of differentiating sets on the basis of geological characteristics rather than just geometry.

Figure 4.19 Ground-based radar being used to generate a digital image of cut slopes near Seoul, Korea. Point clouds can be used to measure discontinuity geometry remotely.

4.5 Geophysics

Geophysical techniques are used to identify the disposition of soil and rock units, based on differences in physical properties such as strength, density, deformability, electrical resistance and magnetism. They can sometimes be used successfully to identify cavities such as mine workings or solution hollows and for identifying saturated ground. Geophysics really comes into its own for offshore investigations where drilling is very expensive. Geophysics can provide considerable information on geological structure and rock and soil mass quality, which is relevant to engineering design, although such techniques are rarely used by themselves but as part of a wider investigation involving boreholes. Many engineering geologists and geotechnical engineers have both good and bad experience of engineering geophysics. Darracott & McCann (1986) argue that poor results can often be attributed to poor planning and the use of an inappropriate technique for the geological situation. More specifically, key constraints are:

- penetration achievable
- resolution
- signal-to-noise ratio, and
- lack of contrast in physical properties.

When geophysics works well, the results can be extremely useful and the method cost-effective. The main options and constraints are set out in BS 5930: 1999 and Clayton (1995).

4.5.1   *Seismic methods*

Seismic refraction techniques, using an energy source ranging from a sledgehammer to explosives, can be useful on land and in shallow water for finding depth to bedrock, for example, to identify buried channels that could otherwise only be proved by numerous boreholes or probes. Large areas can be investigated quite cheaply and quickly. The method works best where there is a strong contrast in seismic

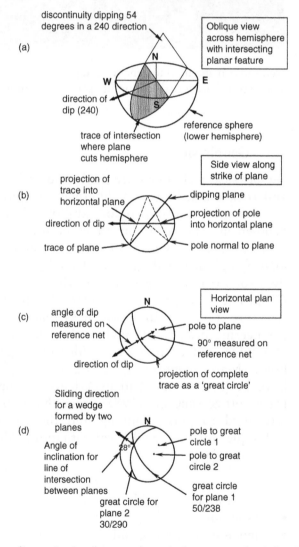

Figure 4.20 Representing discontinuity data as great circles or as poles (after Hencher, 1987).

velocity between the overlying and underlying strata and some knowledge of the geological profile, preferably from boreholes. Otherwise, results will be ambiguous. Where weak (low velocity) strata underlie stronger materials, these may not be identified by seismic survey. Wave velocity (compressive and shear) can be interpreted directly in terms of rock mass quality, deformation modulus and ease of excavation, as reviewed comprehensively by Simons *et al.* (2001). Seismic reflection is a key technique in offshore investigations.

4.5.2   Resistivity

Resistivity is another cheap and rapid method that can prove very effective, particularly in identifying groundwater (low resistance) and voids (high resistance). The technique has been used successfully in the investigation of landslide profiles, in particular for identifying water-bearing strata at depth. Figure 4.21 shows the results of a resistivity survey in Hong Kong to identify underground stream channels as zones of high resistance (voids), which it did extremely well (Hencher *et al.*, 2008).

4.5.3   Other techniques

There are a host of other techniques reported in the literature, with various success rates. Ground-based radar can be useful for finding shallow hidden pipes, etc. Other techniques such as magnetic and micro-gravity rely on particular physical properties of the rock or feature being searched for. Both have been used for locating old mine

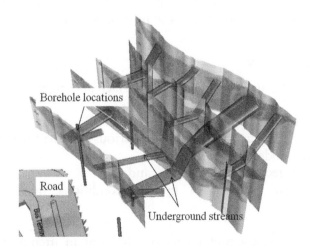

Figure 4.21 Digital image interpretation of resistivity surveys across hillside above Yee King Road, Hong Kong. Tubular features of low resistivity are interpreted as underground streams (Hencher *et al.*, 2008).

shafts – because the brick lining might have a magnetic signature and the void is low gravity. Generally, such techniques are used as a first pass across a site to identify any anomalies, which are then investigated more fully using trial pits, trenches and boreholes. For such investigations, percussive holes, as used for forming holes for quarry blasting (no coring), can be very quick and relatively cheap – the presence of voids is indicated by lack of resistance to drilling and loss of flushing medium. The voids can later be examined using TV cameras, periscopes or sonic devices to try to quantify size and shape. For many reasons, such surveys are not always successful and therefore are not to be relied upon to give a definitive answer (Clayton *et al.*, 1995). Sewell *et al.* (2000) demonstrate the usefulness of marine magnetic and gravity surveys for identifying geological structures.

4.5.4   *Down-hole geophysics*

As with seismic reflection, down-hole geophysics is used routinely in oil and gas exploration, in mining and in sophisticated GI linked to nuclear waste disposal studies. Tools can be used to determine minor stratigraphic contrasts and rock properties. These tools are less used for engineering, with the exception of rock joint orientation (using cameras and geophysical tools) and sometimes for identifying clay-rich layers. These tools are discussed below, together with logging and description.

4.6   Sub-surface investigation

Methods and techniques for sub-surface investigation are dealt with in many publications, including BS 5930 (BSI, 1999), Clayton *et al.* (1995), GCO (1987), Hunt (2005) and Mayne *et al.* (2001).

4.6.1   *Sampling strategy*

There are usually four main objectives in sub-surface investigation:

1. to establish the geological profile
2. to determine engineering properties for the various units within the eventual ground model
3. to establish hydrogeological conditions, and
4. to monitor future changes in ground conditions through instrumentation.

At many sites, it is best to use preliminary boreholes in an attempt to establish the geological profile accurately. This will require sampling over the full depth and with sufficient boreholes to establish lateral and vertical variability. If recovery is low, then boreholes may need to be repeated; it is often the pieces of core that are not recovered that are the most important, because they are also the weakest. It is wise to include a clause in specifications for the GI contractor, setting out a minimum acceptable recovery, to encourage diligent work. A good driller can generally achieve good recovery in almost any ground, providing he has the right equipment and adjusts his method of working to suit the ground conditions. If he does not have suitable equipment (or flushing medium), then that might be the fault of the engineer who specified the investigation, rather than the contractor, and this may need rectification by issuing a variation order to the contract.

Once the preliminary geological model has been established adequately at a site, then additional boreholes can be put down as necessary to take samples for testing or to carry out *in situ* testing and to install instruments for monitoring changes such as response of water table to rainfall. The same approach (sample first to prove the geological model and to identify any geological hazards, followed by a second phase for testing and instrumentation) should be used for any investigation where geological features may be important. This can only be judged by a competent engineering geologist aware of both the local geological conditions and the factors that will control the success or otherwise of the particular civil engineering project.

In practice, boreholes are often put down using a strategy of intermittent sampling and *in situ* testing within a single borehole, which means that the full ground profile is not seen. This can be cost-effective for design when the site is underlain by relatively uniform deposits and where the ground profile is already well-established from previous investigations. The danger is that site-specific geological features might be missed yet prove important for the project.

4.6.2   Boreholes in soil

There are many different tools that can be used to investigate soils and many of these are described by Clayton *et al.* (1995). In the UK, the most commonly used machine for investigating soils is the shell and auger, otherwise known as the cable-percussive rig, as illustrated in Figure 4.22. Such rigs are very manoeuvrable and can be towed behind a field vehicle or winched to the point where the hole is to be put down. They can cope with a wide range of soils, which makes for their popularity in the UK, where mixed glacial soils are common. The hole is advanced by dropping a heavy shell (Figure 4.23). Material between sampling points is usually discarded, although it should be examined and recorded by the drilling contractor and disturbed bulk samples are taken in bags, if specified for the contract. All samples, of course, should be sealed and labelled. If boulders are encountered in the soil profile, these are broken up with a heavy chisel dropped down the hole. Engineers usually specify alternate undisturbed samples for

Figure 4.22 Shell and auger rig in action, Leicester, UK. Casing, used to support the hole, is standing out of ground and a shell is being dropped down hole to excavate further. In the foreground is a U100 sampling tube attached to a down-hole hammer, ready for placing down hole and taking a sample once the hole has been advanced to the required depth. Leaning against the wheel is one of the drillers and also a trip hammer for SPT testing – also awaiting use at appropriate depths and changes in strata.

laboratory testing and *in situ* strength tests at perhaps 1.5 m intervals or changes in strata. The standard penetration test (SPT) is commonly used to measure strength, as discussed below under *in situ* testing. Vane tests might be carried out rather than SPTs, especially in clay soils. USA practice for investigating and sampling soils is described by Hunt (2005). One cheap and quick way of sampling/testing is to use wash boring, whereby the hole is advanced by water jetting as rods are rotated. SPT tests, and possibly other samples, are taken at intervals. None of these methods gives continuous sampling, so geological detail may be missed.

(a)

(b)

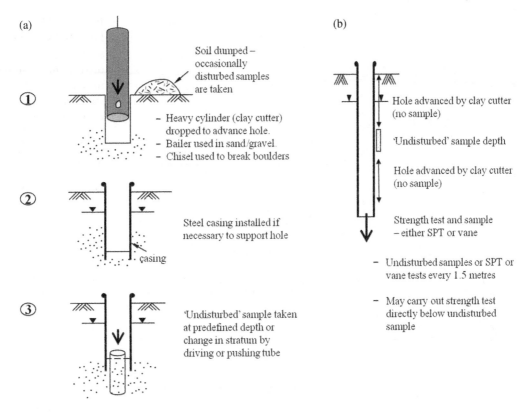

①

Soil dumped –
occasionally
disturbed samples
are taken

– Heavy cylinder (clay cutter)
  dropped to advance hole.
– Bailer used in sand/gravel.
– Chisel used to break boulders

②

Steel casing installed if
necessary to support hole

casing

③

'Undisturbed' sample taken
at predefined depth or
change in stratum by
driving or pushing tube

Hole advanced by clay cutter
(no sample)

'Undisturbed' sample depth

Hole advanced by clay cutter
(no sample)

Strength test and sample
– either SPT or vane

– Undisturbed samples or SPT or
  vane tests every 1.5 metres

– May carry out strength test
  directly below undisturbed
  sample

Figure 4.23 (a) Methodology for shell and auger advancement of boreholes. (b) Sampling strategy.

Undisturbed samples are usually taken using a relatively thin walled sampler of diameter 100 mm (U100), and much of the published empirical relationships that are relied upon by designers are based on tests on samples achieved in this way. This sampling method does not, however, meet the more stringent requirements of Eurocode 7 for class 1 sampling and testing, because of fears over disturbance. This is rather naïve in that it implies that thinner sampling tubes can take an undisturbed sample, which is not the case. Any sample taken from depth, squeezed into a tube and then extruded at the laboratory, will inevitably be disturbed to some degree. Further disturbance occurs during preparation of samples for laboratory testing and initial loading and saturation, as expressed schematically in Figure 4.24 and investigated by Davis & Poulos (1967). The engineering geologist and geotechnical engineer need to be aware of the likely disturbance to any tested samples and take due care in interpretation. Furthermore, the scaling up of results from laboratory to project scale requires careful consideration because it must include the effect of mass fabric and structure, including fractures and discontinuities. This is discussed further in Chapter 5.

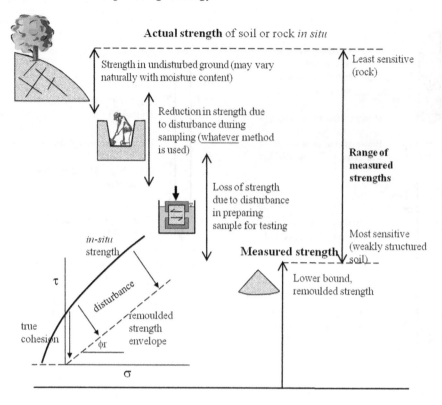

Figure 4.24
Potential sources of
sampling
disturbance leading
to much lower
strengths being
measured in the
laboratory
compared to those
in situ.

4.6.3   Rotary drilling

Rotary drilling is used in all rocks but can also be used to obtain good samples in weaker materials, including colluvium (mixed rock and soil), weathered rock and soil. In weaker ground, a similar investigation strategy is often adopted as for soils, whereby sections are cored followed by SPT tests, although as for soils there is the risk that important geological features may be missed.

A drilling rig rotates a string of drilling rods whilst hydraulic cylinders apply a downward force. At the lower end of the drilling string there is a hollow annulus bit, usually coated with diamonds or tungsten carbide. As the bit is rotated, a stick of core enters into a core barrel at the bottom of the drilling string. The retained core is prevented from falling out as the barrel is brought back to the surface, by some form of core-catching device. Air, water, mud or foam is used to cool the bit and carry rock cuttings back to the surface (Figure 4.25). Where cored samples are not required over a particular length of hole, it can be advanced more quickly using rock roller bits, down-the-hole hammers and water jets, as used in much oil and gas drilling.

At the most basic level, a single-barrel well-boring rig can be used to take core samples but these are often highly disturbed (Figure 4.26). Most drilling is carried out using double-barrel systems in which the outer barrel rotates around an inner barrel that takes in the core. A

Figure 4.25 Rotary drilling above fatal landslide at Fei Shui Road, Hong Kong. Polymer foam (white) is being used as the drilling flush to try to improve recovery.

Figure 4.26 Sample obtained from single-barrel Russian well drilling rig, El Hadjar steelworks, Annaba, Algeria (see preface). Previous logging of similar samples had interpreted the layering as some kind of varved sequence of silt and sand. Actually, the horizon *in situ* is fairly uniform weathered (grade IV) gneiss (the pale material). The dark-brown silt horizons represent occasions when the Algerian driller, bored with the slow drilling progress from his worn-out bits, raised the drilling string and then dropped it again with some force down the hole, letting in a layer of the silty drilling mud, which then became baked by the heat from the drilling process ... The thickness of the pale layers are an indication of the driller's boredom threshold – generally pretty consistent.

problem with the double-tube system is that the flushing medium flows between the core and the core barrel and can wash away some of the cored material, but it is still used internationally because it is relatively inexpensive and can be mass produced. The problems can be reduced by using a triple-tube system. In this system, the core enters a split inner tube, which does not rotate; the flushing medium flows between the inner tube and an outer tube without touching the core. Such equipment has low manufacturing tolerances so must be bought off the shelf, and the bits are very expensive and only last perhaps 8 to 12 m of coring before they need to be replaced, which precludes its use on many projects.

Usually, the larger the diameter of the core barrel, the better the recovery and quality of sample, and it is prudent to start using a large diameter and reduce diameter as necessary with depth. The wide range of casing, core barrel and drill rod sizes are listed in ASTM (1999), which also discusses good practice. When there is good-quality rock overlying soil material, retrieving the softer material can be a problem. As for soil boring, the hole may need to be cased temporarily during drilling to prevent it collapsing. Drillers generally try to recover about 1.5 m of core per run before pulling all the drill string back to the ground surface and dismantling it all. If recovery is low, then the driller might try to reduce the core run to 1 m or even less, but this does not always produce better results. Other parameters such as thrust, torque and flushing medium may have more influence on recovery, and much depends on the experience, knowledge and attitude of the drilling crew.

Wire line drilling employs large-diameter rods, which effectively support the hole as it advances. After each core run, the core barrel is pulled up the centre of the drill rods, the core extracted, then dropped back down the hole to lock into the bottom of the hole, ready to start drilling again. The cutting bit stays at the bottom of the drill rods and is not extracted with the core barrel. To change the cutting bit, however, the whole drill string has to be removed.

A system that is very commonly used in Hong Kong and elsewhere for sampling weathered rock and mixed rock and soil is a Mazier core barrel. This has a soil cutting shoe which is spring loaded and extrudes in advance of an outer rock cutting bit when cutting through relatively weak soil-like material (Figures 4.27 and 4.28). As conditions get harder, the soil cutter is pushed back and the outer coring bit takes over. This system, especially where combined with polymer foam flush, has been shown to produce good recovery of material in weathered and mixed materials (Phillipson & Chipp, 1982). The sample is taken in a plastic tube, which is later cut open so that the sample can be examined, described and tested (Figure 4.29). Drilling contractors will not open tubed samples without instruction to do so, and, in practice, geotechnical engineers sometimes order Mazier samples (from the office) but then never get round to opening and examining the samples, which is poor practice. The author was recently involved in an arbitration where 20 boreholes had been put down with alternate Mazier sampling in soft clays and then SPTs. The project was then designed on the basis of the SPT data alone and went badly wrong, ending in arbitration. The samples had not been opened up for examination or testing. A similar system to the Mazier, used in the USA, is the Dennison sampler (Hunt, 2005).

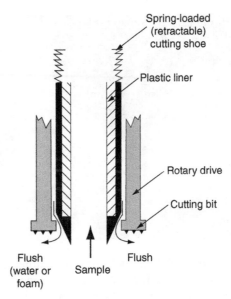

Figure 4.27
Principles of Mazier
sampling.

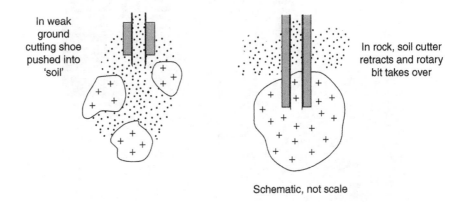

4.7 *In situ* testing

Many parameters are obtained for design by laboratory testing, as discussed in Chapter 5, but the potential for disturbance is obvious, as discussed earlier, especially for granular soil that disaggregates when not confined. There are therefore many reasons for attempting to test soil and to a lesser extent rock *in situ*. Most tests are conducted in boreholes, but some are conducted by pushing the tools from the ground surface or from the base of a borehole to zones where the soil is relatively undisturbed. A self-boring pressuremeter, suitable for clay and sand, drills itself into the ground with minimal disturbance before carrying out a compression test at the required level.

The SPT is probably the most commonly used *in situ* test, whereby the number of blows to hammer a sample tube into the ground is

Figure 4.28 Mazier sampler with nicely recovered weathered granite – the right side third is stained with iron oxides. Spring-loaded cutting shoe is seen extending from the rock cutting bit outside. When the material strength becomes too high for the cutting shoe (exceeds spring stiffness), the outer bit takes over the cutting.

Figure 4.29 Mazier sample plastic tube being cut for examination.

recorded. Soft soils are penetrated easily, hard soils and weak rocks with more difficulty. The SPT data can be interpreted in terms of shear strength and deformability (Chapter 5) and for making predictions of settlement directly (Chapter 6). The split spoon sampler used for the SPT is a steel tube with a tapered cutting shoe. It is lowered down the borehole, attached to connecting rods, and then driven into the ground by a standard weight, which drops a standard height, as illustrated in Figure 4.30 and shown in action from a rotary drilling rig in Figure 4.31. The number of blows for each penetration of 75 mm is recorded; blows for the first 150 mm are recorded but essentially ignored (considered disturbed); the blows for the final 300 mm are

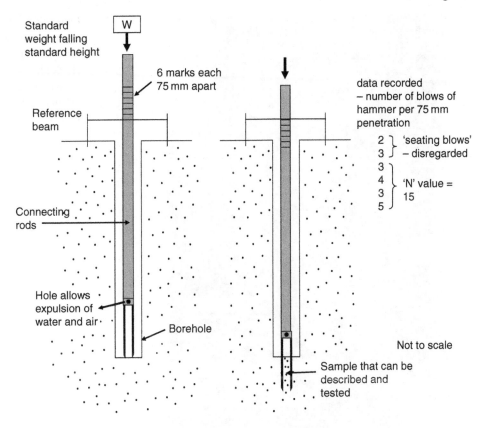

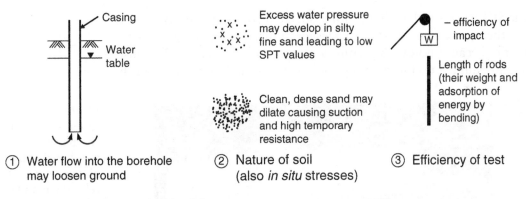

Figure 4.30 Principles and details of the SPT test.

added together as the N-value. Care must be taken in soil that the external water table is balanced, otherwise water may flow in from the bottom of the hole, causing softening and too low an N value. There are various corrections suggested for tests conducted in silty sand and for depth of overburden. Details are given in Clayton (1995).

Figure 4.31 SPT test underway in Hong Kong (1980s). Nowadays, a helmet would be worn.

The SPT test is much maligned for associated errors but nevertheless is still the most common basis for design in many foundation projects, mainly because no-one has come up with anything better. It is also actually quite a useful sampling tool, as illustrated in Figure 4.32. In

Figure 4.32 Split spoon sample of completely weathered granite. Note presence of relict joints and lack of visible disturbance.

the UK, it is normal to stop a test when 50 blows fail to advance the split spoon the full 300mm and instead to record the penetration achieved for the 50 blows. Depending on the ground conditions and sample retrieved, it might be valid to extrapolate the blow count to an equivalent N-value *pro rata*. Overseas, it is common practice to continue the test for 200 blows or more in weathered rock, and designs are often based almost solely on such data, which is rather questionable practice in a profile that might comprise a heterogeneous mix of harder and softer materials. Tests carried out in this way may damage equipment and are tedious for the drilling contractor, who might well be tempted to cut corners if no-one is supervising. The interpretation of SPT testing in weak and weathered rock is discussed in more detail in Chapter 5.

The vane test involves rotation of a cruciform steel tool at a slow rate within the soil (Figure 4.33). The test is especially suitable for soft clay where SPTs are inappropriate because of the indeterminate nature of pore pressure changes brought about by rapid loading. The vane test is assumed to give a direct measure of undrained shear strength for the shape sheared by the rotating tool but interpretation can be difficult, especially in bedded soils.

The static cone penetrometer is a conical tool (like an SPT) that is pushed rather than driven into the ground, usually from a heavy lorry (Figures 4.34 and 4.35). The end force on the cone tip, and drag on the sides of the tool, are measured independently and can be interpreted in terms of strength and deformability. Clay, being cohesive, grips the side proportionally more than sand or gravel, so the ratio between end resistance and the side friction can be used to interpret the type of soil

Figure 4.33 Field vane used for measuring strength of clay down borehole or sometimes pushed from the ground surface. Once at test location, the vane is rotated to measure shear strength of the cylinder of soil defined by the vane geometry. To the left is a sleeve used to protect the vane during installation.

Figure 4.34 Electric static cone penetrometer with piezometric ring. Forces on the cone tip are measured independently from the force on the shaft section above. A combination of all three measurements (including water pressure) gives a good indication of soil type as well as strength characteristics.

Figure 4.35 Heavy lorry being used to conduct static cone penetrometer tests.

as well as strength. A further refinement (piezocone) allows water pressures to be monitored as the cone is pushed in, which again can help in interpreting the soil profile.

Large-scale direct shear tests are sometimes carried out in the field (Figure 4.36), in the hope that scale and disturbance effects might be reduced. In reality, lack of control in the testing process, as well as questions over representation of samples, however large, often outweighs any advantages. The derived data are generally less reliable than those from a series of laboratory tests, which themselves would need very careful interpretation before use at the mass scale, as discussed in Chapter 5.

Figure 4.36 In situ direct shear test in trial trench, Hong Kong.

Small-scale deformability tests down boreholes include the use of inflated rubber packers in soil (pressuremeter) or the Goodman jack in rock where two sides of the borehole are jacked apart. All such tests are very small relative to the mass under consideration and need to be interpreted with due care as to their representativeness. Deformation at project scale is better predicted from loading tests involving large volumes. The inclusion of very high capacity Osterberg jacking cells set within large diameter, well-instrumented bored piles, as discussed in Chapter 6, gives the prospect of deriving much more representative parameters (e.g. Seol & Jeong, 2009). In practice, most rock mass parameters tend to be estimated from empirical relationships derived from years of project experience together with numerical modelling, rather than small-scale tests, as discussed in Chapter 5.

Field tests are really the only option for measuring hydraulic conductivity (also for oil and gas). Simple tests include falling or rising head tests in individual boreholes, whereby water is either added to or pumped out of a hole and then the time taken for water to come back to equilibrium measured. For realistic indications of behaviour at field scale, however, larger-scale pumping tests are required. Even then, water flow is often localised and channelled so tests may not always be readily interpreted.

4.8　Logging borehole samples

Data from ground investigations are generally presented in a report comprising factual data as well as an interpretation of conditions (if the GI contractor is requested to do so). One of the important jobs for an engineering geologist is to examine and record the nature of samples retrieved from boreholes. The data from individual boreholes is usually presented in a borehole log, which provides a record not only of the ground profile but many details of how the borehole was carried out. In the oil industry, where the hole is advanced by a rock-roller bit or similar destructive method, logging is done by examining small chips of rock carried in the flushing mud (well logging); in civil engineering, we generally have rather better samples to examine.

Logging is generally conducted using a checklist approach and employing standard terminology to allow good communication, for example, on the apparent strength of a sample. Such standardisation can, however, result in over-simplification and lack of attention to geological detail. The task might be delegated to junior staff who might not have the experience and training to fully understand what they are examining. In addition, GI contractors will not routinely describe all features of samples recovered, partly because they want to avoid disturbing the samples before the client/design engineer has made a decision on which samples he wishes to select for laboratory testing. Several examples of borehole and trial pit logs are provided in Appendix D. The examples prepared by GI contractors in the UK and Hong Kong demonstrate good practice, whereby the whole process of drilling a hole, testing down the hole and sampling are recorded. The materials encountered are described following standard codes and normal practice. Given the limitations discussed above, designers and investigators may need to examine samples and core boxes themselves and not rely on those produced by the contractor. In Appendix D, examples are given of logs prepared by engineering geologists who have the responsibility for the overall site investigation. These are supplementary to the logs produced by the GI contractors. The Australian example is from an intensive investigation of a failing slope that was threatening a road. There is considerable attention to detail, especially regarding the nature of discontinuities and far more so than in the contractor's logs. In practice, even this level of logging may be inadequate to interpret the correct ground model, and selected samples and sections of core will need to be described in even more detail by specialists, perhaps employing techniques such as thin-section microscopy, radiometric dating and chemical analysis. In all cases and at all levels, logs should be accompanied by high-quality photographs with scales included.

As discussed in Appendix C, guidance on standardised terminology is given in BS5930: 1999, in the GEO guide on rock and soil description (GCO, 1988) and the ISRM guidance on rock mass description (ISRM, 1981). There are many different standards and codes of practice in use worldwide – USA practice is far removed from that in the UK, as is that for Australia, China, New Zealand, Japan and Korea, which leads to confusion, particularly as similar terminology is often used to mean different things. A consequence of this fuzzy standardisation is that when projects go wrong geotechnically, as they sometimes do, then legal arguments often hinge on incorrect or misinterpretation of terminology. The engineering geologist needs to do his homework before practising in any region.

Another criticism made earlier regarding field mapping, but equally applicable to logging, is that standard guides and codes to rock and soil description tend to comprise a series of limited classifications that one has sometimes to force on an unwilling rock mass. For example, rock masses, as exposed in quarries, can seldom be simply described as widely or closely jointed, but loggers are required to apply such classifications to core samples. In the author's opinion, it is far better to concentrate on recording factual data, which can then be interpreted as the overall ground model becomes clearer. An example of oversimplified rock classification terminology is given in Box 4-8 with reference to the term aperture. The problem is that by using such terms it is implied that the feature has properly been characterised, which is not the case. De Freitas (2009) discusses the same point and also notes that many terms and indeed measured values such as porosity are lumped parameters and therefore rather insensitive and uninformative.

Box 4-8 Defining aperture: an example of poor practice by geotechnical coding committees

This example is used to illustrate the inadequacy of current geotechnical standards for soil and rock description to convey an accurate or realistic representation of the true nature of the geological situation.

Mechanical aperture is the gap between two rock discontinuity walls (three-dimensional) and a very important characteristic with respect to fluid flow and grouting. It is expressed in most codes and standards as a one-dimensional scale of measurement, in the same way as joint spacing. The various attempts at revising description of aperture over 25 years (leading to the current BS/Eurocode 7 requirements discussed later) have simply reinvented the measurement scales and terminology but have failed to address or inform users about the fundamental difficulties in measuring and characterising this property.

What is aperture?

It is the mechanical gap between two walls of a rock discontinuity such as a joint or a fault. An example of a small section of joint with a gaping aperture (because the block has moved down slope and dilated over roughness features) is shown in Figure B4-8.1, which is a photograph of a section of sheeting joint in granite from Hong Kong.

Figure B4-8.1 Part of sheeting joint with gaping aperture where seen. Evidently, away from the exposure the aperture is tight and the rock walls are in contact. Example is near Sau Mau Ping, Hong Kong.

In the second example, of a fault exposed at a beach, also in Hong Kong (Figure B4-8.2), it is not quite so easy; there is a groove along the feature but the astute geologist might interpret this as preferential erosion. Some authors advise measuring aperture using feeler gauges. Others have attempted to characterise aperture volumetrically by injecting resin or liquid metals.

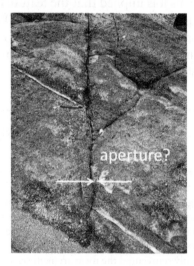

Figure B4-8.2 Minor fault exposed on beach, Peng Chau Island, Hong Kong.

Does it matter?

It is an extremely important property of the rock mass, controlling fluid flow and also related to shear strength. The problem is it is a very complex and unpredictable characteristic, as is the associated fluid flow. A single joint can be locally tight and impermeable, whilst elsewhere can be open allowing huge volumes of water to flow, as discussed by Kikuchi & Mito (1993). Investigation and characterisation can be a nightmare – if a borehole hits a conductive section, then high permeabilities will be measured and an installed instrument will be responsive to changes in water pressure, but this is literally a hit or miss business, as evidenced by many examples in investigations associated with nuclear waste (e.g. Thomas & La Pointe, 1995). The author has the experience of working in a deep tunnel 150m below the sea, where over one section, the rock was highly jointed but dry, but elsewhere, at the same level, there was a steady inflow through what was apparently intact rock. Clearly, it is not just local aperture that matters, but the characteristics of the full fracture network and its connectivity leading to the point of observation. It is an important area for research and for observation linked to geochemical and structural studies together with an appreciation of coupled mechanisms (e.g. Olsson & Barton, 2001; Sausse & Genter, 2005). Without getting to grips with the concept of channelised flow on rock joints and through joint networks, it may be impossible to ever make a safety case for nuclear waste disposal, with all the corollaries, i.e. no nuclear power, global warming and the end of civilisation. Well, perhaps slightly overstated, but not that much.

Apart from the natural variability of fracture networks, are there any other considerations?

Yes. Most rock joints are sampled in boreholes where aperture simply cannot be measured. Furthermore, it is very unlikely that any borehole sample would be representative of the discontinuity at any great distance. Down-hole examination with cameras and periscopes can be used to examine borehole walls, but again there is a problem with sampling and representativeness. In exposures such as quarries or tunnels, exposure is better but there is a question of disturbance – blasting, stress relief and block movement and whether observations at one location are relevant to the rock mass as a whole.

So what advice is given in recommended methods and standards?

1978 ISRM. The discussion on aperture is very useful. Its importance is recognised and many of the difficulties in measurement and interpretation are highlighted.

For description purpose and where appropriate, apertures are split into closed, gapped and open features, each subdivided into three. It is advised that:

a. modal (most common) apertures should be recorded for each discontinuity set
b. individual discontinuities having apertures noticeably wider or larger than the modal value should be carefully described, together with location and orientation data, and
c. photographs of extremely wide (10–100cm) or cavernous (>1m) apertures should be appended.

1999 UK BS5930 (BSI, 1999). Says little about aperture other than noting that it cannot be described in core. Five classes are introduced, which use some of the same terms as ISRM but with different definitions.

2003 INTERNATIONAL STANDARD ISO 14689–1 (BSI, 2003) (for Eurocode 7 users). Provides a new mandatory terminology for one-dimensional measurement that differs from that of BS5930: 1999 and ISRM (1978), as illustrated in Table 4 B8.1 (see below).

Table 4 B8.1 Terms for the description of aperture.

Aperture size term	ISRM 1978[1]	BS5930 1999	ISO 14689–1: 2003
<0.1mm	Very tight	Very tight	Very tight
0.1–0.25mm	Tight	Tight	Tight
0.25–0.5mm	Partly open		Partly open
0.5–2.5mm	Open	Moderately open	Open
2.5–10mm	Moderately wide	Open	Moderately wide
10–100mm	Very wide	Very open	Wide
100–1,000mm	Extremely wide		Very wide
>1,000mm	Cavernous		Extremely wide

[1] In detail, there is further confusion in that ISRM also defines a term *wide* for *gapped* features >10mm; the other terms above, also for apertures >10mm are for *open* features but the difference is not fully obvious.

Hydraulic aperture vs. mechanical aperture

For completeness, it is worth emphasising here that even if we could measure mechanical aperture meaningfully, the actual associated flow characteristics of the rock mass (hydraulic aperture) would be very difficult to estimate or predict. It clearly makes sense to observe and characterise rock masses as best we can, with respect to openness of the fracture network, but hydraulic conductivity can only be measured realistically using field tests, as discussed elsewhere, and even these are often open to different interpretations (e.g. Black, 2010).

Conclusions

After 25 years to digest the ISRM discussion and intensive international experience on research in measuring gaps in discontinuities and associated fluid flow, especially with respect to nuclear waste disposal investigations, the requirement for site investigation in Europe is a new set of linear measurements that are inconsistent with previous ones. No mention is made of the difficulty of characterising aperture in this way. Meanwhile, the New Zealand Geotechnical Society (2005) has produced yet another classification for aperture, which uses a selection of the same terms as in the above table but defined differently (e.g. wide = 60mm to 200mm) and introduces a new set of classes for the middle range: very narrow, narrow, moderately narrow.

Apologies

Apologies for being so critical, but it seems to this author that many codes and classifications over-simplify geological description and constrain/stifle good practice. This is especially so where it is mandated that some particular but fundamentally inadequate terminology shall be used. Unfortunately, inexperienced geotechnical engineers and engineering geologists are led to believe that such codification adequately deals with description and characterisation of the feature, which is not the case.

4.9   Down-hole logging

Down-hole logging technology has largely come from the oil industry and partly from mining. At the simplest level, a TV camera or borehole periscope is lowered down an uncased borehole and used to identify defects or to examine discontinuities. A borehole can be pumped dry of water and observations made of locations of water inflow, although this might need to be inferred from temperature or chemical measurements (Chaplow, 1996). Borehole impression packers were introduced in the 1970s and can be used to measure the orientation of discontinuities. Using an inflatable rubber packer, paraffin wax paper is pressed against the walls of the borehole and when retrieved, the traces of indented joints are clearly visible (Figure 4.37). Dip of the joints is easily determined from the geometry of the borehole but measuring direction relies upon whatever device is used to orientate the packer and, from experience, this can be a major source of error. It is good practice when using the impression packer to specify over-lapping sections of measurement down the hole (by perhaps 0.5 m) so that consistency can be checked. In one borehole we found a 70 degree difference between consecutive sections, resulting from the packer being deflated before the compass had set in position – the contractor was asked to redo the work. A more modern tool is the Borehole Image

Figure 4.37 Impression packer. Paraffin wax paper has been pushed against the walls of the borehole by a rubber inflatable packer. A series of pale-grey traces can be seen, which represent a set of fairly planar joints dipping at about 70 degrees. Direction is obtained from a compass set in glue at the base of the packer. Other options for orienting devices now include flux gate magnetometers and gyroscopes.

Processing System (BIPS), which gives a continual visual record of the borehole wall (Kamewada *et al.*, 1990). The tool is lowered down the borehole and a video camera takes a 360 degree image millimetre by millimetre down the hole through a conical mirror (Figure 4.38). Despite modern instrumentation for this tool, whereby azimuth can be measured by magnetic flux gates or gyroscopes, studies have revealed errors of up to 20 degrees in this measurement (Döse *et al.*, 2008). Care must also be taken in interpretation of discontinuities logged in boreholes, especially if boreholes are all vertical. There will be obvious bias to the measurements – steep joints will be undersampled in vertical boreholes. As an example, during the Ching Cheung Road landslide investigation (Halcrow Asia Partnership 1998a), BIPS measurements were taken in vertical boreholes and

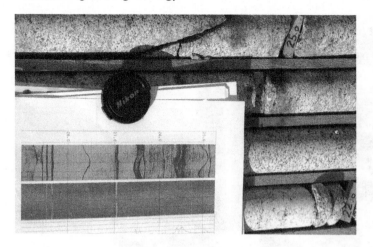

Figure 4.38
Output from BIPS down-hole discontinuity orientation device, being used during logging of rock core, Taejon Station, South Korea.

indicated a completely different style of jointing to those measured in exposed faces (essentially along horizontal scan lines). The data are presented in Figure 4.39 and it can be seen that the borehole data essentially defined a girdle of joints at 90 degrees to the main pole concentration that was measured from the horizontal scan line data. Both sets of data were required to provide the correct geological picture.

Other down-hole tools include resistivity and gamma ray intensity (even in cased holes) which, whilst often useful for oil exploration and coal mining, generally have rather limited application to civil engineering, other than possibly for locating clay-rich horizons.

4.10   Instrumentation

Instrumentation is used to establish baseline ground conditions at a site, most commonly in terms of natural groundwater fluctuations. It is also used to monitor changes at a site brought about by construction activities such as excavation or blasting. Instrument systems need to be designed carefully so that they are reliable; there needs to be built-in redundancy for instruments that may fail or become damaged by site works or by vandalism. Incoming data must be readily interpretable if some action is to be taken as a consequence. Instruments are often used during the works to check performance against predictions. Displacements and water levels can be monitored and compared to those anticipated. First (ALERT) and second (ALARM) level trigger conditions can be defined with prescribed action plans. Data can be sent remotely to mobile phones or by email to engineers who have the responsibility for safety and the power to take action such as closing a road or evacuating a site. Other instruments that might be

Figure 4.39
Comparison
between
discontinuity data
recorded by BIPS
(vertical drillhole)
and from surface
mapping
(horizontal scan
lines). After
Halcrow Asia
Partnership
(1998a).

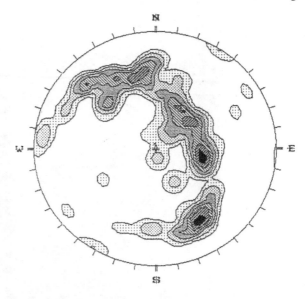

Discontinuities measured in borehole (vertical)

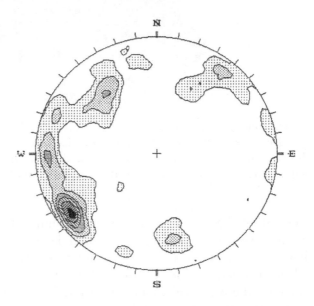

Discontinuities measured in scanlines

employed during a large construction project include sound and vibration meters, especially where blasting is to be carried out.

Piezometers are commonly installed as part of ground investigations to measure water pressures. Detailed information on these and other instruments are given by Dunnicliff (2003). The simplest device is an open-tube standpipe with a porous tip, installed in a

Figure 4.40 Standpipe piezometer tip about to be placed in borehole. Another has already been installed at a deeper level. It is not very good practice to install more than one piezometer in a borehole, because of potential leakage between the different horizons being monitored, but can work providing great care is taken in installation. Portsmouth dry dock, UK.

sand pocket within the borehole, as shown in Figure 4.40. There are also push-in versions available. The water level in the stand-pipe is dipped, perhaps on a weekly or monthly basis, using a mechanical or an electronic device lowered down the hole; for an electric dipmeter, the water closes a circuit to activate a buzzer. To measure high rises in water level between visits by monitoring personnel, Halcrow plastic buckets can be installed on fishing line with a weight at the bottom of the string, at perhaps 0.5 m intervals down standpipes. The buckets are pulled out of the hole when the site is visited – the highest one that is filled with water indicates the maximum level of water (Figure 4.41). At a more sophisticated level, standpipes can be set up so that readings are taken automatically at regular intervals using pressure transducers (divers) or through an air bubbler system (Pope *et al.*, 1982). Data can be recorded on data loggers that can be set up to transmit

Figure 4.41 Halcrow buckets retrieved at Yee King Road landslide investigation (Hencher *et al.*, 2008). These are unusual in that they contain sediments (from turbulent flows down the borehole). Normally, they would just contain water (or not), indicating the highest level that the water has risen in the borehole between inspections. Left side bucket is attached by fishing line to lead weights used to lower the buckets down the borehole.

information by telemetric systems. Other instruments include pneumatic or vibrating wire piezometers that respond very quickly to changes in pressure (Figure 4.42). Because they require almost no water flow to record change of pressure (unlike a standpipe), they can be grouted in place in the borehole and several instruments can be installed in the same hole, which can save cost (Vaughan, 1969; Mikkelsen & Green, 2003).

Instruments that are used to measure displacement include strain gauges, tilt meters, inclinometers and extensometers. They can be mechanical or electrical, for example, using vibrating wire technology. Figure 4.43 shows the end of an extensometer anchored deep behind the working face of a large copper mine in Spain and fitted with lights and a claxon horn to give warning if

Figure 4.42 Pneumatic piezometer being used to take measurements of rapidly changing water pressures during pile driving. Only small volume changes are necessary to measure pressure changes, so readings could be taken every ten seconds or so. Water pressures measured went off scale at about three times overburden pressure (Hencher & Mallard, 1989), Drax Power Station, Yorkshire, UK.

Figure 4.43 Extensometer with claxon and flashing lights used to warn workers at Aznalcollar mine, Spain, of danger from moving slope.

Figure 4.44 Exhumed inclinometer tubing. Four grooves inside (ridges outside) are guides for the wheels on the inclinometer instrument. The device with arms is a spider, which becomes fixed in position against the walls of a borehole whilst the tube can pass up or down inside. It is magnetic and a probe down hole can locate it and measurements can be made of settlement (as well as inclination).

the anchored point moves towards the mining area. Other instruments used to monitor performance at that site included deep inclinometers and a Leica total station, whereby numerous targets on the slope surfaces were surveyed remotely and automatically on an hourly basis, with the data sent to the site office (Hencher *et al.*, 1996). An inclinometer is a tubular torpedo (with wheels), which is lowered down a grooved tube set into a borehole or built into embankment fill. Figure 4.44 shows a section of inclinometer casing with the two sets of orthogonal grooves for the wheels. The torpedo (Figure 4.45) is first lowered down aligned by the first set of grooves, then removed and lowered down the second set of grooves. The section on the figure also has magnetic spiders with magnets, through which the tube can slide and can therefore be used to monitor vertical settlement where the tube is installed in fill. Strain gauges within the torpedo measure tilt, which is recorded against depth. The orthogonal measurements can be resolved to give the true direction and amount of displacement.

4.11   Environmental hazards

4.11.1   General

Site investigation needs to include a review of the potential environmental hazards as well as the immediate ground conditions. There may be risk from natural landslides and rockfall threatening the project, potential for natural subsidence or collapse (say in

Figure 4.45
Inclinometer torpedo about to be lowered down grooved tube. Tuen Mun Highway, Hong Kong.

areas underlain by salt deposits, old mine workings or karst), coastal erosion, wind, rain or earthquakes (Bell, 1999). As noted earlier, for some locations there are published hazard maps, but such maps cannot usually be relied upon on a site-specific scale. It is up to the site investigation team to identify the potential hazards for the project throughout its life (maybe 50 to 100 years) and to quantify these. In some cases, such an assessment might lead to a decision not to proceed with a project. Elsewhere, the hazard can be dealt with by careful design, and the main example of so doing is the hazard of earthquakes.

4.11.2 Natural terrain landslides

Landslides from natural terrain (rather than man-made slopes) are a hazard in most mountainous regions and can range from minor rock and boulder falls to massive landslides which involve >20 million m³ of rock and occur on average every three or four years worldwide (Evans, reported by Eberhardt *et al.*, 2004). Landslides like the one that destroyed Yungay, Peru, in May 1970, and killed about 20,000 people, are very difficult to predict and impossible to engineer. All one can do is identify the landform, the degree of risk and perhaps monitor displacements or micro-seismicity, with a plan to evacuate people and close roads if necessary.

Smaller and more common natural terrain landslides can be predicted and mitigated to some degree by engineering works. The starting point is generally historical records of previous landslides, such as incidents on active roads through mountainous regions. These may allow areas of greatest hazard to be identified and some prioritisation of works. It should be noted, however, that small rockfalls at one location can be indicative of much larger and deep-seated landslides, and minor incidents should be reviewed in this light. Where there is good historical air photograph coverage, sources of landslides can be identified and these correlated to susceptibility maps prepared using geographical information systems (e.g. Devonald *et al.*, 2009). Typical factors that might be linked to probability of landslide occurrence include geology, thickness of soil, vegetation cover, slope angle, proximity to drainage line and catchment area. Once a best fit has been made linking landslide occurrence to contributing factors, maps can be used in a quantitative, predictive way. Consequence of a landslide depends on location relative to the facility at risk (e.g. road, building), volume, debris run-out, possibility of damming a watercourse and eventually impact velocity. From studies in Hong Kong (Moore *et al.*, 2001; Wong, 2005), it is apparent that the greatest risk is generally from channelised debris flows (outlets of streams and rivers) and to facilities within about 100 m of hazardous slopes (the typical limit of debris run-out in Hong Kong). A broader discussion is given by Fell *et al.* (2005). A decision can be made on the resources that are justified to mitigate the hazard, once one has determined the level of risk (which can be quantified in terms of risk to life). There are many options, including barriers and debris brakes in stream courses and catch nets, especially for rockfall and boulder hazards. In some cases, a decision might be made to stabilise the threatening natural terrain using drainage, surface protection, netting and anchors, as for man-made slopes, dealt with in Chapter 6.

4.11.3   *Coastal recession*

Coastal recession is a common problem and rates can be very rapid. For example, parts of the Yorkshire coast are retreating at up to 2 m per year (Quinn *et al.*, 2009). Many studies have been carried out on mechanisms, but the harsh fact is that many properties and land near the coast are at risk and many houses have to be abandoned. Coastal protection measures can be designed successfully but these sometimes fail in a relatively short time and, constructing works at one location, can have consequences for others along the coast, as suspected for the damage to the village of Hallsands in Devon, which had to be largely abandoned (Tanner & Walsh, 1984).

4.11.4   *Subsidence and settlement*

An excellent review on ground subsidence – natural and due to mining, is given by Waltham (2002). Ground subsidence occurs naturally due to lowering of the water table from water extraction, oil and gas extraction, shrinkage of clay, and dissolution of salt deposits, limestone and other soluble rocks (e.g. Cooper & Waltham, 1999). Subsurface piping can occur associated with landslides in any rocks, including granite (Hencher *et al.*, 2008). The results can be dramatic, with sudden collapses of roads or even loss of buildings. Care must therefore be taken to consider these possible hazards during site investigation.

Underground mining dates back thousands of years in some areas (e.g. flints from chalk) and on a major scale for hundreds of years. Consequently, there are very incomplete records. In desk study, the first approach will always be to consult existing records and documents, but wherever there is some resource, such as coal, that might have been mined, the engineering geologist needs to consider that possibility. Investigations can be put down on a pattern, specifically targeted at the suspected way that mining might have been carried out (pillar and stall or bell pit, for example). Air photograph interpretation will often be useful and geochemical analysis of soil can give some indication of past mining activities.

4.11.5   *Contaminated land*

Many sites around the world are severely contaminated, often because of man's activities. This means that if the site is to be used for some new purpose, it may need to be cleaned up to be made habitable. Similarly, when constructing near or through possibly contaminated land, this needs to be investigated and the contamination mitigated, possibly by removing the contaminated soil to a treatment area. Barla & Jarre (1993) describe precautions for tunnelling beneath a landfill site. Guidance on investigation is given in BSI (2001), CIRIA (1995) and many other sources of information are given by the AGS (Appendix A). Sometimes the contamination is dealt with at site. Desk study can often identify projects where there are severe risks because of previous or current land use. Industrial sites such as old gas works, tanneries, chemical works and many mines are particularly problematical. Severe precautions need to be taken when dealing with such sites and works will probably be controlled by legislation.

4.11.6   *Seismicity*

4.11.6.1   *Principles*

Design against earthquake loading is an issue that needs to be considered in many parts of the world, depending upon the importance of the project and risks from any potential damage. In some locations, because of inherently low historical seismicity (UK) or severity of other design issues (e.g. typhoon wind loading in Hong Kong), seismicity might be largely ignored for design other than for high-risk structures like nuclear power plants. Elsewhere, seismicity needs to be formally assessed for all structures and taken into account for design.

4.11.6.2   *Design codes*

Many countries have design codes for aseismic design and these are generally mandatory. Nevertheless, it is often prudent to carry out an independent check and in particular to consider any particular aspects of the site that could affect the impact of an earthquake. For example, the local soil conditions might have the potential to liquefy. These issues are considered in more detail in Chapter 6.

Design codes, where well written and implemented, reduce the earthquake risks considerably. The USA, for example, has a high seismic hazard in some areas but fatalities are few and this can be attributed to good design practice and building control. China also has a high seismic hazard in some areas, but earthquakes commonly result in comparably large loss of life, which might be attributed to poor design and quality of building. Structures can be designed to withstand earthquake shaking, and even minor improvements in construction methods and standards of building control (quality of concrete, walls tied together, steel reinforcement, etc.) can prevent collapse and considerably reduce the likely loss of life (Coburn & Spence, 1992).

4.11.6.3   *Collecting data*

The first stage is to consider historical data on earthquakes, which are available from many sources, including the International Seismological Centre, Berkshire, and the US Geological Survey. These historical data can be processed statistically using appropriate empirical relationships to give probabilistic site data – for example, of peak ground acceleration over a 100 or 1,000-year period. This can be done by considering distance from site of each of the historical earthquake data or linked to some source structure (such as possible active faults). Dowrick (1988) addresses the process well, and some guidance is presented in Chapter 6. In some cases, estimates are made of the largest earthquake that might occur within the regional tectonic regime and similar regimes around the world, to derive a maximum credible event. This postulated worst case could be used by responsible authorities for emergency planning and is also used for some structures – a safe-shutdown event for a nuclear power station design.

4.12   Laboratory testing

Generally, a series of laboratory tests are specified for samples recovered from boreholes, trial pits and exposures, often employing the same GI contractor who carried out the boring/drilling. Geotechnical parameters and how to measure or estimate them are addressed in Chapters 5 and 6.

4.13   Reporting

The results of site investigation are usually presented as factual documents by the GI contractor – one for borehole logs, a second for the results of any laboratory testing. In addition, specialist reports might be provided on geophysics and other particular investigations. These reports may include some interpretation, perhaps with some cross sections if the contractor has been asked to do so, but such interpretation may be rather general and unreliable, not least because the GI contractor will not be aware of the full details of the planned project.

Generally, it is up to the design engineer to produce a full interpretation of the ground model in the light of his desk study, including air photo interpretations and the factual GI (that he has specified). This might be done supported by hand-drawn cross sections and block diagrams – which should ensure that the data are considered carefully and should enable any anomalies and errors to be spotted. There is a tendency now to rely upon computer-generated images, with properties defined statistically to define units (e.g. Culshaw, 2005; Turner, 2006), which might reduce the chance that key features of the model are properly recognised by a professional.

5   Geotechnical parameters

'Putting numbers to geology'

Hoek (1999)

5.1   Physical properties of rocks and soils

For civil engineering design, it is necessary to assign physical properties to each unit of soil or rock within a ground model. These include readily measurable or estimated attributes such as unit weight, density and porosity. Other parameters that are often needed are strength, deformability and permeability. In the case of aggregates (rock used in construction for making concrete) and for armourstone, important attributes are durability and chemical stability.

5.2   Material vs. mass

Most tests and measurements are made on small-scale samples in the field or the laboratory and need to be scaled up according to theoretical or empirical rules, to include for geological variability, fabric and structure. For example, a soil mass might be made up of a mixture of strong boulders in a matrix of weak, soil-like material, and this mix has to be accounted for in assigning parameters for engineering design. Mass strength, deformability and permeability of rock masses are controlled largely by the fracture network, rather than intact rock properties; the permeability of intact rock might be 10^{-11} m/sec, which could be thousands of times lower than for the fractured rock mass.

5.3   Origins of properties

5.3.1   Fundamentals

The strength of soil and rock (geomaterials) is derived from friction between individual grains, from cohesion derived from cementation

filling pore spaces and from inter-granular bonds such as those
formed by pressure solution (Tada & Siever, 1989). The strength
and deformability of soil is also a function of the closeness of
packing of the mineral grains. Densely packed soil will be forced
to dilate (open up) during shear at relatively low confining stresses as
the grains override one another and deform, and the work done
against dilation provides additional strength. The same principles
apply to rough rock joints or fractured rock masses. Different minerals may also have fundamentally different properties – some are
more chemically reactive and may form strong chemical bonds in
the short term, some are readily crushed or scratched, whilst others
are highly resistant to damage or chemical attack. Some, such as talc
and chlorite, are decidedly slippery and if present on rock joints can
result in instability.

The huge range of properties in soil and rock and how these
evolve with time is illustrated by a single sample in Figure 5.1. The
left-hand picture shows a graded series of sediments. The sand
horizons become finer upwards, as is typical of sediments deposited
from a river into a lake. At the top of the sample, there is a second
sand horizon that has been deposited onto the underlying sediment.
This has deformed the underlying sediments, producing a loading
structure, which shows that the soil was in a very soft state at the
time of formation. Contrast this with the rear of the same sample
showing conchoidal fractures in what is actually extremely strong
rock. The conversion from soft mud to rock has occurred over a
long time but has occurred naturally and, in practical geotechnical
engineering, we encounter and need to deal with the full range of
materials, transitional between these end members.

a)

b)

Figure 5.1
(a) Graded,
probably seasonal
bedding with clear
evidence of soft
sediment
deformation. (b)
Rear of the same
sample with
conchoidal
fractures indicating
the strength of this
rock (probably of
the order of
300 MPa).

5.3.2 *Friction between minerals*

Strength at actual contact points between grains of soil or rock is largely derived from electrochemical bonds over the true area of contact, which is only a very small proportion of the apparent cross-sectional area of a sample. At each contact between grains, elastic deformation, plastic flow and dissolution may take place, spreading the contact point so that the actual contact area is directly proportional to normal load. The attractive force over the true area of contact gives rise to frictional behaviour (Hardy & Hardy, 1919; Terzhagi, 1925; Bowden & Tabor, 1950, 1964). Bowden & Tabor, in particular, established that the area of asperity contact changed linearly with normal load for metals by measuring electrical resistance across the junctions. Power (1998) carried out similar tests using a graphite-based, rock-like model material (Power & Hencher, 1996).

The lower-bound friction angles for dry samples of quartz and calcite is reportedly about 6 degrees but higher when wet (Horn & Deere, 1962). The opposite behaviour was reported for mica and other sheet minerals. Perhaps linked to Horn & Deere's observations, mineral species that reportedly give higher friction values when wet are the same minerals that commonly form strong bonds during burial diagenesis through dissolution and authigenic cementation (Trurnit, 1968). It is possible that the presence of water allows asperity contacts to grow in these minerals, even in laboratory tests. Conversely mica, chlorite and clay minerals are rarely associated with pressure solution bonding and inhibit pressure solution and cementation of quartz (Heald & Larese, 1974). Some authors have questioned whether Horn & Deere's data are valid because of possible contamination and natural soil does not exhibit the same phenomena (Lambe & Whitman, 1979), but there is other evidence that basic friction of rock-forming minerals can be so low. Hencher (1976, 1977) used repeated tilt tests on steel-weighted, saw-cut samples of sandstone and slate to reduce the sliding angle from about 32 degrees to almost 12 degrees, which is approaching the low values of Horn and Deere. The reduction in strength was attributed to polishing (Figure 5.2).

5.3.3 *Friction of natural soil and rock*

Whilst basic friction the lower bound of minerals, originating from adhesion at asperities, might be of the order of 10 degrees or even lower, friction angles even for planar rock joints and non-dilational soil are often greater than 30 degrees yet the additional resistance (above basic) is still directly proportional to normal load. This additional frictional component varies with surface finish of planar rock joints and can be reduced by polishing (Coulson, 1971) or by reducing the angularity of sand (e.g. Santamarina & Cho, 2004). Figure 5.3 shows results from two series of direct shear tests on saw-cut and ground surfaces of granite. As

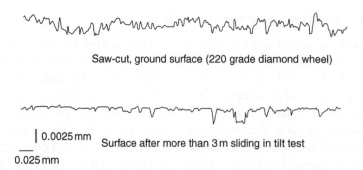

Saw-cut, ground surface (220 grade diamond wheel)

| 0.0025 mm   Surface after more than 3 m sliding in tilt test
—
0.025 mm

Figure 5.2 Ground and polished saw-cut surface of Delabole Slate at high magnification (top) and following repeated sliding tests (bottom).

shown in Figure 5.2, at a microscopic scale such ground and apparently flat surfaces are still rough. Each data point in Figure 5.3 is taken from a separate test with the sample reground beforehand. The upper line (inclined at 38 degrees) is the friction angle measured for moderately weathered (grade III) rock; the lower line inclined at 32.5 degrees is for slightly weathered (grade II) rock. The reason for the higher strength for the more weathered surface is because the surface finish is slightly rougher, the weathered feldspars being preferentially plucked from the surface during grinding. The key observation, however, is the precision of the frictional relationships – an increase in strength that is directly proportional to the level of normal load. Scholtz (1990) reviews the origin of rock friction and concludes that the additional strength is derived from deformation and damage to small-scale textural roughness. It is quite remarkable that this interlocking, non-dilational component still obeys Amonton's laws of friction.

The third contact phenomenon is dilation. Additional work is done against the confining normal load during shear as soil moves from a

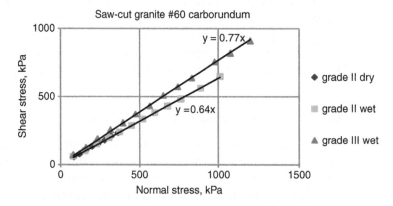

Figure 5.3 Perfect linear, frictional relationships between shear strength and normal stress for saw-cut and ground surfaces of rock. The upper line (stronger) is for moderately weathered granite, the lower for stronger, slightly decomposed rock. This paradox is explained by the fact that in the grade II rock the various mineral grains are of similar scratch resistance and therefore the surface takes a better polish during grinding than the more heterogeneous grade III rock.

Figure 5.4
Measured strength
envelope with
apparent cohesion
and friction, which
can be corrected to
a basic friction line
(non-dilational).

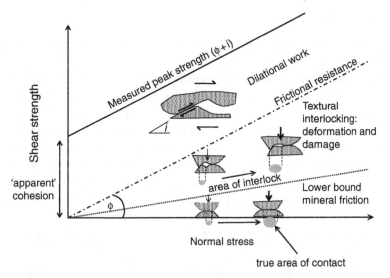

dense to a less dense state or as a rock joint lifts over a roughness feature. If the raw strength data from a test are plotted against normal stress, then the peak strength envelope may show an intercept on the shear strength axis (apparent cohesion), albeit that the peak strength envelope may be very irregular, depending upon the variability of the samples tested. If corrections are made for the dilational work during the test, in many cases the strength envelope will be frictional: the strength envelope passes through the origin. At very high stresses, all dilation will be constrained and the soil or rock asperities will be sheared through without volume change. These concepts are illustrated schematically in Figure 5.4.

5.3.4   *True cohesion*

Rocks and natural soil may also exhibit true cohesion, due to cementation and chemical bonding of grains. For a rock joint, it is derived from intact rock bridges that need to be sheared through. This additional strength, evident as resistance to tension, is essentially independent of normal stress and proportional to sample size. This is discussed further below.

5.3.5   *Geological factors*

In Chapter 1 (Figure 1.5), the concept of a rock cycle was introduced whereby fresh rock deteriorates to soil through weathering and then sedimented soil is transformed again into rock through burial, compaction and cementation. Clearly, at each stage in this cycle the geomaterials will have distinct properties and modes of behaviour.

5.3.5.1   *Weathering*

In fresh igneous and metamorphic rocks, the interlocking mineral grains are linked by strong chemical bonds. As illustrated in Figure 5.5, there is almost no void space, although there may be some tiny fluid inclusions trapped within mineral grains. As weathering takes place close to the Earth's surface and fluids pass through the rock, it develops more voids as minerals decompose chemically and weathering products such as clay are washed out. The bonds between and within individual grains are weakened. Figure 5.6 illustrates how rock that starts off with a dry density of about 2.7 Mg/m^3 (typical of granite) becomes more and more porous so that by the completely

Figure 5.5 Thin section through granite, illustrating tightly interlocking fabric. Width of view approximately 20 mm.

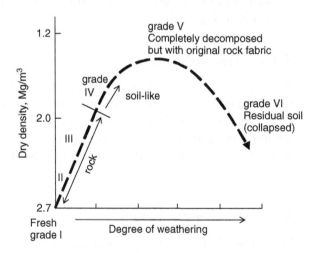

Figure 5.6 Change of dry density in weathered granite. The lowest value is for grade V, completely decomposed material, at which stage the density can be as low as 1.2 despite still having the appearance of granite (fresh state 2.7). At that stage, the material is prone to collapse to a denser, reworked, grade VI state. Based on Lumb (1962).

decomposed stage, the dry density may be reduced by more than 50% if weathering products have been washed out. The final stage is collapse to residual soil and an increase in density. Weathering is discussed in detail in Chapter 3.

Geotechnical properties at the material scale are linked quite closely to density empirically and, therefore, degree of deterioration from the rock's fresh state. Fresh granite might have a uniaxial compressive strength of perhaps 200 MPa but by the time the rock is highly decomposed the strength is reduced to 10–15 MPa and when completely decomposed perhaps 10–15 kPa. Where the rock is relatively strong, then properties and behaviour will be dominated by contained fractures; for most projects, the point at which material strength begins to dominate design decisions is where the rock can be broken by hand.

At the mass scale in weathered profiles, strength and deformation might be affected by the presence of strong corestones of less weathered rock in a weakened matrix, and the problem of characterisation is similar to that of mixed soils and rock such as boulder clay or boulder landslide colluvium, as discussed later.

Permeability in fractured rock or in weathered profiles can be extremely variable and difficult to predict, with localised channel flow providing high permeability. Elsewhere, accumulations of clay or general heterogeneity in the profile can prevent and divert water flow. The complexities of flow through weathered rock profiles and difficulties in measuring permeability are discussed in Chapters 3 and 4.

5.3.5.2 *Diagenesis and lithification (formation of rock from soil)*

As discussed in Chapter 3, soil is transported by water, wind or gravity from the parent rock. During the process of transportation, the sediment is sorted in size. Some soils such as glacial moraine and colluvium remain relatively unsorted. Sediments tend to be continually deposited over a very long period of time, for example, in river estuaries, and each layer of sediment overlies and buries the earlier sediment. The underlying sediment is compacted and water squeezed out. This is termed burial consolidation and is a very important process governing the strength and deformability of sediments. Grains become better packed, deformed and may form strong chemical bonds with interpenetration and sutured margins. Voids may be infilled with cement precipitated from soluble grains in the sediment (authigenic cement) or from solutions passing through the sediment pile, as illustrated in Figures 5.7 and 5.8. Many clay oozes initially have a very high percentage of voids, with the mineral grains arranged like a house of cards. With time, overburden stress and chemical changes cause the flaky minerals to align and the porosity (or void ratio) to decrease markedly, as illustrated in Figure 5.9. Burland (1990) has expressed the rate at which void ratio is reduced with burial depth as a normalised equation although there are

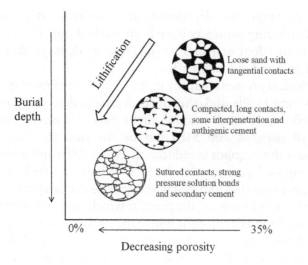

Burial depth

0%   ←————————————   35%

Decreasing porosity

Figure 5.7 Compaction and cementation of granular soils with burial leading to increased strength, reduced deformability and lower permeability.

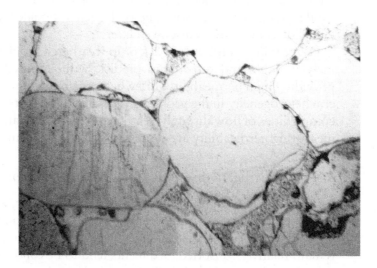

Figure 5.8 Thin section of aeolian sandstone with rounded grains of quartz, interpenetration of grains and flattened surfaces where in contact, with some pressure solution, plus authigenic cementation of grains by silica and iron oxides. As a result of these diagenetic processes, the material has been turned from loose sand into a strong rock. Triassic Sandstone, UK. Large grains about 5 mm in diameter.

often departures from this behaviour in natural sediment piles, due largely to cementation (Skempton, 1970; Hoshino, 1993). The changes in property (especially strength and deformability) that ensue from burial, compaction and consolidation are discussed in Section 5.5. At some locations, the upper part of the sediment pile is considerably stronger than might be anticipated from its shallow burial level because it has become desiccated on temporary exposure above water level. Where soils are uplifted and upper levels eroded, or otherwise loaded, and then that load removed (e.g. by the melting of a glacier), then the strength and stiffness will be relatively high and the soil is termed

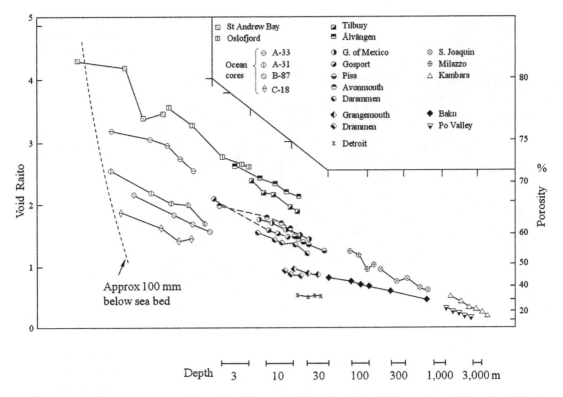

Figure 5.9 Compression curves for naturally consolidated and partially cemented clay (modified from Skempton, 1970).

overconsolidated. In the case of sand, the history of burial compaction can result in an extremely dense arrangement of the sand particles that cannot be replicated in the laboratory. Such locked sands, with grains exhibiting some interpenetration and authigenic overgrowths, not surprisingly, have high frictional resistance and dilate strongly under shear (Dusseault & Morgenstern, 1979).

5.3.5.3   Fractures

Natural fractures occur in most rocks close to the Earth's surface and in many soils once they begin to go through the processes of burial and lithification. Figure 5.10 shows a quarry face where discontinuities dominate mass geotechnical parameters such as deformability and permeability. Vertical joints in relatively young glacial till are shown in Figure 5.11. Fractures will often dominate fluid flow through the mass, as well as mass deformability and strength. They need special consideration and characterisation, as addressed in Chapters 3 and 4 and discussed later.

5.3.5.4   Soil and rock mixtures

Many soils such as glacial boulder clay and colluvium comprise a mixture of finer soil and large clasts of rock, and these need special

Zone of intense jointing

Zone of sparse jointing

Figure 5.10
Predominantly
vertical jointing
(probably
combined cooling
and tectonic during
emplacement) in
granite. Mount
Butler Quarry,
Hong Kong.

Figure 5.11
Vertical joints
developed in
boulder clay. Robin
Hood's Bay, North
Yorkshire, UK.

Figure 5.12
Options for slope
stability analysis.
After Hencher &
McNicholl, 1995.

Option	Schematic diagram	Approach for defining parameters and analysis
1. Treat as uniform (continuum)	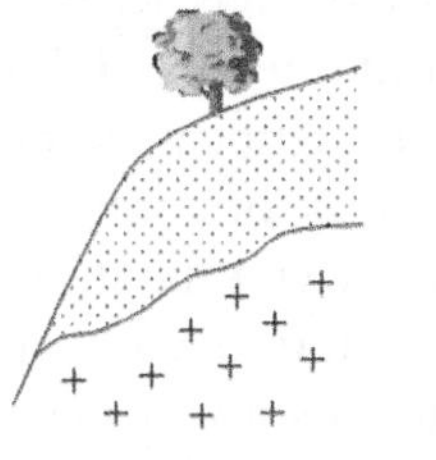	• parameters from laboratory or *in situ* tests taken to be representative of zone
2. Treat as uniform but weakened by discontinuities (continuum)	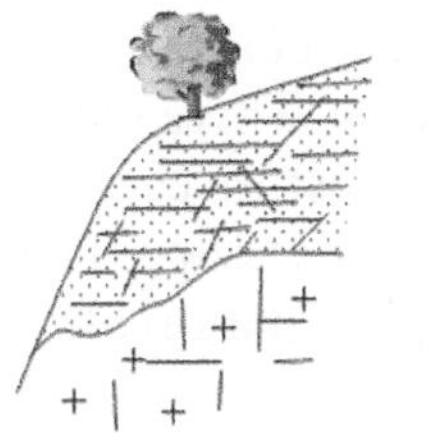	• allowance made for influence (but not control) of discontinuities on mass properties (e.g. Hoek-Brown)
3. Treat as heterogeneous (continuum)	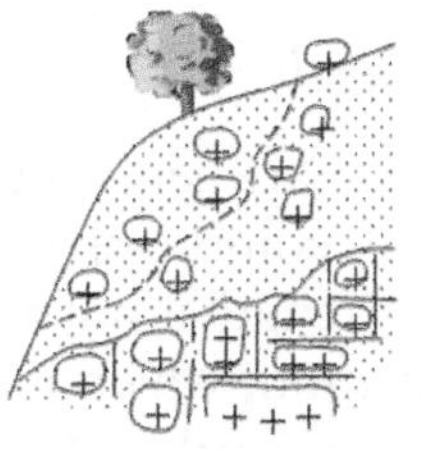	• consideration given to influence of strong inclusions with deviated failure paths
4. Treat as discontinuous due to structural control	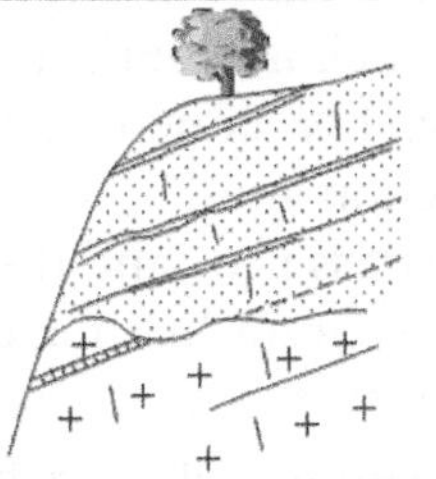	• discontinuity controlled

consideration in terms of their properties. Weathered rocks can similarly comprise mixes of weak and hard materials but there is also the added complication of relict rock fabric and structure. The overall nature of the mass will strongly affect the options for engineering assessment, as illustrated for slopes in Figure 5.12. Geotechnical parameter determination for such mixed deposits is considered in Section 5.8.

5.4   Measurement methods

Methods of testing soil and rock are specified in standards such as BS 1377 for soil in the UK (BSI, 1990), BS 5930 for several field tests (BSI, 1999) and ASTM, more generally in the USA. The International

Society for Rock Mechanics provides guidance on many field and laboratory tests (Ulusay & Hudson, 2006). Recommendations for the same test sometimes differ, for example regarding sample dimensions and testing rate, so care has to be taken that an appropriate method is being adopted and referenced. Furthermore several different techniques or different equipment can sometimes be used ostensibly to measure the same parameters but inevitably with different results. For example, small strain dynamic tests may give very different values for soil stiffness compared with large-scale loading tests but each might be appropriate to some aspect of numerical analysis and design within a single project (Clayton, 2011). It should also be remembered that, however much they are standardised, all tests on soil and rock are experiments. There will be many variables, not least the geological nature and moisture content of the sample to be tested, so interpretation is always necessary. Further judgement is required before attempting to apply small-scale results at the larger scale (e.g. Cunha, 1990).

5.4.1   *Compressive strength*

Intact rock, clay and concrete are generally classified in shorthand by their unconfined (or uniaxial) compressive strength (UCS) as discussed in Chapter 4. Compressive strength is not a relevant concept for purely frictional materials such as sand, which must be confined to develop shear resistance. Indicative UCS values for various materials are presented in Table 5.1; fresh rock is often considerably stronger than the highest strength concrete. For concrete, UCS is used as a quality assurance test on construction sites.

In a UCS test the axial stress is σ_1 and the confining stresses (σ_2 and σ_3) are zero. Despite the apparent loading condition, the sample does not actually fail in compression but either in tension or in shear or in some hybrid mode. If the sample contains adverse weak fabric such as incipient joints or cleavage, then the sample will fail at lower strength than it would without the flaws. UCS is really essentially an index test used especially in rock mass classification. In practice strength can often be estimated quite adequately using index tests such as hitting with a geological hammer (see Box 5-1). UCS can also be measured using point load testing, which is quick and easy, but correlation with UCS from laboratory testing may be imprecise. The Schmidt hammer is sometimes used to estimate strength using standard impact energy to measure rebound from a rock or concrete surface. It is sensitive to surface finish and any fractures behind the impact location will cause low readings. It is also insensitive to strength over about 100 MPa. It is generally unsuitable for testing rock core – its main use in engineering geology is as an index test to help differentiate between different degrees of weathering as discussed in Chapter 4.

Table 5.1 Indicative unconfined compressive strengths for some rock, soil and concrete.

Material	Uniaxial Compressive Strength, UCS MPa	
Natural rock and soil		
Fine-grained, fresh igneous rock such as dolerite, basalt or welded tuff, crystalline limestone	>300	Rings when hit with geological hammer
Grade I to II, fresh to slightly weathered granite	100–200	Difficult to break with hammer
Cemented sandstone (such as Millstone Grit)	40–70	Broken with hammer
Grade III, moderately weathered granite	20–40	
Chalk Grade IV highly weathered granite	5–30	Readily broken with geological hammer Weaker material broken by hand
Overconsolidated clay	0.6–1.0	Difficult to excavate with hand pick
Very stiff clay-rich soil	0.3–0.6	Indented with finger nail
Concrete		
High-strength concrete (e.g. Channel Tunnel liner)	50–100	
Typical structural concrete	30–50	
Shotcrete in tunnel	20–40	

Box 5-1 To test or not to test?

Many ground investigations are wasteful in that they do not target or identify critical geological features, and laboratory tests are commissioned without real consideration of whether or not they will be useful.

Example 1

Figure B5-1.1 shows the formation level (foundations) for the Queen's Valley Dam, Jersey, which was completed in 1991. The dam was to be an earth dam, which exerts relatively low stresses on its foundations, compared to a concrete dam such as an arch or gravity dam. With a maximum height of 24 m and an assumed unit weight of 20 kN/m^3, the bearing pressure might be of the order of 500 kPa. The author, who was mapping the foundations, was asked to select samples of core to be sent to the laboratory for uniaxial compressive strength testing.

Rock over much of the foundation was rhyolite that was extremely difficult to break by geological hammer and had an estimated compressive strength of more than 300 MPa. The rhyolite, however, contained numerous incipient fractures (Figure B5-1.2), which would mean that the mass strength was somewhat lower and, more significantly, would cause samples to fail prematurely in the laboratory. The author argued that if the samples were sent to the laboratory, the reported result would simply be scattered with a range from 0 to 300 MPa and what would that tell us that we didn't already know? The allowable bearing pressure for rock of this quality (Chapters 6) would be at least five times the bearing pressure exerted by the dam. In the event, the samples were still sent off to the laboratory for testing (because they had already been scheduled by the design engineers) and the money was duly wasted.

Figure B5-1.1 View of left abutment of Queen's Valley Dam, Jersey, UK, under construction.

Figure B5-1.2 Extremely strong rhyolite. Hammer and clipboard for scale.

Example 2

The Simsima Limestone is the main founding stratum in Doha, Qatar, and is found extensively across the Middle East. It is a highly heterogeneous stratum including calcarenite, dolomite and breccia. The rock is often vuggy and re-cemented with calcite. RQD can be very high, with sticks of core a metre or more in length without a fracture; elsewhere the RQD is zero. An example is shown in Figure B5-1.3.

The properties of the stratum are clearly important for design of foundations and for other projects such as dredging, as discussed in Chapter 3. UCS test data tend to be very scattered, in part because the integral flaws in many samples lead to early failure. If a strongly indurated sample with few flaws is tested,

Figure B5-1.3 Example of core through Simsima Limestone (courtesy of Karim Khalaf, Fugro, Middle East).

then it can give UCS strength of 60 or 70 MPa (higher than structural concrete). Samples of inherently weaker material (as could be estimated from scratchtesting) or containing vugs or other flaws, will fail at much lower strengths. A typical range of data is given in Figure B5-1.4. If smaller intact pieces of dolomitised limestone are point load tested selectively, they will, of course, err towards the higher strength of the rock mass. As a consequence, conversion factors from point load test to UCS for this rock are usually taken empirically as 8 to 9 (Khalaf, personal communication). Data converted in this way are included in Figure B5-1.4. For more uniform rocks elsewhere in the world, conversion factors of about 22 are more commonly applied (Brook, 1993). If such a factor were to be used for the Simsima Limestone, then it would imply strength for the intact limestone, without flaws, up to about 200 MPa.

Given this very wide range of possible strengths, it would seem unwise simply to rely on a statistical testing campaign for characterising the rock mass. Far better to try first to characterise the rock geologically into units based on the strength of rock materials and then mass characteristics including flaws, degree of cementation and degree of fracturing. In this case, index tests (hammer, knife), combined with visual logging and selective testing of typical facies, are likely to give a far better indication of mass properties than UCS testing alone. To obtain parameters for the large scale (say foundations) then *in situ* tests such as plate loading and perhaps seismic tests would help, as would full-scale instrumented pile testing. Where rock mass strength is very important, as for the selection of dredging equipment, then it would be very unwise to take UCS data at face value (as a statistical distribution). As for many tests, there are numerous reasons why values measured in the laboratory might be unrepresentative of conditions *in situ*, often too low, and considerable judgement is required if the parameters are critically important.

Lesson: compressive strength of most rocks can often be estimated adequately by hitting with a hammer and the use of other index tests; if a hard blow by a hammer cannot break the material, then its strength probably exceeds that of any concrete structure to be built upon it. Where strength is critical, as in the selection of a tunnelling machine or choice of dredging equipment, then any test data must be examined

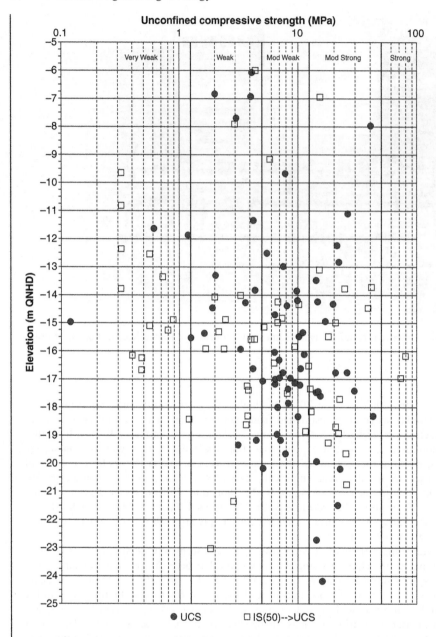

Figure B5-1.4 Typical UCS data from Simsima Limestone (courtesy of Karim Khalaf, Fugro, Middle East).

critically. If laboratory test samples contain flaws such as discontinuities, then measured intact strength may be too low. Of course, at the mass scale, the flaws and joints will be extremely important but their contribution cannot be properly assessed by their random occurrence and influence on laboratory test results.

5.4.2 Tensile strength

Although rocks actually usually fail in tension rather than compression, tensile strength is rarely measured directly or used in analysis or design, compressive strength being the preferred parameter for rock mass classifications and empirical strength criteria (see later). Tensile strength of rock and concrete is relatively low, typically about 1/10th of UCS. It is because of the weakness of concrete in tension that reinforcing steel needs to be used wherever tensile stresses are anticipated within an engineering structure.

5.4.3 Shear strength

Shear strength is a very important consideration for many geotechnical problems, most obviously in landslides where a volume of soil or rock shears on a slip surface out of a hillside. It is also important for the design of foundations and in tunnelling (Chapter 6). There are two main types of test used to measure shear strength in the laboratory – direct shear and triaxial testing. There are also many other *in situ* tests used to measure shear strength parameters, either directly (e.g. vane test) or indirectly (e.g. SPT and static cone penetrometer tests), and these have been introduced in Chapter 4.

For persistent (continuous) rock discontinuities, direct shear testing is the most appropriate way of measuring shear strength. Details of testing and interpretation are given in Hencher & Richards (1989) and Hencher (1995). Because of the inherently variable roughness of different natural samples, dilation needs to be measured and results normalised, as discussed later. If this is not done then, in the author's opinion, the tests are usually a total waste of time. The details of a shear box capable of testing rock discontinuities and weak rocks with controlled pore pressures is described by Barla *et al.* (2007).

Direct shear tests are also carried out on soil and are much easier to prepare and conduct than tests on rock discontinuities, although the stress conditions are not fully defined in the test, which can cause some difficulties in interpretation (Atkinson, 2007). This is one reason why triaxial testing is preferred for most testing of soils. Other advantages are that factors like drainage and pore pressure measurement can be carefully controlled. A disadvantage is that the soil may well become disturbed during trimming and preparation for the test as well as during back saturation and loading/unloading, but that is a problem for all testing. In a triaxial test, the cylindrical sample is placed inside a cell and then an all-around fluid pressure applied (σ_3). This remains the constant minimum principal stress throughout the test. Some tests are carried out drained, in that water is allowed to seep out of the sample as it is compressed; in others drainage is prevented, the water pressure changes as the sample is loaded and can be measured. In some tests the sample is initially loaded and consolidated to a required effective stress in an attempt to simulate the field condition. Once the sample is in equilibrium, it is gradually compressed axially whilst the confining stress remains constant. The process is illustrated graphically using Mohr stress circles in Figure 5.13. Note that within the sample, the angle between σ_1 and σ_3 is 90 degrees, but in the Mohr circle presentation, this stress field is expressed as a hemisphere (180 degrees). The hemisphere represents the stress state on any plane drawn through the sample. The test proceeds from the state where $\sigma_1 = \sigma_3$, then σ_1 is increased (hemispheres grow towards the right) until the sample eventually fails. Normal stress on any plane through the sample is measured on the horizontal axis, shear stress on the vertical axis. The stress normal to a vertical plane through the sample is σ_3 and the shear stress is zero; the normal stress on a horizontal plane through the sample is σ_1, the shear stress zero. These planes are known as principal planes. For a plane inclined at 10 degrees (shown as 20 degrees graphically within the Mohr circle) the normal stress on that plane is σ_{10} and at 45 degrees it is σ_{45}, with the corresponding shear stress (τ), as indicated. At failure, the shear plane through the sample will be developed at some angle ($\theta/2$ degrees) to the horizontal, expressed as θ in the Mohr circle graph. The Mohr stress circle representing the stress state at that stage is shown in Figure 5.14 for a single test. Further tests would be carried out on other similar samples at different confining stresses and used to define a strength envelope (a line joining the stress states at which all samples failed). Usually the envelope for a set of samples can be defined in terms of friction (gradient of line) and apparent cohesion, c, which is the intercept on the shear stress axis at zero normal stress (Figure 5.4).

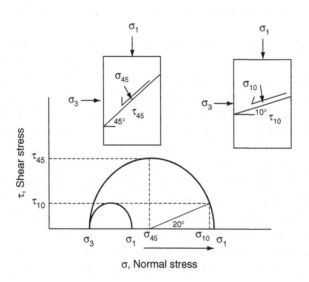

Figure 5.13 General representation of stress conditions in an individual sample.

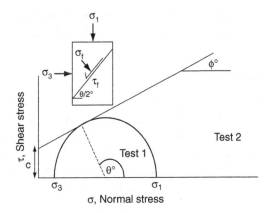

Figure 5.14
Mohr circle at shear
failure.

5.4.3.1   *True cohesion*

The nature, origin and even existence of cohesion – strength at zero normal load – causes considerable debate and confusion. This is partly because it can be either apparent (the result of dilation during a test and varying with confining stress) or a real physical entity and due to cementation, grain bonding or impersistence of discontinuities in the rock mass. Quite often both factors contribute to the measured strength in the same test, for example, if shearing intact rock. In artificially prepared samples of remoulded soil there is no true cohesion and apparent cohesion is a function of the density of packing of the soil grains relative to the confining stress. A theory of critical state soil mechanics has been developed for such soil that links shear strength to deformation characteristics (Roscoe *et al.*, 1958; Schofield, 2006). Burland (2008) however notes the importance of geological history to natural soils, with the development of bonding and fabric leading to true cohesional, non-dilational and stress-independent strength. While Burland was really discussing relatively young soils, it has been demonstrated earlier (Figure 5.1) how, with time, true cohesion can become very high and far outweigh the contribution of friction to shear strength. Conversely, as rock is gradually weathered it is primarily the cohesional strength that is lost – friction remains essentially constant.

5.4.3.2   *Residual strength*

After high shear displacement, cohesion is lost, and shearing continues at a residual friction level. This is non-dilational friction but in nature can be lower than the critical state – also non-dilational – because of change in structure with, for example, flattening and alignment of particles in a clay or the development of highly polished shear surfaces. Such strengths can be very low (sometimes of the order of 7 degrees for montmorillonite clay-rich rocks) and very significant, especially for landslides (see discussion of Carsington Dam failure in Chapter 7). To test residual strength, torsional ring shear boxes are used, in which an annulus-shaped sample is prepared and then rotated until a constant low strength is obtained.

5.4.4   *Deformability*

Young's Modulus (E) is expressed as stress/strain (with units of stress) and is a key parameter for predicting settlement of a structure or deformation in a tunnel and needs to be defined at a mass scale. For soil, samples are consolidated in oedometers and measurements taken of deformation against time. The main derived parameters are m_v, which is an inversion of E, i.e. strain/stress, and C_c, which is a measure of rate of consolidation. For normally consolidated clay that has been simply buried by overlying sediment, there will be a gradual improvement in strength and stiffness with depth, as illustrated for natural soils in Figure 5.9. Soil that has been overconsolidated because of its geological history will exhibit relatively high stiffness up to the loading level corresponding to its earlier pre-consolidation stress state. Once that pressure is exceeded, the stiffness will revert to the natural consolidation curve. At very small strains, overconsolidated clay can be much stiffer than at higher strain levels, and this can be important for realistic modelling of excavations (Jardine *et al.*, 1984; Clayton, 2011). Geophysical testing can be used to interpret stiffness parameters from velocities of wave propagation through soil, and values are again on the high side compared to static tests at relatively high strains (Mathews *et al.*, 2000). The same is true of rock masses – interpretation of compressional or shear velocities tend to give higher stiffness values than do static loading tests, and this probably reflects the low strain nature of loading from transient dynamic waves (Ambraseys & Hendron, 1968). Because of the difficulties in determining E at the rock mass scale from first principles or testing, it is common to rely on empirical published data as discussed at 5.6.3.

5.4.5   *Permeability*

Permeability is an intrinsic parameter of soil and rock, relating to rates of fluid flow through the material and strictly varies according to the fluid concerned – e.g. oil, water or gas. It has dimensions of area (L^2). In hydrogeology and geotechnical engineering, the term permeability is generally used interchangeably with hydraulic conductivity and is the volume of water (m^3) passing through a unit area (m^2) under unit hydraulic gradient (1m head over 1m length) in a unit of time (per second). This reduces to m/s. For low permeability rock suitable for a nuclear waste repository, the permeability, k, might be 10^{-11} m/s. For an aquifer of sandstone suitable for water extraction, it might be 10^{-6} m/s and for clean gravel 10^{-1} m/s. Typical values for other soils are given in BS 8004 (BSI, 1986).

In some soil such as alluvial sand, the material permeability could be similar to that of the mass, so laboratory testing might be relevant, but for many ground profiles water flow will be localised and involve natural pipes, fissures and open joints or faults. Field tests are then generally necessary to measure mass-scale permeability, as outlined in Chapter 4. Large-scale pumping tests from wells with observational boreholes at various distances can give reliable parameters for aquifer behaviour but localised testing in boreholes, as specified in BS 5930 (BSI, 1999), can be unreliable (Black, 2010). As discussed in Chapters 3 and 6 and illustrated in examples in Chapter 7, localised geological features often control fluid flow through the soil or rock mass, so testing must be linked to relevant geological and hydrogeological models.

5.5 Soil properties

5.5.1 Clay soils

As Skempton (1970) showed (Figure 5.9), for clay soil deposited offshore at rates of perhaps 2 m per thousand years, consolidation behaviour due to self-weight is fairly well defined. As the porosity diminishes and water is squeezed out, so strength increases and deformability reduces, even in the absence of other diagenetic processes. Hawkins *et al.* (1989), for example, show a consistent linear increase in shear strength with depth over 20 m at a test site in Bothkennar, Scotland, based on vane tests. Cone test data from the same site are very similar to other sites in the UK, confirming the trend. Similar results have been achieved from other sites worldwide, with a typical relationship:

$$s_u = 10 + 2.0d$$

Where s_u = undrained shear strength, kPa and d = depth below ground, m.

Elsewhere, values can be somewhat lower; for the Busan Clay in Korea, the gradient is closer to 1.0 times depth (Chung *et al.*, 2007). Nevertheless the trend is similar so for design in soft to firm clay it is usual practice to carry out a series of vane tests down boreholes or cone penetrometer soundings, and then try to define a relationship of increasing strength with depth that can easily be input to numerical simulations. Relationships are published both for shear strength and modulus of clay interpreted from SPT tests, and these are reviewed in Clayton (1995) although the SPT is less appropriate for clay than for granular soils. Most of the values obtained from field tests are necessarily undrained and expressed as a value of apparent cohesion with no frictional component. Undrained shear strength of clay can also be obtained from undrained tests in the laboratory and is estimated during field description using index tests like resistance to finger pressure or in a rather more controlled way using a hand penetrometer, as discussed in Chapter 4. Undrained strength is useful for assessing the fundamental behaviour of clay empirically, for example, in designing foundations (Table 6.1). It is also used for numerical analysis in soils of low permeability immediately after or during construction. Conversely drained conditions apply where excess pore pressures have dissipated following construction or where they dissipate relatively rapidly during construction. For design of structures in clay under drained conditions, effective stress parameters are required – friction and possibly some cohesion where there has been some geological bonding. These parameters are generally obtained from triaxial testing, in which pore pressures are monitored and corrected for throughout the test (e.g. Craig, 1992). Effective stress parameters can also be interpreted from *in situ* piezocone penetrometer soundings (Chapter 4).

Laboratory tests are relied upon for characterising natural clay far more than for any other soils, because reasonably undisturbed samples can be taken and the small grain size relative to testing apparatus means that scale effects are not evident. An exception is in settlement analysis, where it is found that standard oedometer tests give lower stiffness than larger-scale plate load tests or are evident from back analysis of the construction of a structure. Specialised testing is necessary to simulate low strain deformation (e.g. Atkinson, 2000).

As noted earlier, for some active and ancient landslides, the strength along the slip plane through clay/mudstone is reduced below the critical state friction angle to a residual friction angle well below 20 degrees, even for clay of relatively low plasticity such as kaolinite or illite (Skempton *et al.*, 1989). Such low values can be measured in the laboratory using ring shear boxes and back-analysed from landslide case histories.

Clays include some groups of very problematical soils. Quick clays are clay and silt size but mostly detrital materials (rock flour produced by glacial scour), weakly cemented by salt, which can become disturbed and then flow, sometimes to disastrous effect. The Rissa, Norway, landslide in 1978 was filmed, flowing rapidly across flat ground, indicating the sensitivity of such materials. Other clays such as black cotton soils swell and shrink dramatically with changes in moisture, which causes damage to roads and other structures. The clay mineral group smectite (montmorillonite/bentonite) is most commonly associated with volume change and is typically identified by X-ray testing. Its presence is also indicated from high liquid limits and high plasticity indices in Atterberg limit tests (Chapter 3). These clay minerals can have very low shear strengths. Starr *et al.* (2010) describe a creeping major rock slope failure where the rock is smectite-rich and for which the operating residual friction angle was only about 7 degrees as established by numerical back-analysis and confirmed from laboratory tests.

5.5.2   *Granular soil*

The behaviour of granular soil such as silt, sand and gravel can be examined in the laboratory but for design, geotechnical parameters are generally determined by *in situ* testing, because of the difficulties of a) obtaining and transporting undisturbed samples and b) the problems of scale effects in testing samples of large grain size.

The most common test for characterising silt, sand and gravel is the SPT, as discussed in Chapter 4. Measured resistance needs to be corrected for various influences, including overburden pressure and the silt content of sand. Resistance may be affected by water softening in the base of a borehole. Details are given in Clayton (1995). SPT N-values are used to infer a range of properties, including density (unit weight), friction angle and deformability which are then used for the design of many types of structure, including foundations, retaining walls and slopes. CPT tests can also be used in this way and are particularly useful for the design of offshore structures.

5.5.3   *Soil mass properties*

Usually, properties of intact soils of sedimentary origin are assumed to be representative of the larger soil mass layer or unit. This can be an over-simplification in that even quite recent soils can contain fractures and systematic joints and many are layered with different layers having different properties. In the latter case, permeability parallel to bedding might be orders of magnitude higher than at right angles to bedding, and there are many geotechnical situations where such a condition would be important. McGown *et al.* (1980) discusses origins of fractures in soil and how they might be dealt with when assessing geotechnical properties. London Clay, for example, contains many fissures that can be interpreted using structural geological techniques (Fookes & Parrish, 1969). Chandler (2000) describes the significance of bedding parallel flexural-slip surfaces extending at least 300 m in London Clay. Similar features are discussed by Hutchinson (2001).

5.6   Rock properties

5.6.1   *Intact rock*

5.6.1.1   *Fresh to moderately weathered rock*

Fresh to moderately weathered rock, by the definitions adopted here (Chapter 3 and Appendix C), cannot be broken by hand at the intact sample scale, as in a piece of core. That being so, it has an unconfined compressive strength of at least 12.5 MPa and is definitely rock-like in that it could carry most structures without failing (presumed bearing capacity of at least 1 MPa according to Table 6.1) and will not fail in a man-made slope, in the absence of discontinuities, almost irrespective of height and steepness.

The strength of fresh rock is a function of its mineralogy, internal structure of those minerals (cleavage), grain size, shape and degree of interlocking, strength of mineral bonds, degree of cementation and porosity. Some rocks have intact strength approaching 400 MPa – these might include quartzite, welded tuffs and fine- and medium-grained igneous rocks such as basalt and dolerite. Corresponding intact moduli can be as high as 1×10^6 MPa (1×10^3 GPa) (Deere, 1968).

Compressive strength is measured most accurately using very stiff servo-controlled loading frames, whereby, as the rock begins to fail, so the loading is paused to limit the chance of explosive brittle failure. Such test set-ups allow the full failure path to be explored, which can be important in underground mine pillars where, despite initial failure in one pillar, there may be sufficient remnant strength after load is transferred to adjacent pillars, so that overall failure of the mine level does not occur. For most civil engineering works, UCS values measured by less sophisticated set-ups are adequate. Nevertheless, the specification for UCS testing is onerous, particularly regarding test dimensions and flatness of the ends of samples. If these requirements are not followed, local stress concentrations can cause early failure. If samples are too short, then shear failure might be inhibited. As noted in Box 5-1, there are alternative ways of estimating UCS that might be adequate for the task at hand.

UCS is the starting point for many different empirical assessments of rock masses, including excavatability by machinery such as tunnel boring machines. Other parameters that might need to be quantified include abrasivity and durability. Appropriate tests are specified in the ISRM series of recommended methods (Ulusay & Hudson, 2006).

Intact rock modulus is rarely measured for projects and is not usually an important parameter for design. An exception is in numerical modelling of fractured rock mass, e.g. using UDEC (Itasca), where this parameter is required, but for this purpose, values are typically estimated from published charts or even selected to allow the model to come to a solution within a reasonable time. Models tend to be insensitive to the chosen parameter.

5.6.1.2   Weathered rock

Intact weathered rock has true cohesion from relict mineral bonding. In some cases there may be secondary cementation, especially from iron oxides and the redistribution of weathering products within the rock framework. At the strong end of grade IV where it can just be broken by hand UCS might be about 12.5 MPa and cohesion of about 3 MPa might then be anticipated (Hencher, 2006). In practice, such high values have never been reported. Ebuk, who tested a range of

Figure 5.15
Peak strength
envelopes for
grades IV, V and VI
granite (based on
Ebuk, 1991). It is
highly likely that
Ebuk (and others)
have not carried out
or reported tests on
stronger grade IV
materials (or if so,
the author has not
seen them).

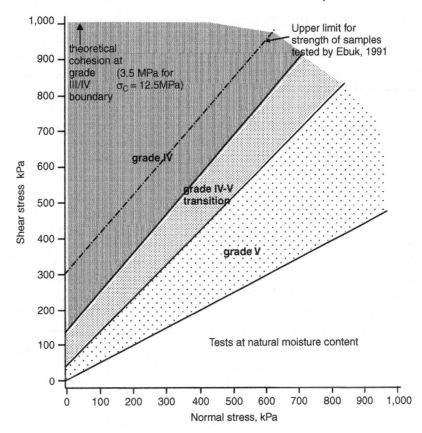

weathered rocks in direct shear, measured a maximum cohesion of 300 kPa for grade IV samples (Figure 5.15) but may have only been testing the weaker range of grade IV.

For design, parameters for weathered rock are often estimated from SPT N-value data. Tests are often continued to 100 or even 200 blows, which is questionable practice for many reasons, not least damage to equipment. In terms of rock mass modulus, E, a typical relationship adopted for design is:

$$E = 1.0 \text{ to } 1.2N \text{ MPa} \quad \text{(Hencher \& McNicholl, 1995)}$$

For foundation design, parameters such as side friction and end bearing are also often estimated from empirical relationships linked to SPT data. Full discussion of practice in Hong Kong is given in GEO (2006).

5.6.2 *Rock mass strength*

The presence of discontinuities in many rocks means that intact rock parameters from the laboratory are inappropriate at the field scale.

Therefore, many attempts have been made to represent the overall strength of the rock mass using simple Mohr-Coulomb parameters, friction and cohesion, based on overall rock quality, using classifications such as those presented in Appendix C. For example, using the Rock Mass Rating (RMR) of Bienawski (1989), 'poor rock' would be assigned cohesion 1–200 kPa with friction angle 15–25 degrees, and 'good rock' would be assigned cohesion 3–400 kPa with friction angle 35–45 degrees.

A rather more flexible and geologically realistic approach is to use the Hoek-Brown criteria (Hoek & Brown, 1997; Brown, 2008), which is linked to a Geological Strength Index (GSI) for rating overall rock mass conditions such as 'blockiness' and the roughness or otherwise of discontinuities. The GSI chart is presented and discussed in Appendix C. Given a GSI estimate, the uniaxial compressive strength for the rock blocks and a constant, m_i, which differs for different rock types and has been derived empirically from review of numerous test data (Hoek & Brown, 1980), one can calculate a full strength envelope for the rock mass. A program, RocLab, is downloadable from https://www.rocscience.com and allows values for cohesion and friction to be calculated but it needs to be checked that these relate to the appropriate stress level for the problem at hand. For example, Figure 5.16 shows a steep cut slope in weathered tuff. The question is whether it needs to be cut back or otherwise reinforced or supported. The rock mass is severely

Figure 5.16 Cut slope through weathered volcanic tuff.

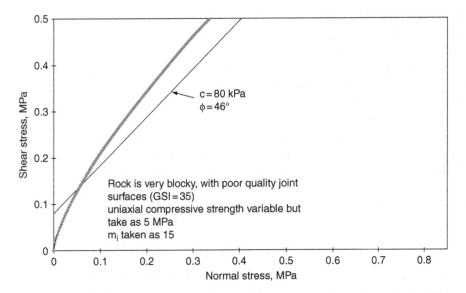

c = 80 kPa
φ = 46°

Rock is very blocky, with poor quality joint
surfaces (GSI = 35)
uniaxial compressive strength variable but
take as 5 MPa
m_i taken as 15

Figure 5.17 Strength envelope for slope in Figure 5.16, based on Hoek-Brown criteria (see text).

weathered. There are corestones of very strong tuff but these are separated and surrounded by highly and completely weathered materials
that are much weaker. There are many joints and some of these have
kaolin infill. In this case, there are no structural mechanisms for translational failure along daylighting joints, and it is a clear candidate for
where a Hoek-Brown/GSI approach might help the assessment. From
the GSI chart, one might best characterise the mass as 'very blocky' with
'poor' joint surfaces. The rock type is tuff, so the m_i value is 15 (for
granite it would be 33). The difficult parameter is intact strength. In this
case, the corestones have UCS values in excess of 100 MPa, but for this
assessment I have taken into account the strength of the weakest material making up this slope and, on balance, an average of 5 MPa is
considered conservative. Using a spreadsheet from Hoek *et al.* (1995)
modified for low stress conditions, the curve shown in Figure 5.17 is
obtained. On that basis, for a potential slip surface at a depth of about
10m (vertical stress say 0.27 MPa), appropriate strength parameters
might be c = 80 kPa and phi = 46 degrees, as shown. Carvalho *et al.*
(2007) discuss the assessment of rock mass strength where the intact
rock has relatively low uniaxial compressive strength in more detail.

5.6.3 *Rock mass deformability*

Rock mass modulus is very difficult to predict with any accuracy, and
measurements in boreholes or even by large *in situ* tests need to be
considered critically and certainly should not be used directly in design
without due consideration of the rock qualities of the zone tested
(including relaxation) vs. the larger mass volume. Back calculations

have been made from large projects, including dams and tunnels, and these data provide the main database for prediction (e.g. Gioda & Sakurai, 2005). Generally, poor quality, highly fractured rock (up to RMR = 50) will have a rock mass modulus increasing from soil-type values of perhaps 500 MPa to about 20 GPa with decreasing fracture spacing and increasing intact compressive strength. As the rock mass quality improves, so the modulus increases markedly, up to values of 60 GPa or so for good-quality rock with RMR = 80. Many authors have attempted correlations between a variety of rock mass classifications (RMR and Q especially) and rock mass modulus, but with considerable scatter. This is perhaps not surprising given the inherent difficulties of 1) trying to represent an often complex, heterogeneous geological situation as a single quality number and 2) the non-uniform loading conditions of any project vs. the measurement system (deficiencies of data).

Hoek & Diederichs (2006) carried out a detailed review and proposed optimised equations linked to the GSI classification. The best-fit equation obtained was:

$$E_{\mathrm{mass}}(\mathrm{MPa}) = 10^5 (1-D/2) / \left(1 + e^{((75+25D-GSI)/11)}\right)$$

where GSI is as taken from the chart in Appendix C (Table C11). The factor $D = 0$ for undisturbed masses, 0.5 for partially disturbed and 1.0 if fully disturbed. Hoek & Diederichs present a more refined version of this equation using site-specific data for intact strength and modulus, but in many situations the rock mass will not be uniform, so considerable judgement is necessary anyway. Richards & Read (2007) tried applying the Hoek-Diederichs equations to the Waitacki Dam in New Zealand, which was founded on greywacke, and found that the mass modulus was considerably underestimated for a judged GSI of 20, but examination of their data shows how sensitive any prediction is on the GSI adopted. As discussed elsewhere, features like joint spacing and continuity are extremely difficult to measure and characterise and very risky to extrapolate from field exposures because of variations with weathering and the structural regime. This all reinforces the need for considerable judgement and engineering geological expertise in establishing ground models, and caution when applying any empirical relationships.

Large-scale pile loading tests can provide data on rock mass deformation (Hill & Wallace, 2001). They found that published correlations based on RMR and Q classifications overestimated the *in situ* modulus for deep foundation design by up to one order of magnitude, but this was only a significant consideration where the Rock Mass Rating was below 40 (poor and very poor rock masses), and in such cases site-specific testing might be required. As discussed in Chapter 6, the increasing use of Osterberg-type jacks embedded in large-diameter bored piles will no doubt provide very useful data in the future for assessing deformability of rock masses and this, combined with sophisticated numerical modelling, is allowing refinements to the empirical approaches currently in use.

5.7   Rock discontinuity properties

5.7.1   *General*

The majority of rocks, and some soil masses near the Earth's surface, contain many discontinuities and these dominate mass properties, including strength, deformability and permeability. Discontinuities include bedding planes, cleavage, lithological boundaries, faults and joints. The origins, nature and development of discontinuities are discussed in detail in Chapter 3. For the rest of this discussion, I will discuss joints but this is generally relevant to other discontinuities. Many joints are initiated geologically as incipient weakness directions and only with time do they develop as full mechanical discontinuities, as illustrated in Figure 5.18 and discussed by Hencher & Knipe (2007). In this figure, the incipient cleavage in the slate below the unconformity with the Carboniferous Limestone generally has cohesion almost as high as the rock orthogonal to that cleavage direction. Nearby, however, cleavage and bedding has opened up due to exposure and

Figure 5.18
Variable development of cleavage and bedding features as mechanical discontinuities. Horton in Ribblesdale, West Yorkshire, UK.

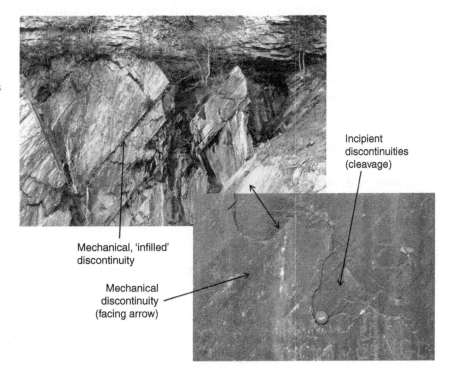

Incipient
discontinuities
(cleavage)

Mechanical, 'infilled'
discontinuity

Mechanical
discontinuity
(facing arrow)

Figure 5.19 Well-defined daylighting discontinuities, clearly only stable due to impersistence (cohesion), Taiwan.

weathering to form persistent joints with zero cohesion. Also shown is a bedding-parallel surface that is infilled with soil – actually a sedimentary feature. At intermediate stages, before rock joints become full mechanical fractures, sections of incipient fractures are cohesive and will contribute strongly to shear strength and shear stiffness along the discontinuity plane. This is illustrated in Figure 5.19. The persistence and shape of rock joints are very challenging parameters to measure or even estimate. Rawnsley (1990) tried to relate joint properties such as style and persistence to geological origin. He concluded, after studying numerous rock outcrops of wide geological age, that whilst persistence can be typified at the scale of joint sets, it is far less predictable at smaller scales (Rawnsley *et al.*, 1990). Zhang & Einstein (2010) review the situation and make some suggestions based on measurement, modelling and theory (see also discussion of DFN modelling in Chapter 3).

5.7.2 *Parameters*

The main properties of rock joints that need to be measured or estimated are shear strength, normal and shear stiffness and permeability/hydraulic conductivity. These properties depend on the geometry of the joints, including roughness, the nature, strength and frictional properties of the wall rock and any infill between the walls, and their tightness or openness.

Shear and normal stiffness of rock joints are not parameters that are normally required for civil engineering design but are needed as inputs when carrying out numerical simulations of jointed rock masses where each joint is modelled discretely using software such as UDEC. Guidance is given in the UDEC manuals (Itasca, 2004). Permeability of joints depends on their openness, tortuosity and connectivity. It is a very difficult but important subject area, especially for nuclear waste disposal considerations and tunnel inflow assessments (Black *et al.*, 2007).

5.7.3   *Shear strength of rock joints*

When dealing with rock slopes, often any discontinuity that appears that could be persistent, is treated as so (ignoring potential cohesion from rock bridges). This is a conservative thing to do (see discussion in Chapter 6) but there is little alternative. It is generally agreed that the shear strength of persistent joints is derived from some basic frictional resistance offered by an effectively planar natural joint, plus the work done in overriding the roughness features on that joint. This is expressed by the following equation (after Patton, 1966):

$$\tau = \sigma \tan(\phi_b^\circ + i^\circ)$$

where τ is shear strength, σ is normal stress, $\phi_b{}^\circ$ is a basic friction angle for a planar joint and i° is a dilation angle that the centre of gravity of the sliding slab follows during shear, i.e. the deviation from the direction that the shearing would have followed if the plane had been flat and sliding had occurred along the mean dip direction of the joint. Despite the apparent simplicity of the Patton equation, derivation of the parameters can be difficult, especially for judging the effective roughness angle. The available international standards and codes deal with this inadequately.

5.7.3.1   *Basic friction, ϕ_b*

Basic friction of natural joints can be measured by direct shear testing on rock joint samples, but samples taken from different sections of the same joint and joint set can be highly variable, particularly in terms of roughness. Furthermore, it is found that any rough rock joint sample will give different values for peak strength, depending on the direction of shear under the same normal load. Tests need very careful set up, instrumentation, analysis and interpretation, if they are to make sense. A series of tests on different samples of a joint will often yield very wide scatter, which is meaningless without correcting for sample-specific dilation, as described by Hencher & Richards (1989) and Hencher (1995). Dilation reflects work being done in overriding asperities. The dilation angle measured during a shear test will vary, especially according to the original roughness of the sample and the stress level. It is test-specific, will vary throughout a test and with direction of testing. It is not the same as the dilation angle, $i°$, which needs to be assessed at field scale, although the mechanics are the same. To avoid confusion, the laboratory-scale dilation angle measured during a test is here designated $\psi°$, whereas the field-scale dilation angle to be judged and allowed for in design is $i°$, as defined by Patton (1966).

Typically, because of the complex nature of shearing, with damage being caused to some roughness asperities whist others are overridden, the dilation angle, $\psi°$, is difficult to predict for an irregularly rough sample, although numerous efforts have been made to do so with some limited success (e.g. Kulatilake *et al.*, 1995; Archambault *et al.*, 1999). In practice, rather than trying to predict dilation, which will be unique to each sample, stress level and testing direction, it is a parameter that needs to be measured carefully during direct shear tests so that corrections can be made to derive a normalised basic friction angle for use in design. Figure 5.20 shows the result from the well-instrumented first stage of a direct shear test on a rough interlocking joint through limestone. The measured strength throughout the test includes the effect of the upper block having to override the roughness as the joint dilates and work is done against the confining pressure. The dilation curve in Figure 5.20 superficially appears fairly consistent, but if one calculates the dilation angle over short horizontal increments, from the same data set, it is seen to be much more variable and strongly reflects the peaks and troughs of the measured shear strength throughout the test (compare Figure 5.21 with Figure 5.20).

These instantaneous dilation angles can be used to correct (normalise) the shear strength incrementally throughout the test, using the following equations:

$$\tau_\psi = (\tau \cos\psi - \sigma \sin \psi)\cos \psi$$

$$\sigma_\psi = (\sigma \cos \psi + \tau \sin \psi)\cos \psi$$

where τ_ψ and σ_ψ are the shear and normal stresses corrected for positive dilation caused by sample roughness. The signs are reversed where compression takes place. By making such corrections, the basic friction angle can be determined for the natural joint surface. In practice, experience shows that for a system measuring to an accuracy of about $\pm0.005\,mm$, analysis over horizontal displacement increments of about $0.2\,mm$ generally gives accurate dilation angles, even for a rough tensile fracture (Hencher, 1995). By comparison, if one were to use the average dilation angle throughout the test, as implied in the ISRM Suggested Method (ISRM, 1974), this would not allow the variable shear strength to be understood and might lead to serious errors in determining basic friction values.

Tests can be run multi-stage, in which the same sample is used for tests at different confining stresses, which is very cost-effective, given

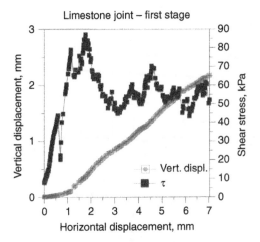

Figure 5.20 Results from single stage of direct shear test on rough induced tensile fracture through limestone. Upper curve shows very spiky shear stress against displacement. The lower line shows vertical vs. horizontal displacement (dilation) throughout this stage of the test. The line has a fairly consistent gradient.

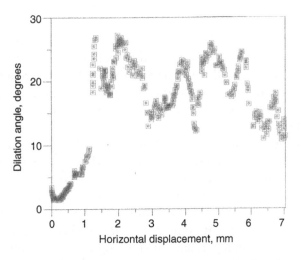

Figure 5.21 Detailed analysis of dilation curve from Figure 5.20 calculated over 0.2 mm horizontal increments. The revealed underlying spikiness in the dilation curve matches that of the shear strength curve in 5.20 and is clearly the cause of variable strength. Details of how the dilation can be corrected for to reveal the underlying basic friction are given in the text.

the difficulties of obtaining and setting up samples. At each stage, the normal load is generally increased (or decreased for experimental reasons) and then the sample sheared until peak strength plus a few mm. Tests must be properly documented, however, with photographs, sketches and profiles, so that any variable data can be explained rationally (Hencher & Richards, 1989). Generally, it is found that

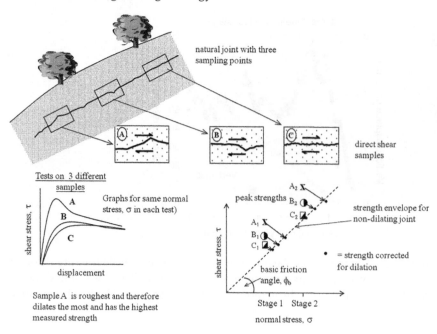

Figure 5.22
Methodology for
selecting a series of
samples of rock
joint, testing and
correcting to yield a
basic friction angle
for the naturally
textured rock joint
(after Hencher
et al., 2011).

tests on a series of samples from the same joint set (with similar surface mineralogy and textures) provide a reasonably well-defined dilation-corrected strength envelope, as illustrated in Figure 5.22. That strength is frictional (obeys Amonton's laws) and comprises an adhesional component plus a non-dilational damage component that varies with texture and roughness.

Barton (1990) suggested that the dilation-corrected basic friction angle might be partly scale-dependent, as assumed for the asperity damage component in the Barton-Bandis model (Bandis *et al.*, 1981), but further research using the same testing equipment as Bandis but with better instrumentation, indicates that this is unlikely (Hencher *et al.*, 1993; Papaliangas *et al.*, 1994). Dilation-corrected basic friction is independent of the length of the sample. Scale effects do need to be taken into account in design but as a geometrical consideration when deciding on an appropriate field scale $i°$ value.

Many silicate rocks are found to have a basic friction $\phi_b \approx 40$ degrees (Papaliangas *et al.*, 1995), and Byerlee (1978) found the same strength envelope ($\tau = 0.85\sigma$) for a large number of direct shear tests on various rock types where dilation was constrained by using high confining stresses. Empirically, it seems to be about the highest value for basic friction achievable for natural joints in many silicate rocks and applicable specifically to joints that are forced to dilate during shear or where dilation is suppressed because of the high normal load. Conversely, much lower friction angles can be measured for natural joints

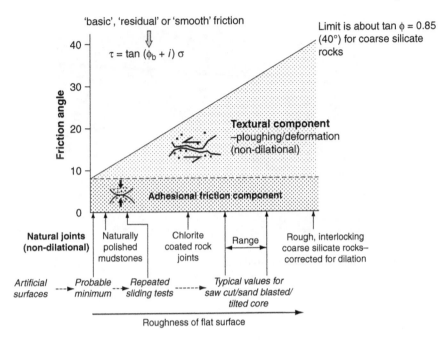

Figure 5.23 Concept of basic friction for a rock joint (after Hencher *et al.*, 2011).

where they are planar and where the surface texture is very fine, polished or coated with low-friction minerals, as illustrated by a case example in Box 5-2. The author has measured values of only 10 to 15 degrees for naturally polished joint surfaces through Coal Measures mudstones of South Wales, UK, and such low values are lower than measured for saw-cut surfaces through the parent rock. The range of variation for basic friction, measured for natural joints with different surface textures and for artificially prepared (saw-cut and lapped) joints, is indicated in Figure 5.23.

Box 5-2   Yip Kan Street landslide – an example of use of direct shear testing.

The Yip Kan Street landslide occurred in July 1981 on a dry Sunday night. It mainly involved large blocks of rock of up to $10 \, m^3$, which slid on persistent joint planes dipping at only 22 degrees out of the slope (Hencher, 1981b). The total failure volume was estimated to be $1,235 \, m^3$. The 8 m high, near-vertical slope was cut in very strong, slightly decomposed, coarse-grained igneous rock (quartz-syenite). The upper part of the slope was in saprolite. The failure occurred next to a construction site where blasting had been carried out recently, before the failure but not over the weekend. There had been intense rainfall a week before the failure. The slope had been deteriorating in the days preceding the failure, with cracks in chunam cover in the weathered part having been repaired five days before failure.

Figure B5-2.1 Failure plane with debris cleared off.

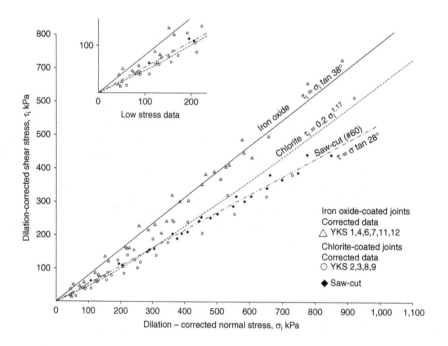

Figure B5-2.2 Shear strength data, Yip Kan Street rock slope failure.

Because of the low angle of sliding, it was decided to investigate in some detail. Blocks were collected – both matching discontinuities and mismatched. It was noted that some blocks were coated with red iron oxides and others with green chlorite (a hard, thin coating). Each sample was carefully described and then tested multi-stage in a Golder Associates direct shear box. At each stage, the test proceeded until peak strength was reached and then for another mm or two, following which the normal stress was increased, without resetting the sample in some cases. For some tests, complete runs of about 15 mm shear displacement were conducted and in one test the sample was tested at the highest stress level first, which was then reduced in stages incrementally. Samples were photographed, roughness measured and damage described carefully. For reference, a series of tests were conducted on saw-cut samples, ground with grade 60 carborundum powder.

Results from the tests are presented in Figure B5-2.2. Tests on natural joint surfaces were corrected for dilation incrementally. It can be seen that the saw-cut surfaces gave a friction angle of about 28 degrees, which is about what might be expected.

The tests from natural joints fall into distinct groups. The data from joints coated with iron oxides define a friction angle of 38 degrees, which is the same as one finds for many weathered rocks (Hencher *et al.*, 2011). The data for the chlorite-coated joints were much lower, however, and unexpectedly so. At low stress levels especially, values were very low, below that of the saw-cut joints, as can be seen from the inset figure and about the same as the angle of dip of the planes along which the failure took place ($\phi \cong 20$ degrees at the lowest stress levels). Field-scale roughness was measured at 5 degrees using a 420 mm diameter plate and 9 degrees using a base plate of 80 mm. It was concluded that the failure was progressive, probably having been exacerbated by blasting and previous rainfall and that the initial movements overcame the field-scale roughness. The eventual failure was explained by the presence of persistent chlorite-coated joints with inherently low frictional resistance (Brand *et al.*, 1983).

5.7.3.2 *Roughness*

Roughness at the field scale will often be the controlling factor for the stability of rough or wavy persistent joints and for engineering design must be added to the basic friction, ϕ_b, of the effectively planar rock joint, as determined from corrected shear tests. Roughness is expressed as an anticipated dilation angle, $i°$, which accounts for the likely geometrical path for the sliding slab during failure (deviation from mean dip). There are two main tasks for the geotechnical engineer in analysing the roughness component: firstly, to determine the actual geometry of the surface along the direction of likely sliding at all scales (Figure 5.24) and secondly to judge which of those roughness features along the failure path will survive during shear and force the joint or joints to deviate from the mean dip angle. This is the most difficult part of the shear strength assessment, not least because it is impossible to establish the detailed roughness of surfaces that are hidden in the rock mass. Considerable judgement is required and has to be balanced against the risk involved. Hack (1998) gives a good review of the options, and the difficulties in exercising engineering judgement are discussed in an insightful way by Baecher & Christian (2003).

In practice, the best way of characterising roughness is by measurement on a grid pattern in the way originally described by Fecker & Rengers (1971), adopted in the ISRM Suggested Methods (1978) and described in Richards & Cowland (1982), although spatial variability may be an important issue; the important first-order roughness represented by major wave features may vary considerably

Figure 5.24
Characterising
discontinuity
roughness using
plates of different
diameter. Skipton
Quarry, West
Yorkshire, UK.

from one area to another, as of course also might the mean dip of the plane. At one location, a block might be prevented from sliding by a wave in the joint surface causing a reduction in the effective down-dip angle along the sliding direction; elsewhere, a slab of perhaps several metres length may have a dip angle steeper than the mean angle for the joint as a whole because it sits on the down-slope section of one of the major waves. Defining the scale at which roughness will force dilation during sliding, rather than being sheared through, requires considerable judgement. Some assistance is provided by Schneider (1976) and by Goodman (1980) who indicate that for typical rough joint surfaces, where slabs are free to rotate during shear, as the length of the slab increases (at field scale), the dilation angle controlling lifting of the centre of gravity of the upper block will reduce. The problem cannot be finessed by improved analytical methodology. There is no substitution to careful engineering geological inspection, investigation, characterisation of the ground model and judgement based on experience of similar joints and geological settings, and an appreciation of the fundamental mechanics controlling the potential failure.

5.7.4 Infilled joints

The two walls of a joint might be separated by a layer or pockets of weaker material which may reduce shear strength. A similar situation arises from preferential weathering along a persistent

Figure 5.25 Cut slope at Rhuallt, North Wales, UK. Traversing the slope is a very persistent narrow stratum of weak clay which, combined with cross-cutting faults, provided the mechanism for major rock failure in this otherwise excellent-quality rock mass.

joint. The effect of the infill is a function of the relative height of roughness asperities in the wall rock vs. the thickness of weaker material (Papaliangas *et al.*, 1990). If persistent and the infill is of low strength, the consequences can be serious. Cut slopes on the A55 at Rhuallt, North Wales (Figure 5.25), failed by sliding on bedding-parallel thin clay infilled discontinuities with faults acting as release surfaces (Gordon *et al.*, 1996). The mechanism had not been anticipated from ground investigation prior to the failure, which involved more than 185,000 m^3 of rock.

In some slopes, incremental movement may take place over many years before final detachment of a landslide and, following each movement, sediment may be washed in to accumulate in dilated hollows on the joint (Figure 5.26). The presence of such infill might cause alarm during ground investigation but in many cases is confined to local down-warps and probably plays little part in decreasing shear strength, other than in restricting drainage (Halcrow Asia Partnership, 1998b). It may, however, be taken as a warning that the slope is deteriorating and approaching failure.

5.7.5 Estimating shear strength using empirical methods

Various empirical criteria have been proposed for estimating shear strength of rock joints, based on index tests and idealised joint shapes. The most widely used is that proposed by Barton (1973). Frictional resistance for saw-cut or other artificially prepared planar surfaces is taken as a lower bound. The limiting value is typically 28.5 to 31.5 degrees according to Barton & Bandis (1990). An additional

Figure 5.26
Patchily infilled
sheeting joints
following
intermittent
displacement prior
to failure. Details
are given in
Hencher (2006).

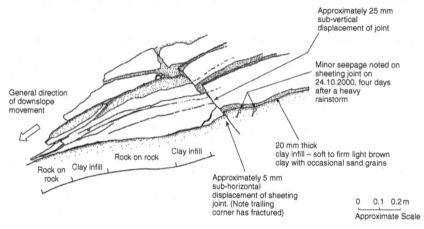

component is then added to account for roughness using a Joint
Roughness Coefficient (JRC) usually judged from standard profiles
and ranging from 0 to 20. This can be difficult in practice (Beer *et al.*,
2002). JRC is then adjusted for the strength of the rock asperities vs.
stress conditions and for scale. Details are given in Brady & Brown
(2004) and Wyllie & Mah (2004).The criterion can be incorporated
within numerical software for modelling rock mass behaviour such as
UDEC (Itasca, 2004). The contribution to shear strength from small-
scale roughness is measured or estimated from standard shape profiles (Joint Roughness Coefficient), although this can be difficult in
practice (Beer *et al.*, 2002). Larger-scale roughness (waviness) then
must be accounted for, over and above JRC, and scale corrections
applied.

   An important point that needs to be emphasised is that dilation-
corrected basic friction parameters from direct shear tests on natural

joints should not be used interchangeably in empirical equations as this could lead to an overestimation of field scale strength by perhaps 10 degrees in many cases.

5.7.6 *Dynamic shear strength of rock joints*

There is some evidence that frictional resistance for rock joints is dependent on loading rate, and this may be significant for aseismic design and for understanding response to blasting. For a block of rock sitting on an inclined plane, given a value for static friction, one can calculate the horizontal acceleration necessary to initiate movement and when the block should stop, given a particular acceleration time history, as illustrated in Figure 5.27. This type of calculation is the basis of the Newmark (1965) method of dynamic slope stability analysis, which is used to calculate the distance travelled, as discussed in more detail in Chapter 6. Hencher (1977) carried out a series of experiments and found that initiation of movement was generally later than anticipated (or did not occur), implying greater peak frictional resistance than predicted from static tests. The effective friction for initiation increased with the rate of loading (Figure 5.28). The implication is that if the loading is very rapid and reversed quickly (as in blast vibrations), shear displacement might not occur, despite the supposed critical acceleration being exceeded. However, once movement was initiated, Hencher found that the distance travelled was higher than anticipated from static strength measurements and interpreted this as reflecting rolling friction and the inability of strong frictional contacts to form during rapid sliding. Hencher (1981a) suggested that for Newmark-type analysis, residual strength should

Figure 5.27
Transient loading
of block on a plane.

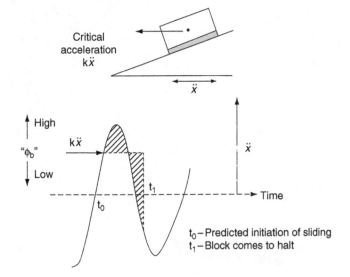

t_0 – Predicted initiation of sliding
t_1 – Block comes to halt

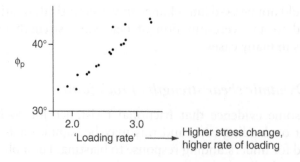

Figure 5.28
Relationship
between phi
(dynamic) and rate
of loading. The
higher the rate of
application of load
(frequency), the
greater the initial
strength. Data from
Hencher (1977,
1981a).

be used for calculating displacements. Recent work confirms low slid-ing friction angles post-failure (Lee *et al.*, 2010).

5.8 Rock-soil mixes

It has long been recognised that mixes of soil and rock, such as illustrated in Figure 5.29, can often stand safely at steeper angles than if the slope were comprised only of the soil fraction. From testing on soils together with theoretical studies, the point at which the hard inclusions start to have a strengthening effect is about 30% by volume.

Figure 5.29
Cutting through
boulder colluvium.
East of Cape Town,
South Africa.

5.8.1 Theoretical effect on shear strength of included boulders

Hencher (1983d) and Hencher *et al.* (1985) report on the back-analysis of a landslide involving colluvium containing a high percentage of boulders, in which an attempt was made to estimate dilation angles on the basis of the coarse fraction percentage estimated in the field and measurements taken from idealised drawings. These estimated field dilation angles were added to the strength for the matrix, determined from laboratory testing. West *et al.* (1992) took this further and identified several ways that included boulders might influence shear strength, based on physical modelling and back analysis of slopes (these are illustrated in Figure 5.30). Factors envisaged included: boulders preventing failure along an otherwise preferred failure path, failure surface forced to deviate around a boulder, and a failure zone incorporating the boulder. Triaxial tests reported by Lindquist & Goodman (1994) similarly concluded that boulders increase the mass strength. Additional review is provided by Irfan & Tang (1993).

Practical methods for addressing the strength of mixed soils and rocks remain difficult. One of the main problems is that such masses can be highly heterogeneous and difficult to characterise realistically. The other is that whilst trends of increasing mass strength with percentage of rock clasts and boulders are clear, general rules have not yet been formulated. Further advances will probably be by numerical

Figure 5.30 Mechanisms of failure through a mixed rock and soil slope. After West *et al.* (1992).

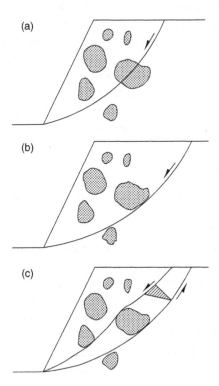

modelling and could be done using PFC3D (Itasca). Whilst the largely intractable geological characterisation nature of the problem would remain, the problem could probably be resolved parametrically in a similar way and with a similar level of success for prediction as the Hoek-Brown model for fractured rock masses.

5.8.2   Bearing capacity of mixed soil and rock

Mixed soil and rock deposits include sedimentary deposits like colluvium and glacial boulder clay, but also some weathered rocks. As for assessing shear strength, there are considerable difficulties for sampling and testing and there can also be significant problems for construction (e.g. Weltman & Healy, 1978). The conservative position for design is to take the strength and deformability of the matrix as representative of the mass, but allowance might be made for the included stiffer and stronger clasts by rational analysis, perhaps backed up by numerical modelling.

5.9   Rock used in construction

Crushed rock and quarried or dredged sand and gravel are important materials used in making concrete and construction generally, perhaps as fill. Rock is also used as armourstone, for example, in protecting earth dams from wave action or for forming harbours. It is also cut and polished as dimension stone to be used as kitchen work surfaces or as cladding on the outside of prestigious buildings. Engineering geologists are often required to identify sources of aggregate, either from existing quarries but sometimes from new borrow areas in the case of sand for reclamation or new quarries for a remote project such as a road. Some of the properties that are important for their use are the same as in much of geotechnical design: strength, unit weight and porosity, but there are other properties that need to be tested specifically.

5.9.1   Concrete aggregate

For concrete, the aggregate must be sound, durable and chemically stable. Materials to be avoided include sulphates (e.g. gypsum and pyrites), clay and some silicate minerals such as opal and volcanic glass, which can cause a severe reaction and deterioration of the concrete if present in the wrong proportions (see case example of Pracana Dam in Chapter 7). Tests are available and should be used to ensure that the aggregate being sourced is suitable. These include mortar bar tests whereby a test mix of concrete is formed and observed to see if it expands with time. Other factors might include the need for light- or heavy-weight concrete, fire resistance and overall strength. Concrete mix design for a large project may require a research programme to optimise the aggregate specification and type of cement to use. For smaller projects or where the demands are less onerous, then cost may be the controlling factor; aggregates and quarries have place-value, which is a matter of the quality of aggregate at a particular quarry together with the costs of transport to the project site. A useful review of the factors to be considered in specifying concrete aggregate is given by Smith & Collis (2001).

5.9.2 *Armourstone*

Armourstone is used to protect structures primarily from wave action and is often made up of blocks of rock of several tonnes. Generally, the rock must be durable and massive. If it softens or discontinuities open up with time, then the function is lost. Massive crystalline limestone often works well, as do many igneous and metamorphic rocks. Usually, durability (and availability and cost) is all-important but see the case history of Carsington Dam in Chapter 7 where the choice of limestone as riprap contributed to adverse chemical reactions and environmental damage. Weak or fractured rocks are obviously not appropriate. For many coastal defence works in the east of England, large rock blocks are brought by barge from Scandinavia because of a lack of suitable local rock. CIRIA (2005) provides useful guidance. Where suitable rock is not available then concrete tetrapod structures known as dolosse are used in the same way, piled on top of one another and interlocked, to protect coasts and structures by dissipating wave energy.

5.9.3 *Road stone*

Aggregate is used in road construction in many different ways – as general fill or in the sub-base, as drainage material and in the wearing course. There are many different standard tests to be applied in road construction, and these are described in Smith & Collis (2001). The most demanding specification is for wearing course material. Rock must be strong and durable but also must resist polishing as it is worn by traffic. This requires the rock used to comprise a range of different minerals that are strongly bonded but wear irregularly. Rocks like limestone are generally unsuitable (the polished stone value, PSV, is too low). Rocks like Ingleton granite, which is really an arkose, have excellent properties and therefore very high place values – worth quarrying and transporting large distances – even from a National Park.

5.9.4 *Dimension stone*

Dimension stone is quarried to be used directly in building, construction or even sculptures. Typical rocks quarried in this way include marble, granite and slate for roofs. Rocks are generally chosen for their colour and appearance – the quarry at St Bees headland, Cumbria, UK (a fairly ordinary sandstone), was re-opened temporarily in the 1990s to provide rock for shipping to New York to repair buildings faced with sandstone carried by ships as ballast in the 19th Century – because of its appearance. Dimension stone must also be resistant to wear, frost and chemical attack. This can be difficult to determine from direct testing, so experience of the long-term performance of a particular rock from a particular quarry may be the best clue.

6 Analysis, design and construction

6.1 Introduction

In Chapter 2, a brief introduction was given to civil engineering practice and types of structure. This chapter provides more detail so that the engineering geologist can better understand the requirements of projects, in terms of site investigation, design and construction issues.

6.2 Loads

Most civil engineering projects involve either loading the ground, say from the weight of a new building, or unloading because of excavation of a slope or in a tunnel. Load changes can be permanent or temporary, static (due to weight) or dynamic (due to blasting, for example). A further important consideration for most geotechnical problems is the self weight of the ground and other *in situ* stresses.

6.2.1 Natural stress conditions

At any point in the Earth's crust, the stresses can be resolved into three orthogonal directions. These are termed the maximum, intermediate and minimum principal stresses and depicted σ_1, σ_2 and σ_3, respectively. By definition, the planes to which the principal stresses are normal are called principal planes and the shear stresses on these planes are zero. An important point regarding rock engineering is that all unsupported excavation surfaces are principal stress planes because there are no shear stresses acting on them (Hudson, 1989). One of the principal stresses will always be perpendicular to the Earth's surface (Anderson, 1951) and is generally vertical.

For projects close to the Earth's surface, such as cut slopes or foundations, natural stresses include self weight, weight of included water and buoyancy effects below the water table, which reduces the total stress to an effective stress (weight of soil minus water pressure), as illustrated in

Box 6-1. As the rock or soil is compressed under self weight, it tries to expand laterally and a horizontal stress is exerted. This is termed the Poisson effect. Typically, in a soil profile at shallow depths (tens of metres), the *in situ* horizontal stress (σ_h) due to self weight will be between about 0.3 (in loose sand) and 0.6 times (in dense sand) the vertical gravitational stress. The value 0.3 to 0.6 is called the coefficient of earth pressure at rest. In normally consolidated clay, the value is about the same as for dense sand: 0.6. For most rocks, the Poisson's ratio is slightly less than 0.3. Most continental rocks weigh about 27 kN/m³, so at a depth of 500m the total vertical stress can be anticipated to be about 13.5 MPa, and horizontal stresses (σ_h) about 4 MPa.

Box B6-1   Example stress calculations

Generally, stresses are estimated by calculating the total weight of a vertical column of soil based on unit weight measurements. Effective stress is estimated by subtracting measured or estimated water pressure from the total stress due to the bulk weight of the soil or rock (including contained water).

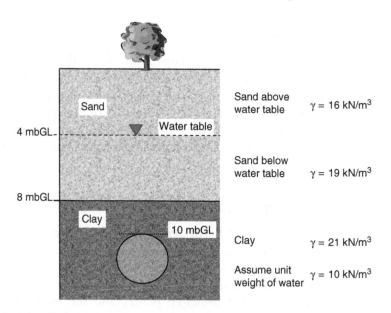

Figure B6-1.1 Soil profile with tunnel to be constructed with crown at 10mbGL

In Figure B6-1.1, a ground profile is shown with sand overlying clay. The water table (upper surface of saturated ground) is 4m below ground level (mbGL).

The unit weight (γ) of the damp sand above the water table is 16 kN/m³; the unit weight below the water table, sand plus pores full of water (γ_{sat}), is 19 kN/m³. The underlying saturated clay has unit weight γ_{sat} = 21 kN/m³. The unit weight of fresh water, γ_w, is about 9.81 kN/m³ (10 is generally a near-enough approximation given other assumptions).

We wish to estimate the vertical stress at the crown of a tunnel to be constructed at a depth of 10 mbGL. As shown in Figure B6-1.2.

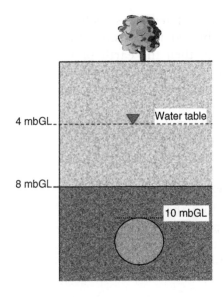
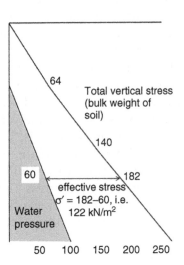

Figure B6-1.2 Total and effective stresses(vertical)

At depth	Total vertical stress $(\sigma_v)kN/m^2$	Water pressure (u) kN/m^2	Effective vertical stress $(\sigma')kN/m^2$
4m	4×16 = 64	0	64
8m	64 + (4×19) = 140	4×10 = 40	140 − 40 = 100
10m, at tunnel crown	140 + (2×21) = 182	6×10 = 60	182 − 60 = 122

Therefore, before tunnel construction, the estimated vertical effective stress at the tunnel crown is 122 kN/m^2. During construction, due to seepage into the tunnel the water table would be lowered or this might be done deliberately to excavate 'in the dry' to avoid flowing or ravelling of the soil into the tunnel. If the water pressure dropped, so the effective stress would increase. If the water table was lowered so that water pressure was zero at tunnel crown level, then the effective stress would equal the total stress (= 182 kN/m^2).[1]

At some locations, however, tectonic or topographic stresses can be dominant even very close to the Earth's surface, with horizontal stresses sometimes locked in from a previous geological event and far in excess of that due to gravity and the Poisson

---

[1] Note: the actual stress conditions near a tunnel would be more complex than this calculation. The tunnel would distort the stress field – refer to Muir Wood (2000) or Hoek *et al.* (1995).

effect. As illustrated in Box 6-2, in overconsolidated clays such as London Clay, where the rock has been buried to considerable depth before uplift, erosion and unloading, then the earth pressure at rest can be up to three times the vertical stress. In tectonically active regions, stresses can be higher or lower than lithostatic. Horizontal: vertical stress ratios as high as 15 have been measured in areas where tectonic or thermal stress has been locked in as the overburden has been eroded (Hoek & Brown, 1980). These stresses can adversely affect engineering projects, resulting in deformation in tunnels, rock bursts and propagation of fractures (e.g. Karrow & White, 2002). In mountainous terrain, principal stress trajectories will follow the topography so that the maximum principal stress runs parallel to steep natural slopes, and this leads to spalling off of the rock parallel to the natural slope (Chapter 3) and valley bulging at the toe of the slope.

Box B6-2   Variations from lithostatic stress conditions

Whereas in many areas of the Earth's crust, stress conditions can be estimated reasonably well by calculating the weight of the soil/rock overburden to give vertical stress and taking account of Poisson's effect for horizontal stress, considerable variation is found (Hoek & Brown, 1980). In particular, horizontal stresses can be higher or lower than anticipated.

Example 1 Overconsolidated clay

Soils and weak rocks that have gone through a cycle of burial, partial lithification and then uplift and erosion are termed overconsolidated. They typically have lower void ratios (percentage of pores) and are stiffer than would be expected for normally consolidated soils at similar depths of occurrence. They are also sometimes partially cemented, as described in Chapters 1, 3 and 5. Under compression, they demonstrate high moduli up until the original maximum burial stress, at which point they revert to the normal consolidation stress curve, as described in soil mechanics textbooks (e.g. Craig, 1992). Because the stress level has been much higher in geological history, the horizontal stress may have become locked-in as a residual stress and may be much higher than the vertical principal stress, as illustrated in Figure B6-2.1. Craig quotes earth pressure at rest K_0 values up to 2.8 for heavily overconsolidated London Clay. Further discussion of earth pressures and how they relate to geological history is given by Schmidt (1966).

Example 2 Active and ancient tectonic regions

Deviations from lithostatic stress conditions can be anticipated at destructive plate margins, as along the western margins of North and South America where high horizontal stresses are to be expected. Conversely, in extensional tectonic zones the horizontal stresses can be anticipated to be tensile. Variations can also be expected in ancient mountain chains or areas of igneous intrusion where relict horizontal stresses can be very high, resulting in rock bursts and large deformation of structures (e.g. Holzhausen, 1989).

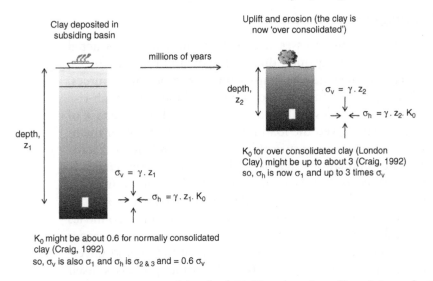

Clay deposited in
subsiding basin

millions of years

Uplift and erosion (the clay is
now 'over consolidated')

depth,
z_1

depth,
z_2

$\sigma_v = \gamma \cdot z_2$

$\sigma_h = \gamma \cdot z_2 \cdot K_0$

K_0 for over consolidated clay (London
Clay) might be up to about 3 (Craig, 1992)
so, σ_h is now σ_1 and up to 3 times σ_v

$\sigma_v = \gamma \cdot z_1$

$\sigma_h = \gamma \cdot z_1 \cdot K_0$

K_0 might be about 0.6 for normally consolidated
clay (Craig, 1992)
so, σ_v is also σ_1 and σ_h is $\sigma_{2\,\&\,3}$ and = 0.6 σ_v

Figure B6-2.1 Stress conditions in overconsolidated soil. Uplift and erosion will result in a reduction in the vertical stress on the soil element but some residual horizontal stress may be retained from its burial history.

Example 3 Topographic stresses

Stress conditions may be strongly affected by local topography exacerbated by geological conditions. At an extreme scale, large-scale mountain structures are ascribed to gravity gliding (e.g. Graham, 1981) and certainly large landslides have ample evidence of compression and tensile zones. Other key examples of the effect of localised topographic stress are sheeting joints (Hencher *et al.*, 2011) and valley bulging (Parks, 1991).

Stress conditions have been measured across the world from instruments, by interpretations of breakouts in deep drillholes for oil and gas exploration, or analysis of earthquakes, and many such data are compiled centrally and are freely available at http://www.world-stress-map. *In situ* stresses are sometimes investigated specifically for projects (Chapter 4) but this is expensive and can be inconclusive because of the small scale and localised nature of tests.

Where stress assumptions prove wrong, the consequences can be severe, as at Pergau Dam, Malaysia, where it had been anticipated that stresses would be lithostatic (i.e. caused by self weight). During construction, open joints and voids were encountered in tunnels together with high inflow of water (Murray & Gray, 1997). It was established that horizontal stresses were much lower than had been expected and this necessitated a complete redesign of shafts and high-pressure tunnels and their linings, at considerable cost. Low horizontal stresses can occur in the proximity of valley sides. Further examples are given later in this chapter.

6.2.2   *Loadings from a building*

A structure will change the stresses in the ground and, in turn, be acted upon by stresses from the ground due to gravity and tectonic forces, wind, snow, earthquakes and perhaps from anthropogenic sources, including blasting and traffic. The loading condition for a high-rise building constructed on piles is illustrated in Figure 6.1. It is the task of the geotechnical team, given the loading conditions from other members of the design team, to ensure that there is an adequate Factor of Safety for the foundations against failure and that settlement is within the tolerance of the structure. The traditional permissible stress approach, involving a lumped Factor of Safety to cover all uncertainties, has been replaced in Europe and some other countries and design codes by a limit state approach, which encourages more rigorous consideration of different modes of failure and uncertainties in each parameter and in the calculation processes itself (Table 2.2).

Total vertical load above pile = Σ [Dead load (including concrete self-weight & imposed dead loads) +Live load + vertical component of Wind load due to structural response from lateral wind force on each floor.].

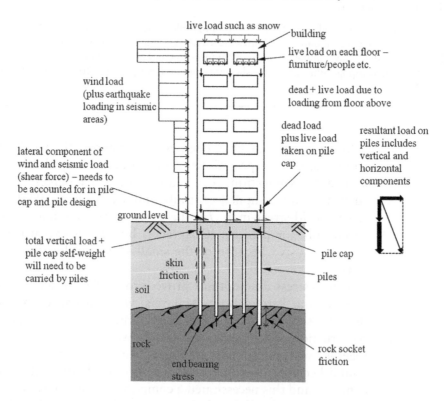

design loading on each pile = total vertical load above each pile + pile cap self-weight

Figure 6.1 Typical loading conditions for a high-rise building to be founded on piles.

6.3   Temporary and permanent works

The engineer's design generally concerns the permanent works – the long-term stability and performance of the finished project. Performance is measured by criteria specific to a project, such as settlement, leakage, durability and long-term maintenance requirements. During construction, there will usually be other design considerations including stability of temporary excavations, disturbance to the groundwater conditions and water inflow to the works. Temporary work design is generally the responsibility of the contractor and his design engineers, perhaps checked by an independent checking engineer. The design of deep temporary excavations can be just as demanding as for permanent works, as illustrated in Figure 6.2. Catastrophic failure of such works is unfortunately common – in recent years affecting such high-profile projects as the International Finance Centre in Seoul, Korea, and the Nicoll Highway subway works in Singapore (Chapter 7). In both cases, the strutted excavations collapsed. Guidance on the design of such structures is given in Puller (2003) and GCO (1990).

In tunnels, during construction there may be a need to stabilise the walls and possibly the working face using rapidly applied techniques, including shotcrete with mesh or steel fibres, steel arches or lattice girders and rock bolts (Hoek *et al.*, 1995). Such measures are generally specified and installed by the contractor, typically agreed with a supervising engineer who may well be an engineering geologist. The engineering geologist will probably be involved in identifying the rock

Figure 6.2 Temporary works for an underground station construction in Singapore. Piles to the left were excavated by a large-diameter drilling rig and then concreted. As excavation has proceeded, the piles have been anchored back into the ground and strutted using systems of waling beams (horizontal, along the face of the piles) and struts, supported where necessary by additional king posts.

mass conditions and identifying any geological structures that might need specific attention, as discussed later. The decisions taken will often have cost as well as safety implications. Usually, measures installed to allow safe working will be ignored when designing and constructing permanent liner support, but in some tunnels there is no permanent lining so the temporary measures also become permanent works. In the latter case, the materials and workmanship will be specified accordingly and as appropriate to the design life of the project. Close supervision will be required on site to ensure that the specified requirements are met and the quality of the works is not compromised.

6.4 Foundations

Foundations are the interface between a building and the ground and transfer loads from the building to the underlying soil and rock. Detailed and practical guidance on foundation design and construction issues is given by Tomlinson (2001). Wyllie (1999) deals specifically with foundations on rock. If ground conditions are suitable, then shallow foundations are used because of cost considerations. These include strip footings beneath the walls of a house (Figure 6.3), pads beneath columns for a steel or concrete-framed structure, or a raft supporting several loading columns and walls.

6.4.1 Shallow foundations

For traditional design involving a single Factor of Safety, which is probably the easiest to understand and still employed as the

Figure 6.3 Concrete strip foundations on weathered limestone for a house, Portugal.

standard approach to design in many parts of the world, the following definitions are used:

Bearing pressure	The net loading pressure: load from structure, divided by the area of the foundation, minus the weight of material removed from the excavation.
Ultimate bearing capacity	The loading pressure at which the ground fails. This is the same as the ultimate limit state in the limit state approach (Eurocode 7).
Allowable bearing pressure	The maximum loading pressure that meets two criteria: 1. An adequate Factor of Safety against failure. 2. Settlement within tolerance of the structure (specific to the particular structure).
Presumed bearing pressure	A net loading pressure considered appropriate for a given ground condition, based usually on local experience and incorporated in building regulations or codes of practice such as BS 8004 (UK) (BSI, 1986) and CP4 (Singapore Standard, 2003).
	Typical values are presented in Table 6.1 and can be used for preliminary design purposes. They allow the practicability of foundation options to be assessed and to select appropriate ground investigation, testing and design methods. Presumed values are only appropriate if the site is approximately level (not, for example, at the top of a steep slope) and where the geology is relatively uniform and isotropic with no lenses or layers of significantly weaker or compressible material within the zone of ground that will be stressed. Such tables are generally very conservative and economies can be made by conducting more detailed characterisation with testing and analysis, although sometimes regulating bodies (building authorities) may be loathe to allow higher values to be used without considerable justification.

In Europe, since 2010, Eurocodes have replaced national standards and should be used for design (BSI, 2004). The ultimate limit state (ULS) is essentially the same as ultimate bearing capacity but with possible failure modes spelt out, including sliding resistance and structural capacity, heave, piping, and so on, which were implicit in the BS 8004 approach as factors that a responsible geotechnical engineer should consider. The serviceability limit state (SLS) of Eurocode 7 is defined as: 'states that correspond to conditions beyond which specified service requirements for a structure or structural member are no longer met', and this equates effectively to the idea of allowable bearing pressure, as far as settlement is concerned, but includes other considerations such as vibration annoyance to neighbours, and so on – again, factors that would usually be considered automatically by experienced and responsible geotechnical engineers when adopting a traditional approach to design.

From Table 6.1 it can be seen that, for rock, the two governing parameters are generally taken to be uniaxial compressive strength (UCS = σ_c) and degree of fracturing. This is expressed in charts presented

Table 6.1   Examples of presumed bearing pressures. These values, which can be used for option assessment, are a selection of more extensive recommendations given in Tomlinson (2001) and BS 8004 (BSI, 1986).

	Examples of rock type (indicative only)		Presumed bearing value (MPa)	
ROCK	Bearing on surface of rock		Strip footings < 3m wide. Length not more than ten times width	
	Strong. Discontinuity spacing more than 200mm		10–12.5	
	Strong. Discontinuity spacing 60–200mm		5–10	
	Moderately strong. Discontinuity spacing 60–200mm		1–5	
	Notes: Figures given are for igneous rocks, well-cemented sandstone, mudstone and schist/slate with flat-lying cleavage/foliation. For other rock types see references quoted. Strength definitions are from BS 5930:1999. Strong rock (σ_c = 50–100 MPa) requires more than one hammer blow to break. Moderately strong rock (σ_c = 12.5–50 MPa) – intact core cannot be broken by hand.			

	Examples of soil type (indicative only)		Presumed bearing value (MPa)	
SOIL	Sand and gravel: foundations at least 0.75m below ground level	SPT N-value	Foundation width	
			<1m	<2m
	Very dense	> 50	0.8	0.6
	Dense	30–50	0.5–0.8	0.4–0.6
	Medium dense	10–30	0.15–0.5	0.1–0.4
	Loose	5–10	0.05–0.15	0.05–0.1
	Clay: foundations at least 1m below ground level	Undrained shear strength (MPa)	Foundation width	
			<1m	<2m
	Hard	> 0.30	0.8	0.6
	Very stiff	0.15–0.30	0.4–0.8	0.3–0.5
	Stiff	0.075–0.15	0.2–0.4	0.15–0.25
	Firm	0.04–0.075	0.1–0.2	0.075–0.1
	Soft	0.02–0.04	0.05–0.1	0.025–0.05

in BS 8004 and similar standards worldwide. For rock such as sandstone or granite with an intact compressive strength of 12.5 MPa (just break by hand), the allowable bearing pressure would also be 12.5 MPa, provided discontinuities are widely spaced apart, reducing to about 10 MPa as discontinuity spacing is about 0.5m and reducing to 2.5 MPa when discontinuity spacing is 150mm. If the fracturing is particularly adverse or includes discontinuities with low shear strength that could combine to form a failing wedge, then this needs specific consideration and analysis, as dealt with by Goodman (1980) and Wyllie (1999).

Variability across the foundation footprint may also be an issue. If there are soft or weathered pockets, these may need to be excavated and replaced with concrete or other suitable material. Karstic conditions with voids at depth that may be particularly difficult to investigate comprehensively can pose particular difficulties for foundation design and construction, as illustrated by a case example in Chapter 7 and discussed by Houghten & Wong (1990). Conversely, if there are particularly strong areas – for example, an igneous dyke through otherwise weak rock in a pad foundation, then this must be accounted for, otherwise the foundation may fail structurally. In all cases, it is essential to check any assumptions from preliminary design as the foundation excavation is exposed. If the ground is worse than anticipated then redesign may be required. In severe cases where, for example, a major fault is exposed unexpectedly, the required change in design may be drastic, but that is the price paid for an inadequate site investigation. Time must be allowed for checking during construction and taking any actions that prove necessary.

For soils, compressibility and settlement is often the main concern and much more so than for rock. The presumed values given in Table 6.1 should restrict settlement to less than 50 mm in the long-term, but estimates may be widely in error and even supposedly sophisticated methods of prediction are often inaccurate. For foundations on granular soils, empirical methods relying on SPT or CPT data tend to be used for predicting settlement. Burland & Burbridge (1985) compiled data for sand and gravel and showed that predictions of settlement are often in error by factors of two or more. Das & Sivakugan (2007) provide an updated review.

For cohesive soil, where relatively undisturbed samples can be taken to the laboratory, oedometer tests are used to determine settlement potential and to predict rate of consolidation. Estimates of settlement can be made, given the thicknesses of the various strata in the ground profile, their compressibility and the stress changes. Details are given in many references, including Tomlinson (2001) and Bowles (1996). For major structures, engineers will often carry out numerical modelling using software such as Plaxis or FLAC, which can be used for sensitivity studies. Such software is also used to predict deformations during different stages of excavation and construction and to determine support requirements.

6.4.2 Buoyant foundations

If the weight of the soil removed from an excavation is the same as the building constructed within the excavation, then no settlement should occur, as illustrated schematically in Figure 6.4. This design concept has been used for many major structures incorporating deep basements which can be utilised for parking spaces. There may be a need to include holding-down piles or anchors in the design to combat any uplift forces. Construction of deep foundation boxes often involves the construction of diaphragm walls using the same techniques as for

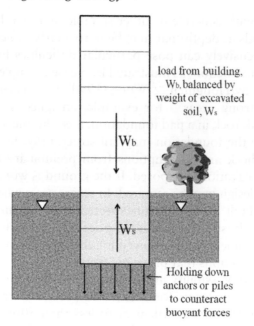

load from building, W_b, balanced by weight of excavated soil, W_s

W_b

W_s

Holding down anchors or piles to counteract buoyant forces

Figure 6.4 Concept of buoyant foundation design. The weight of the building balances the excavated soil so that the net increase or decrease in pressure is minimised.

barrettes, as discussed below. Once the walls are in place, excavation is conducted inside the walls, with either bracing and/or anchorages used to stabilise the works.

6.4.3   Deep foundations

6.4.3.1   Piled foundations

Piles are used to transfer building loads, via pile caps, to deeper levels in the ground profile. There are two main types: driven and bored. Driven piles are hammered into the ground and are also termed displacement piles. Hammering is sometimes done by dropping a large weight on the top of the pile from a crane or using a diesel or hydraulic machine (Figure 6.5). Bored piles are generally constructed using bucket augers, soil grabs and rock roller bits, with heavy-duty rock cutting tools used to grind their way into the underlying rock and to form rock sockets as necessary. Even using the most powerful equipment, formation of sockets can take a very long time, advancing perhaps only 100mm per hour in strong rock, and therefore can be relatively expensive, so designers should be wary of being ultra-conservative in their specification of socket length.

6.4.3.1.1   DRIVEN PILES

Driven piles are generally made of timber, steel or concrete. Figure 6.6 shows concrete piles being manufactured on site in a factory-type operation to allow 20,000 piles to be driven in just 18 months for Drax Power Station completion (Hencher & Mallard, 1989). The purpose-made pile beds were heated to allow rapid curing of concrete,

Figure 6.5 Diesel hammer (centre of photo) being used to drive pre-stressed concrete piles, Drax Power Station, UK. Elsewhere, piles are being pitched into holes formed by auger. In the background, the kentledge can be seen for a proof test on a working pile.

and the piles were pre-stressed to improve their resistance to tensile stresses and to allow the piles to be lifted from their forms quickly. For most sites, piles will be manufactured off site, sometimes as different lengths that are joined together on site to suit requirements. One of the advantages of using driven piles is that an estimate can be made of the driving resistance, given the known energy being used to drive the pile and the penetration into the ground per blow of the hammer. Piles are therefore driven to a set, which is a predefined advance rate (such as 25 mm for 10 blows by the hammer). However, resistance during driving may not always give a very good indication of how the pile will behave under working conditions, because of false sets, generally due to water pressure effects, as described for the Drax operation in Chapter 7.

Figure 6.6 Piles being cast in formers, Drax Power Station, UK. Note lifting eyes cast into the concrete piles, steel plates at end of piles (trapezoidal) and pre-stressing cables, which are to be cut before lifting piles from the casting beds.

Driving resistance and set are, however, part of the process of quality control during construction. Pile driving analysers (PDAs) using accelerometers and other instruments attached to the pile can be used to estimate driving resistance in a more sophisticated way than the traditional method of measuring the quake with a pencil, although the same limitations apply regarding whether or not dynamic behaviour is a reliable indicator of future performance. PDAs are sometimes used after the pile has been installed (both driven and bored piles) to test its capacity, but this can be somewhat of a black art with many assumptions being made and the method is certainly not foolproof or as reliable as full static load tests, as discussed below.

6.4.3.1.2   BORED PILES

Bored piles are excavated as described earlier. Temporary or permanent steel tubes (casing) may be used to prevent collapse of the hole and, if the hole is formed below the water table, often bentonite or some other mud or polymer is used to support the sides of the hole. Once the hole has been completed and cleaned out, then a steel reinforcing cage is introduced and, finally, concreting carried out. Concrete needs to be tremied by a pipe from the surface to the bottom of the hole. This avoids the concrete disaggregating, and the concrete will hopefully displace soft sediment that might have accumulated at the bottom of the bored hole after the final clean out. It will also displace the bentonite slurry or water from the bored pile excavation, so this can be a very messy operation. Despite best efforts, soft toes of sediment will still sometimes occur (perhaps associated with the removal of temporary casing) and sometimes ground movements occur causing necking of piles. Clearly, there is a need for high-quality work and for close supervision. Currently, in Hong Kong, all bored piles are installed with steel tubes attached to the reinforcing cage (Figure 6.7). After concreting, rotary drilling is carried out down one of the tubes, through the concrete and into the underlying natural ground, to prove that the pile is founded as designed and that there are no soft sediments. If there are, then remedial measures such as pressure grouting might be needed. Other tubes installed through the concrete are used to carry out geophysical cross-hole tests (seismic) to check for necking and other construction defects. In severe cases, piles may prove inadequate to carry the loads and remedial works are required. This might not be discovered until the superstructure is constructed. In one extreme case in Hong Kong, two 44-storey tower blocks had to be demolished. Such problems may be put down to workmanship, the inherent difficulties of the operation, poor investigation and design and sometimes fraud (Hencher *et al.*, 2005).

Once the piling is completed, a pile cap is constructed as a reinforced box of concrete that bridges between several piles to support major columns in the superstructure.

Figure 6.7
Reinforcing cage
for bored pile with
included tubes to
allow proof drilling
through the toes of
the completed pile
and cross-hole
geophysical testing
to prove integrity,
Hong Kong.

6.4.3.2 Design

Piles are designed to suit the ground profile. If rockhead is at relatively shallow depth and the overlying soil does not contain boulders that could cause difficulties, then driven piles might be adopted, end bearing onto the rock (Figure 6.8a). At Drax, the piles were driven to found several metres into dense sand overlying sandstone, thereby picking up some skin friction as well as end bearing (Figure 6.8b). If there is no rock, then the piles will need to gain their resistance mostly from skin friction in the soil. For example, the Sutong Bridge across the Yangtze River, China, which is (in 2011) the longest cable-stayed bridge in the world, with a main span of 1,088 m, is founded on bored piles taken to 117 m and relying upon skin friction from alluvial sediments (Figure 6.8c).

Ways to estimate skin friction parameters and end-bearing resistance are given in textbooks such as Tomlinson (2001) and might be governed by standards such as AASHTO (2007), used as the basis for design of the 2nd Incheon crossing completed in 2009. The principles are quite simple: skin friction is calculated as soil shear strength times some adhesion factor multiplied by the surface area of the pile shaft. End bearing is often calculated as an empirical value for the soil or rock quality multiplied by the basal area of the pile. At some sites, the bottom end of the pile is enlarged by under-reaming to increase the end-bearing contribution, although sometimes the difficulty of this operation is hardly justified by the increase in pile capacity that might ensue.

A worked example of pile design to Eurocode 7, using partial factors specified uniquely for the UK (to correlate with traditional design experience), is presented in Box 6-3, based on one

a)

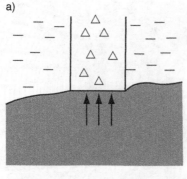

b)

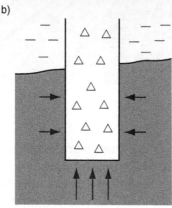

c)

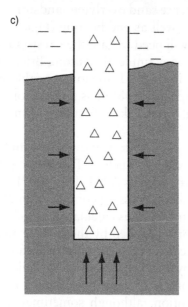

Figure 6.8 Design concepts for piles. a) End bearing, b) end bearing plus skin friction and c) skin friction dominating.

presented by Bond & Simpson (2010). Other countries might use different partial factors and other approaches, as allowed in the Eurocode. In the example presented, the main unknowns – variable live load, shaft resistance and base resistance – are factored up and down as appropriate towards a safe solution. The results

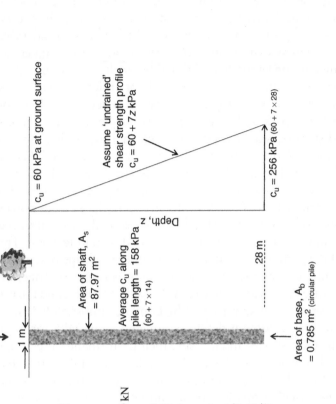

Box 6-3   Pile design (based on Bond & Simpson, 2010)

Shaft resistance, $Rs = 87.97 \times (\alpha \times 158)/\gamma^1$

where α is an adhesion factor (take as 0.5) and γ is a model factor = 1.4

So $Rs = 4,964$ kN

Base resistance, $Rb = 0.785 \times (Nc \times 256)/\gamma$

where Nc is a bearing capacity factor (take as 9) and γ is the same model factor = 1.4

So $Rb = 1,292$ kN

Total design compressive resistance, $Rc = 4,964/1.6 + 1,292/2.0 = 3,749$ kN

With partial factors 1.6 on shaft and 2.0 on end bearing, which are designated in the UK National Annex to BS EN 1997–1 for a bored pile not subject to test. If tested, factors would be 1.4 and 1.7 respectively (see Bond & Simpson, 2010 for discussion).

The factored design resistance, 3.749 MN, is less than the factored compressive load, 4.3 MN, so longer piles would be required.

The traditional, unfactored global FoS would be $(4,964 + 1,292) \times 1.4 / 4,000$, i.e. 2.19, which is lower than would normally be required following traditional design rules (typically 2.5 to 3.0) in the absence of extensive preliminary and working pile tests.

[1]Note that the symbol γ has been used in Eurocodes to represent 'partial factors'. This can lead to confusion in that the symbol γ is more generally used to represent unit weight as in γ_R for the unit weight of rock.

are compared to the FoS, as determined using a traditional approach – best estimate of strength divided by best estimate of loading. It is to be noted from this example that whichever approach, there is considerable judgement and approximation involved. Shear strength is taken as undrained, which is conceptually questionable for the long-term; adhesion factor estimates range from 0.3 to 0.9 for different soils. If an effective stress approach was adopted – as would generally be done for sand and weathered rock – then estimates would be needed of stress conditions and shaft resistance coefficients, which also requires estimation and judgement. Workmanship may also play a key role in whether or not shaft friction will be mobilised and whether the base of a bored pile excavation is properly cleaned out prior to concreting. The use of a partial factors approach does concentrate on where the key unknowns are (rather than geometry and fixed loads) but doesn't take away the need for proper ground characterisation, analysis and design judgement. The fixed nature of the partial factors might seem rather prescriptive to cover all soil, rock and founding situations. Selection of parameters, adhesion and shaft resistance factors are reviewed well in GEO (2006), and the use of Eurocode 7 for design is summarised by Bond & Simpson (2010).

A site-specific way to obtain design parameters, especially for large projects, is to install test piles and measure their performance at perhaps 2.5 times the design load of the working piles. Test piles are often instrumented along their length using strain gauges so that the actual resistance being provided by the ground can be measured throughout the full profile, and these parameters can be used in the design of other piles. Traditionally, piles are loaded from the top using kentledge of concrete blocks or steel (Figure 6.9). Jacks are used to push the pile into the ground whilst the kentledge provides the reaction. One of the difficulties of this is that much of the support comes from the upper soil at early stages of the test, and there is little idea of how the toe is

Figure 6.9 Pile test set up with kentledge. Donghai Bridge, China. Figure courtesy of Leonard Tang, Halcrow.

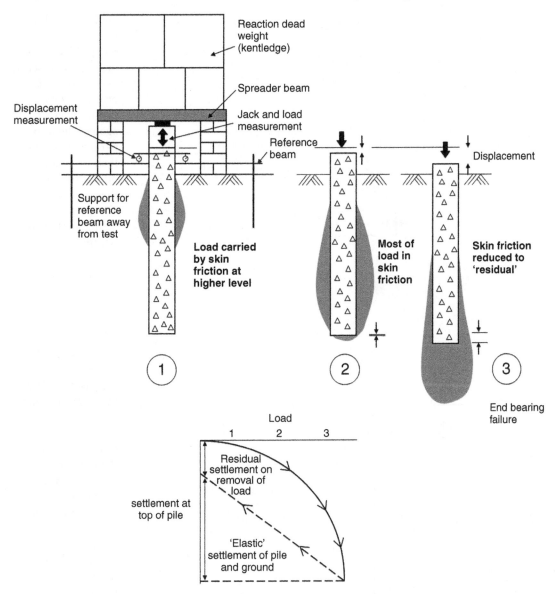

Figure 6.10 Typical set-up for pile load test. At early stages (1), most of the ground resistance will come from skin friction at shallow depths. End bearing is not mobilised until later stages (2) and (3) of the test (depending on the configuration of the pile and ground profile). The rate of settlement increases as the ground resistance becomes fully mobilised and there will be some permanent displacement (residual settlement) once the pile is unloaded.

performing until a test approaches failure (Figure 6.10). Recently, a system has been introduced where Osterberg cells are incorporated into the pile construction at depth and then expanded against the test pile, both upwards and downwards (Figure 6.11). The end-bearing resistance below the cell is balanced by the skin friction from the soil above the cell. This system was used for the Incheon Bridge design, using

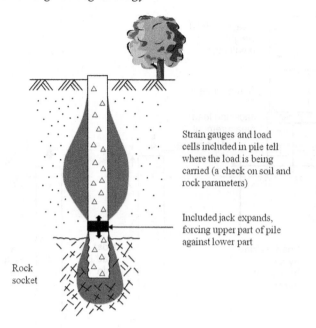

Strain gauges and load cells included in pile tell where the load is being carried (a check on soil and rock parameters)

Included jack expands, forcing upper part of pile against lower part

Rock socket

Figure 6.11 Pile test using an integral jack or set of jacks. This set-up allows the end bearing part of the pile to be jacked against the upper parts (skin friction). If strain gauges are built into the pile, then a good interpretation can be made of ground parameters.

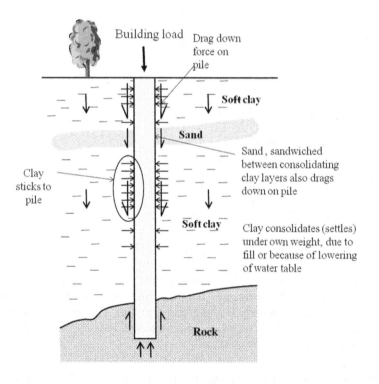

Building load   Drag down force on pile

Soft clay

Sand

Sand, sandwiched between consolidating clay layers also drags down on pile

Clay sticks to pile

Soft clay

Clay consolidates (settles) under own weight, due to fill or because of lowering of water table

Rock

Figure 6.12 The concept of negative skin friction. Where the ground around a pile or group of piles settles significantly (cm), then the ground will cause a drag down force on the pile. At the same time, the upper parts of the pile cannot provide positive skin resistance.

up to 5 cells in a single 3 m diameter pile to generate forces of over 30,000 tonnes (Cho *et al.*, 2009b). The obvious advantages include the fact that no reaction is required at the ground surface, but a limitation is that the forces upwards must be balanced by those downwards, which would be difficult to achieve where the pile is mostly end bearing.

An additional aspect to be considered in pile design is possible future settlement of the ground around the pile due to self weight, earthquake liquefaction or perhaps groundwater extraction, which can result in a drag-down force on the pile, known as negative skin friction. This is illustrated in Figure 6.12. The potential for negative skin friction is generally a matter of engineering judgement based on the ground profile and perceived future usage of the site and applied as a nominal additional load to be carried by the piles.

6.4.3.3   Proof testing

Proof tests are typically carried out on one in a hundred piles or so. The test pile should be selected by the supervising engineer, after construction and with no pre-warning to the contractor so that he does not exercise special care in its construction. Full loading tests are carried out with kentledge or some other reaction system such as ground anchors and should be taken up to loads of perhaps 1.5 times the working load for the pile. The displacement during the test (partly elastic deformation of the pile) and residual settlement after the test is completed are used as criteria of whether the tested pile and its neighbours are acceptable (Figure 6.10). If not, then additional piles may need to be installed and the existing piles down-rated. There may be time or space restrictions (such tests are very expensive and time consuming) and the contractor might urge the use of dynamic pile analysers as an alternative way of proving acceptability. As noted earlier, such tests are often unreliable and may give no measure of end-bearing resistance. Specialist tests are used to determine pile integrity, for example, by using a vibrator to take the pile through a series of frequencies so that its response can be measured. Resonance will indicate the length of the responding section, which will help in deciding whether or not the pile is broken.

6.4.3.4   Barrettes

Barrettes, like piles, are deep foundations but constructed in excavated trenches using special tools called hydrofraises, often under bentonite to support the sides of the trench. Otherwise, construction is similar to a bored pile, with a steel cage inserted in to the trench prior to concreting. Barrette shapes can follow the geometry of load-bearing walls in the finished structure.

An example of the use of barrettes rather than bored piles is for the International Commerce Centre (ICC) in Hong Kong. The 118-storey building is the tallest in Hong Kong and fourth tallest in the world (in 2011). Granite bedrock is reportedly 60–130 m deep below the building, and the designers decided to use 241 post-grouted rectangular barrettes rather than more traditional end-bearing bored piles (Tam,

2010). The barrettes were cracked by high-pressure water injection down pre-installed pipes whilst the concrete was still at low strength. Once the concrete had reached its 21-day strength, high-pressure grouting was carried out through the cracked path around the barrettes, metre by metre from the base to improve the skin frictional resistance.

6.4.3.5   *Caissons*

Caissons are large box structures formed of steel or concrete and are used as a common solution for bridge foundations offshore. The box is typically constructed onshore then floated and towed to its location where it is sunk. Sometimes caissons are sunk into the ground by driving and digging, elsewhere they just sit on a prepared surface on the sea floor. Different types are illustrated schematically in Figure 6.13. Once the caisson is in place/sunk to the required depth, then it is backfilled with rock and concrete. Caissons are also often used to form sea walls for reclamation schemes, the boxes are formed on land then floated and towed to position where they are sunk onto prepared foundations and then backfilled.

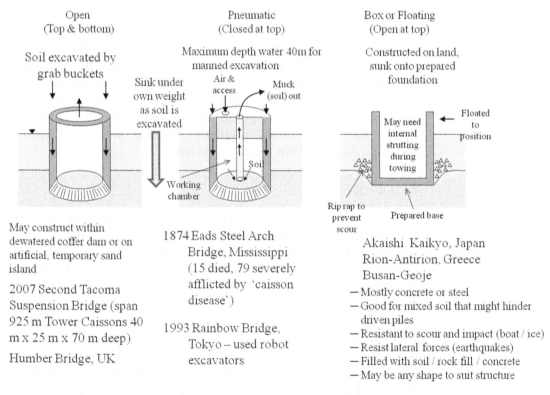

Figure 6.13 Different types of caisson commonly used for large bridge foundations, with examples.

6.5 Tunnels and caverns

6.5.1 *General considerations for tunnelling*

Tunnels will be constructed as part of an overall project, for example, water supply, drainage, rail, road, or in connection with power generation. As a result, there may be little flexibility over route and, therefore, geological and hydrogeological conditions and size and shape of tunnel. It is up to the engineering team to come up with a cost-effective solution.

One factor that will influence the chosen method of construction and lining (or not) are the final finish requirements for road and rail tunnels and whether or not it might carry water under pressure in hydraulic tunnels, as addressed at 6.6.5 below. The main issues for the engineering geologist and design team are likely to be:

– The geology along the route; how this will affect the selected method of tunnelling and any particular hazards such as natural caverns, mining or major faults.
– Stress levels and ratio of vertical to horizontal stress. High stress at depth and the concentrations in stress resulting from perturbation of the stress field by the construction can result in failure of the rock, which might result in spalling in brittle rocks or squeezing in generally weaker rocks (Hoek & Brown, 1980; Hoek *et al.*, 1995).
– Hydrogeological conditions and the risk of unacceptable water inflows and possible flooding; this is always a major issue for undersea tunnels, but can also be a concern under land.
– Existing structures that might be adversely affected by the tunnel during construction, for example, by blast vibrations or undermining as the tunnel passes by. In the longer-term, lowering of groundwater may cause settlement and/or affect water supply boreholes.

As for all geotechnical work, one needs a ground model for design. Because tunnels are often long and may be at great depth, it may be impractical to do more than a rather superficial investigation, relying largely on geological mapping and extrapolation of data, although if a serious obstacle is anticipated, such as a major fault zone, then boreholes might be targeted at that feature using inclined boreholes or even drilled along the line of the tunnel. Alternatively, a small-diameter pilot tunnel might be constructed before the main tunnel – possibly for later use as a drainage or service tunnel – because small diameter tunnels tend to have fewer difficulties (Hoek, 2000). The pilot tunnel essentially works as a large-diameter exploratory borehole.

The ground model needs to include estimates of rock or soil quality along the tunnel drive. For rock, this is often done using rock mass classifications (RMCs) such as Q, RMR or GSI, described in Chapter 4 and Stille & Palmström (2003). This will allow some estimation of support requirements and allow a contractor to choose his method of working and type of machine if a tunnel boring machine (TBM) option is selected (Barton, 2003). The ground model will also be used for hazard and risk analysis, as discussed later, and may sometimes be used as the basis for Reference Ground Conditions in Geotechnical Baseline Reports (Chapters 2 & 4), against which any claims for unexpected or differing ground conditions can be judged. As noted in Chapter 2, however, RMCs may be too coarse to represent geological conditions realistically. They may also be open to different interpretations, so that disputes are difficult to resolve.

6.5.2   *Options for construction*

Up to about a century ago, all tunnels in soil or rock were excavated by hand, using explosives where necessary to break up the rock in advance of mucking out. Nowadays, many are excavated using powerful machines. The main options generally adopted in modern tunnelling and typical support measures are set out in Table 6.2. The method of tunnelling will often be decided on factors including length of tunnel, availability of TBM, local experience and expertise. In South Korea, for example, most rock tunnels, including very long ones, have been constructed in preference by drill and blast rather than TBM. There is a wide variety of tunnel boring machines designed for all kinds of conditions from rock to soft soil. The engineering geologist needs to be able to predict the ground conditions so that the tunnel designers and tendering contractors can select the correct machine. It usually takes a long time to manufacture and launch a TBM with a whole series of ancillary equipment in the following train, and if the machine proves unsuitable, for any reason, it can be a costly mistake. Some machines are designed to be able to cope with mixed ground conditions but can still run into difficulties. Nevertheless, many TBM tunnels proceed well and at much faster rates than hand dug/drill and blast tunnels. The adoption of hazard and risk analysis (BTS, 2003), as discussed at 6.5.8, will help reduce incidents but will not necessarily eliminate hazards entirely.

Table 6.2   Options for tunnelling (after Muir Wood, 2000).

Ground type	Excavation	Support
Strong rock	Drill and blast or TBM	Nil or rockbolts
Weak rock	TBM or roadheader	Rockbolts, shotcrete, etc.
Squeezing rock	Roadheader	Variety depending on conditions
Overconsolidated clay	Open-face shielded TBM or roadheader	Segmental lining or shotcrete etc.
Weak clay, silty clay	EPB closed-face machine	Segmental lining
Sands, gravel	Closed-face slurry machine	Segmental lining

6.5.3   *Soft ground tunnelling*

Soft ground, including severely weathered rock, may be excavated by hand or by tunnel boring machine. For open-face excavation, behaviour can be predicted using classification such as the Tunnelman's Classification of Heuer (1974), which allows prediction of whether the soil will stand firmly whilst the liner is put in place or is likely to ravel, run, flow, squeeze or swell. Behaviour depends on the nature of soil, water conditions and stress levels. For example, un-cemented sand might be expected to flow below the water table, especially at depth. Such empirical predictions are also useful for weathered rocks where the application of conventional soil mechanics principles is questionable (Shirlaw *et al.*, 2000). When tunnelling in soil or in mixed-face conditions, it is the behaviour of the weakest or most mobile material that generally governs the need for, and magnitude of, the support pressure that is needed at the tunnel face.

If the soil is stiff and cohesive, then NATM methods can work successfully, as has been achieved, for example, in the London Clay (van der Berg *et al.*, 2003) and in the Fort Canning Boulder Bed and the Old Alluvium in Singapore (Shirlaw *et al.*, 2000). Where soils are unstable, then various options include grouting, dewatering, freezing or the use of compressed air. All of these are costly, may have severe health and safety implications and restrictions, and take time to install. Nevertheless, such methods are often necessary to recover and restart a tunnel that has encountered a major problem and perhaps collapsed.

Tunnel boring machines used in soft ground are of the closed-face type, as illustrated in Figure 6.14 a and b. Guidance on machine selection and use is given by the British Tunnelling Society (BTS, 2005).

Earth pressure balance (EPBM) and slurry machines use pressurised soil at the cutting face to hold up the ground as the tunnel advances. In an EPBM machine, the broken down soil remains in the plenum chamber behind the cutting head, balanced by pressure in the Archimedes screw, which removes the spoil under the control of the operators. In a slurry machine, which tends to be used in higher permeability soils, bentonite slurry is introduced to the plenum chamber, mixes with excavated soil, which is then removed for separation, disposal and re-use (bentonite) by pipes rather than on a muck conveyor. Permanent concrete lining is formed from precast segments, directly behind the machine, and this liner is used as a reaction to push the TBM forward. TBMs often work well for the specific conditions for which they are designed but also commonly run into problems with the machine getting stuck or running into rock that is either too hard or too soft or too wet for the type of machine (see Table 6.3). Shirlaw *et al.* (2003) report cases of settlement and collapse in Singapore, even using sophisticated EPBMs. Similarly, an EPBM machine was recently stopped by silt breaching the tunnel liner on a contract in the UK. A further example is discussed in Chapter 7. Recovery

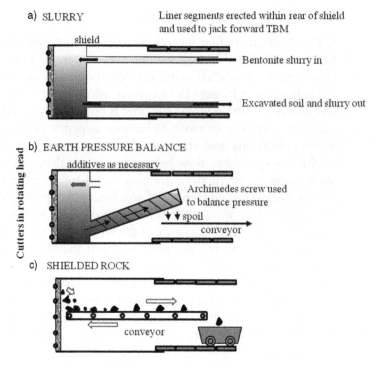

Figure 6.14 Schematic diagrams of shielded TBMs. a) Slurry machine; bentonite slurry is pumped to plenum chamber and mixes with spoil cut at the face. Mixture is removed for separation and treatment before recycling. b) Principles of EPBM. Cut soil (with additives as necessary) is removed by a screw device with the pressures monitored and maintained. c) Single-shield rock TBM. Rock cut from the face is mucked out and TBM pushes forward against the liner erected to the rear of the shield. Other rock TBMs use grippers pushed against the walls of the tunnel and use this as the reaction force for advancing the TBM.

options include freezing the ground and grouting the ground to stabilise it to allow the TBM to be withdrawn (NCE, 19 January 2011).

Where the materials to be excavated include strong and weaker material, this is known as mixed-face conditions. For stability, the major issue concerns relative mobility of the materials rather than just strength. A mixed face of strong boulders and hard clay presents problems in terms of rate of excavation, but generally not in terms of heading stability. However, a combination of strong, stable rock with a more mobile material, such as flowing, rapidly squeezing or fast ravelling material, provides conditions where the overall stability of the heading can be very difficult to control as well as difficult to excavate. Shirlaw *et al.* (2003) provide examples of major inflows resulting from the use of conventional rock tunnelling methods too close to the transition from rock-like to soil-like conditions. Ironically, this particular type of mixed-face condition has become even more problematic with the introduction of modern tunnelling technology.

Table 6.3 Typical problems with TBM tunnels and possible mitigation measures.

Problem	Mitigation
Ground too strong (intact strength and/or lack of discontinuities)	May need to pull TBM back and advance with drill and blast
Ground too weak and collapsing (should have been an earth balance or slurry machine perhaps)	Ground improvement might be necessary in advance of tunnel drive – grouting or freezing
Major faults	Collapse of ground and TBM gets stuck. May need to sink a shaft in front of machine and construct a tunnel back to and around the TBM to free it up. Ground treatment and possible hand construction through fault zone may be required to get the TBM going again
Weak ground and high *in situ* stresses leading to squeezing action on TBM	Can cause huge delays. Ground improvement to strengthen the ground and resist the squeezing pressures
Too much clay for slurry treatment	Can cause delay and necessitate installation of additional treatment plant – extra hydrocyclones, etc
Ground abrasive because of high silica content causing too much wear on teeth, leading to cost and delay	Cost may be prohibitive, necessitating a change of excavation method
Too much water and TBM electrics not protected	Drilling and grouting in advance of machine or possibly ground freezing or compressed air working. Possible change of method to drill and blast or employ different machine with suitable spec
Excess tunnel slurry pressure causes blowout at ground surface	Lower pressure
Pressure too low causes face collapse	Reverse of the above

6.5.4   *Hard rock tunnelling*

The main options are drill and blast, a roadheader excavating machine or to use a TBM that may be either open (without a protective shield) or shielded.

6.5.4.1   *Drill and blast/roadheaders*

Generally, drill and blast tunnels are more flexible than TBMs and allow difficult ground conditions to be understood and overcome, but they may be much more time consuming unless a number of access points can be found to allow operations to proceed from several faces at the same time.

Holes are drilled in the face, and explosives placed in the holes. Issues of tunnel blast design are addressed by Zare & Bruland (2006). The holes are detonated sequentially to break to a free face over micro seconds. The aim is to break the rock to manageable size so it can be excavated (mucked out) readily with machines, without further blasting or hammering. Other aims may be to keep blast vibrations to a minimum and not cause damage or offence to nearby residents, and usually to keep as closely as possible to the excavation shape prescribed by the designers, i.e. minimising overbreak. Typical advances per round are 3 to 3.5 m, sometimes up to 5 m in very good rock conditions. Depending on the size of tunnel and ground conditions, the full face may be blasted in one round or may be taken out as a series of smaller headings – top, or side, that may be supported by sprayed concrete with steel mesh or steel/carbon fibres, rock bolts, and/or steel arches or lattice girders, before the tunnel is advanced. Figure 6.15 shows a tunnel portal following the first blast, with steel arches being erected to protect the tunnel access.

Figure 6.15 After first blast and mucking out, construction of temporary steel arches to protect tunnel portal, Queens Valley Reservoir, Jersey, UK.

Figure 6.16 (a) Convergence in rock tunnel to stable condition. (b) Local failure and ravelling to ground surface.

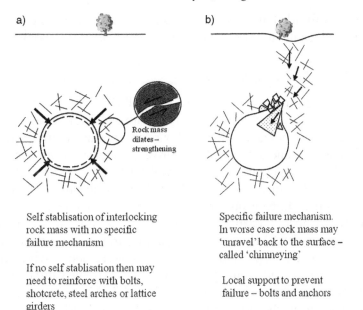

a)

Rock mass dilates – strengthening

Self stablisation of interlocking rock mass with no specific failure mechanism

If no self stablisation then may need to reinforce with bolts, shotcrete, steel arches or lattice girders

b)

Specific failure mechanism. In worse case rock mass may 'unravel' back to the surface – called 'chimneying'

Local support to prevent failure – bolts and anchors

After blasting, and dust and gases have dissipated and safety checks made (e.g. for methane or radon), the broken rock is mucked out and it is the engineering geologist's task to examine and map the geological conditions exposed. The freshly blasted rock may well be unstable, and the geologist should not approach the face until the contractor has carried out all necessary scaling and/or rock support work to make the tunnel safe. The contractor has overall responsibility for site safety and his instructions should be followed at all times in this respect. A decision will then be taken on whether the ground is as expected, if the ground is changing (and probing ahead is required), and the support requirements. Any potential for deteriorating conditions or, for example, a major potential wedge failure, need to be identified quickly so that support measures can be taken. As illustrated in Figure 6.16, often the rock mass is self-supporting. As the tunnel is excavated, the tunnel walls move inwards, the rock mass dilates and generally locks up. If there is an inherent weakness, such as a free wedge of rock or a fault zone, then local collapse can be followed by ravelling failure, which could chimney to the ground surface. In two of the examples discussed in Chapter 7, the situation deteriorated quickly. If conditions are poor and getting worse, then the ground might be supported in advance of the tunnel by an umbrella of spiles or canopy tubes, and/or by pressure grouting.

In suitable rock, other mining approaches may be used, including the use of large roadheaders that cut their way into the rock but do not excavate the full face profile in one operation, unlike a TBM. In a tunnel formed by drill and blast or roadheader, it is possible to examine and record the ground conditions throughout construction and make decisions as to the support required. In a TBM tunnel, little can be told about the ground ahead of the machine without stopping and drilling in front of the face, which disrupts operations and is therefore to be avoided.

6.5.4.2   *TBM tunnels in rock*

The design and use of modern hard rock TBMs is covered comprehensively by Maidl *et al.* (2008). In good rock with high RQD, open TBMs are sometimes used, but generally only for relatively small diameter tunnels. The tunnel advances by jacking forward against grippers that are extended laterally against the tunnel walls. Clearly, if the rock becomes poor quality then there may be problems with the grippers. There is also no way of preventing groundwater ingress other than by grouting, preferably in advance of the machine. In Chapter 7, a case (SSDS) is presented where open-rock TBMs were selected, anticipating good rock conditions with low water inflows, and the operations were halted when inflows became too great and grouting in advance was extremely difficult.

In poorer-quality rock, generally, shielded TBMs are used. A single-shield machine pushes against the liner, as for soil TBMs (Figure 6.14c). In other set-ups there are two shields; the rear shield has grippers and provides the reaction against which the front shield can push forward. The cutter head has discs that rotate as the cutter head itself rotates. The thrust of the machine causes the rock to fail, mainly in tension. A major consideration is the lifetime of the cutting discs before they need to be replaced, as addressed by Maidl *et al.* (2008). A case example in Chapter 7 describes considerable wear in an EPBM used to tunnel through abrasive sandstone.

6.5.5   *Tunnel support*

6.5.5.1   *Temporary works*

Rock tunnelling, in general, relies largely on the rock mass locking up as joints and interlocking blocks of rock interact and dilate during the process of convergence towards the excavation. Good-quality rock often forms a natural arch and no or little support is needed. However, in weaker ground, such as in fault zones, the rock mass cannot support itself, even with reinforcement, and requires artificial support in the form of steel arch ribs, typically encased in shotcrete. Optimising support requirements in weaker ground requires prediction of likely convergence rates, making observations as excavation is undertaken, i.e. observational methods, and then applying support such as rock bolts and/or shotcrete and/or steel arch ribs to control the movement and prevent excessive loosening (Powderham, 1994). In stronger, blocky rock masses, rock movement will be much less, and the purpose of the support is then to prevent loss of loose blocks and wedges, which would destabilise the arch and maybe lead to ravelling failure.

Rock mass classification systems introduced in Chapter 4 are linked to charts allowing decisions to be taken as to the immediate (temporary) support measures required. These are reviewed by Hoek *et al.* (1995). In practice, decisions may often be biased by other considerations such as the materials and equipment at hand and the workers' perceptions of the degree of risk and how well previous support measures have worked. This may of course have cost implications and may also later become a matter of dispute as to what was really necessary, as discussed and analysed by Tarkoy (1991). The importance of good engineering geological records during construction is emphasised. In severe situations such as high stress or intense water inflow, steel lining may be used but even then this sometimes proves inadequate as happened during the construction of the Tai Po to Butterfly Valley water supply tunnel in Hong Kong, where unexpectedly high water pressures buckled the liners (Robertshaw & Tam, 1999; Buckingham, 2003).

6.5.5.2 *Permanent design*

There are two main areas for consideration: firstly, the area around the portal, especially for tunnels that are part of a road or rail system, and, secondly, need for a permanent liner.

6.5.5.2.1 PORTAL DESIGN

The area above the entrance to a tunnel often requires careful engineering to make it safe, both during construction and during operation. The problems are essentially the same as for general slope stability design, as discussed later in this chapter, but the need for long-term inspection and maintenance, whilst maintaining tunnel usage, sets portal design in a rather special category. A canopy is often constructed to protect the portal area from falling rock and other debris, as illustrated in Figure 6.17. Catch nets, barriers (such as gabion walls) and *in situ* stabilisation can

Figure 6.17
Canopy extending out from tunnel liner (being waterproofed), to protect portal area. A55, North Wales.

be used to prevent debris impacting the portal area. Rock and soil masses immediately above the portal area are often covered with steel mesh and shotcrete or similar hard covering and dowelled, nailed or anchored back using post-tensioned bolts and cable anchors. The requirements for designing, protecting and maintaining ground anchorages are set out in national standards and codes of practice such as BS 8081 (BSI, 1989) and BS EN 1537 (BSI, 2000). Despite such standards, things occasionally go wrong, either because of ground conditions or flaws in the anchorage itself, and designers must appreciate the practical difficulties that might be associated with maintenance programmes whilst ensuring safety for the road user. If a major problem is found, then the tunnel might need to be closed or restricted in use whilst the problems are rectified. Several cases of the failure of rock anchorages, even in projects post-dating BS 8081, are discussed in Chapter 7.

6.5.5.2.2   PERMANENT LINERS

The options for permanent tunnel liner design include:

– Unlined (ignoring temporary support measures)
– Unreinforced concrete
– Reinforced concrete
– Steel.

Lined tunnels can be designed to be undrained, in which case the permanent lining must withstand the full groundwater pressure as well as rock loads. Other tunnels are designed to be drained, whereby the outer surface (or extrados) of the arch of the liner is lined with a waterproofing membrane laid onto geotextile sheets, which carry water down to drains and sumps below the tunnel invert. The sumps may need continual pumping, and the whole drainage system needs maintenance over the life of the project. Figure 6.18 shows details of a design, as used in some recent rail tunnels in Hong Kong. After shotcreting the tunnel walls, layers of geotextile (outer) and waterproof membrane (inner) are placed, followed by an inner concrete liner (250 mm thick). Groundwater is thereby channelled via the geotextile to an egg box drainage system in the invert. For any drained lining design, care must be taken that any permanent drawdown in the water table has no adverse affects on structures above the tunnel or on water supply from groundwater sources.

Precast concrete segments are commonly erected as part of a TBM excavation and support process, mainly in soft ground tunnels, but also in some hard rock applications. The segments are manufactured externally and then erected within the shield surrounding the advancing machine and bolted together. If required, segments can be fitted with gaskets to form fully waterproof concrete liners (Figure 6.19). As noted earlier, the installed liner can be designed to provide a reaction to push the TBM forward.

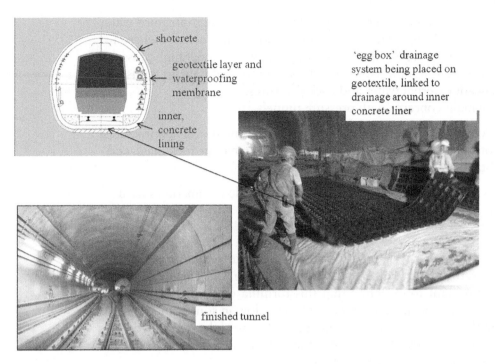

Figure 6.18 Egg box drainage system for a drained tunnel (courtesy of MTRC, Hong Kong).

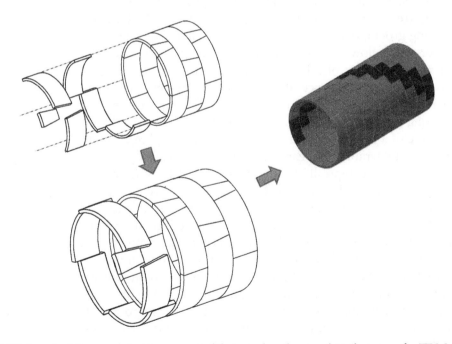

Figure 6.19 Interlocking tunnel segments, prefabricated and erected to the rear of a TBM shield (figure courtesy of Mike King, Halcrow).

One of the most severe design situations is in high-pressure water supply tunnels associated with hydropower constructions where for some operational periods the tunnel carries water under high pressure, but at other times the same tunnels are empty and have to withstand significant external water and rock pressures.

The main concerns with pressure tunnels are:

– Potential damage by hydraulic fracturing (formation of new fractures) or jacking (opening of existing fractures) within the rock mass, and
– Stability, durability and low maintenance.

To avoid hydraulic fracturing, an empirical rule is sometimes used:

$$\frac{D\gamma_R}{H\gamma_W} > 1.25 \qquad \text{(Haimson, 1992)}$$

Where γ_R is unit weight of rock and γ_W is unit weight of water, D is rock overburden at tunnel location and H is the water head. However, it is important to recognise that this formula only considers vertical *in situ* stress. Horizontal stress can be very low in some situations, for example, close to valley sides, and this will control the risk of hydraulic fracture or jacking if water from the tunnel can reach the excavated rock surface at sufficiently high pressure.

Where the confining rock stress, vertical and/or horizontal, is too low, fully welded continuous steel liners are generally used to prevent the high-pressure water from reaching the rock mass. Concrete liners may be used in competent rock but might crack under high internal water pressure if the confining stresses are too low. In such cases, there is a risk of leakage to surrounding ground (with a risk of causing landslides in some situations) and/or water flow into other underground openings. Haimson (1992) presents examples of schemes where the importance of stress conditions and the correct choice of lining only became evident late in the design process, with 'unpleasant consequences'. An important task of the engineering geologist is to ensure that the *in situ* stress conditions along the route of a pressure tunnel are evaluated fully and reported to the design team, preferably at an early stage in project planning.

In certain situations, typically in low pressure headrace tunnels, a concrete liner can be designed with drainage holes to relieve water pressure on the tunnel lining. Consolidation grouting is usually carried out around the tunnel to reduce leakage out of or into the tunnel (depending on the relative internal and external water pressures). Unlined tunnels can be used in good rock conditions and with favourable *in situ* stresses, but there may be higher maintenance requirements and the need to construct rock traps to catch any fallen debris. The proper design of hydraulic pressure tunnels is particularly important as the consequences of failure are usually very severe and costly to repair. A comprehensive summary of the principal design and construction considerations is presented by Benson (1989).

6.5.6   *Cavern design*

Caverns are large-span underground openings and these are used for many purposes, including sports halls, power stations and oil and gas storage (Sterling, 1993). Hydroelectric power caverns and large three-lane road tunnels are typically 20 to 25 m span, but caverns have been constructed successfully in good-quality rock with spans in excess of 60 m (Broch *et al.*, 1996), and natural caves are found with much larger spans.

There is considerable guidance in the literature on approaches to their design and construction (e.g. Hoek & Brown, 1980; GEO, 1992). Many design issues are similar to tunnels but because they are at fixed locations, ground investigation decisions are more straightforward. The other major difference is scale. Whereas many tunnel walls lock up as the rock dilates, and need little support to ensure stability, in a cavern there is more potential for large-scale strain and failure mechanisms to develop. For example, large caverns were required for a proposed high-speed rail station at Taegu, Korea, in strong mudstone. Preliminary numerical analyses were carried out to design permanent concrete liners and bolting support, assuming essentially isotropic rock mass parameters. The design had to be revised when it was realised that the rock structure was strongly anisotropic with bedding mostly horizontal and many near-vertical tensile joints infilled with calcite (Figure 6.20). These joints could allow discrete failure into the crown of the openings, as illustrated by Maury (1993) for mine workings.

Figure 6.20 Rock core from vertical borehole in strong mudstone, Taegu, South Korea. Note near-vertical persistent joint infilled with calcite. This network of joints (two sets orthogonal to bedding) were encountered frequently in preliminary boreholes, and appreciation of their significance led to a) reconsideration of the potential rock loads on the permanent liners and b) additional ground investigation using inclined rather than vertical holes to characterise the rock mass better.

Hoek & Moy (1993) and Cheng & Liu (1993) describe different aspects of the design and construction of the Mingtan pumped storage project in Taiwan and illustrate the need for an integrated approach of geological investigation, modelling, design, observation, construction and instrumentation. An exploration/drainage gallery and two other galleries were used to install long corrosion-protected permanent cable anchors to reinforce the roof arch of the main cavern 10 m below, prior to its excavation. Small loads were applied to the cable anchors, which only took on their full loads as the cavern was excavated.

6.5.7   *Underground mining*

Underground mining is quite different from the formation of caverns and tunnels for civil engineering, although many of the skills required are the same. In mining, the objective is to extract the ore whilst minimising waste rock production. Safety is a prime concern, as it is for civil operations, but mining involves the formation of non-permanent voids, many of which will be allowed to collapse or packed loosely with waste rock, so the fundamental operational concepts are obviously quite different. Rock mechanics of underground mining operations are discussed by Brady & Brown (2004). In terms of geological hazards, of particular concern are flammable and/or noxious gases, including radon, and the control of dust and ventilation is very important. Such matters are generally mandated by national standards on health and safety but still accidents occur regularly worldwide.

A general concern for construction in mining areas is continuing ground settlement or sudden collapse of old workings. These are matters to be considered at the desk study stage of site investigation, as addressed in Chapter 4.

6.5.8   *Risk assessments for tunnelling and underground works*

In Chapter 4, a system was introduced whereby site investigation is conducted or reviewed following a checklist approach whereby firstly geological hazards are considered, then environmental factors and finally hazards associated with the specific type of project or construction method. Tunnels are often particularly risky undertakings because they are so dependent upon geotechnical conditions, which may vary considerably along their length, and it is seldom feasible to carry out as comprehensive a ground investigation as it is for other types of project. Good reviews of tunnel collapse mechanisms and case histories are given by Maury (1993) and GEO (2009) respectively. Consequently, industry has developed several approaches whereby hazards are considered in detail, so that strategies can be prepared to reduce or mitigate the risks. This can be done at the option assessment and design stages and then later as part of the management of construction.

The British Tunnelling Society Joint Code of Practice for Risk Management of Tunnel Works in the UK (BTS, 2003) was prepared jointly by the Association of British Insurers and the BTS and sets out requirements regarding risk assessment and management for any tunnel with a contract price of more than £1 million. In effect, it is mandatory in the sense that without its adoption, no insurance will be forthcoming for an underground project. The Code of Practice sets out how and when risk is to be assessed and managed, and by whom. Risks are to be assessed at the project development stage (design), by the contractor at tender stage and during construction through a risk register.

The Code also requires the ground reference conditions or geotechnical baseline conditions to form part of the contract, but as noted in Chapters 2 and 4, definition of such conditions is not always straightforward. Whilst the intention to avoid dispute is laudable, there may be considerable difficulty in summarising geological and geotechnical conditions succinctly and unambiguously.

6.5.8.1   Assessment at the design stage

The ways that risk can be assessed at investigation and design stages are illustrated by the example of the 16.2 km Young Dong rock tunnel in Korea, as presented in Appendix E-1 and E-2. Given an appreciation of the ground conditions along the route, based on a well-conducted site investigation, the hazards associated with the various options for construction can be considered. Once these have been identified, their likelihood and seriousness can be rated in terms of potential consequence (e.g. programme, cost, health and safety) and methodologies devised for mitigation prior or during construction. Decisions can then be made on how to proceed.

6.5.8.2   Risk registers during construction

During construction, hazards that were anticipated at the design stage may prove real or illusory. New ones will be identified and need to be dealt with. The current way of so-doing is to employ a risk register in which hazards are identified and assigned to individuals in the project team to derive strategies for their avoidance or mitigation. In the BTS Code of Practice (2003), this is identified as a task for the contractor but the register will include risks brought forward from the project development stage. In practice, it may well be the project engineer rather than the contractor who manages the construction risk register, perhaps at monthly meetings held to monitor progress on mitigating each of the identified risks, remove from the register those that have been dealt with, and recognise and assign to individuals any new risks identified during the course of the work. Brown (1999) outlines the risk management procedures adopted for the successfully completed Channel Tunnel Rail Link Project in the United Kingdom, and a list of typical tunnelling hazards to be considered during construction is presented in a table in Appendix E-3.

6.6   Slopes

Landslides cause major economic damage and kill many people each year. Slopes can be split into natural and man-made. The hazards from natural terrain landslides in mountainous regions and at the coast are considered in Chapter 4.

Man-made slopes include cut slopes (cut into the natural hillside) and fill slopes. Fill slopes might simply comprise the excess debris from an adjacent cutting, dumped or compacted onto the adjacent hillside to form an extra carriageway, but can also include sophisticated, high and steep slopes incorporating geotextiles or other materials to strengthen the soil (Figure 6.21). Stability needs to be assessed by engineers and if considered unstable, measures must be taken to improve the stability to an acceptable level.

6.6.1   Rock slopes

Rock slope stability is generally controlled by the geometry of pre-existing, adversely oriented discontinuities, including bedding planes, faults and master joints including sheeting joints. Failure types can be grouped as illustrated in Figure 6.22 and as follows:

Shallow: superficial failures, generally low to medium volume.

Structurally controlled: sliding may occur on one or more intersecting discontinuities that are adversely oriented relative to the slope geometry. Toppling can result because of the presence of unstable columns

Figure 6.21 Construction of reinforced earth embankment for Castle Peak Road widening, Hong Kong.

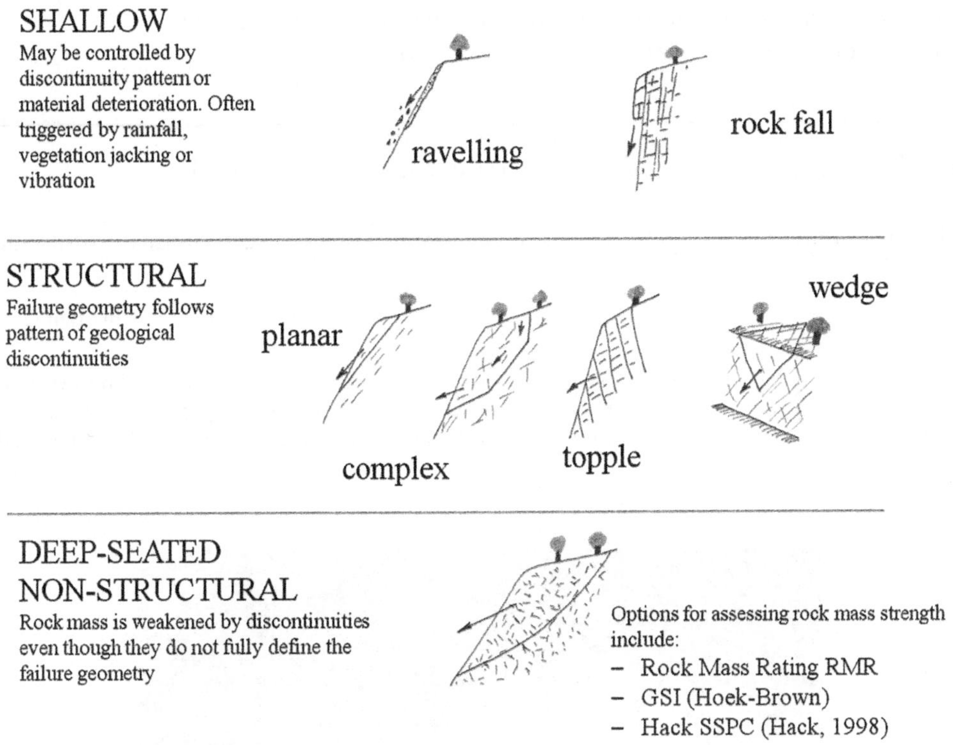

SHALLOW

May be controlled by discontinuity pattern or material deterioration. Often triggered by rainfall, vegetation jacking or vibration

ravelling

rock fall

STRUCTURAL

Failure geometry follows pattern of geological discontinuities

planar

wedge

complex

topple

DEEP-SEATED NON-STRUCTURAL

Rock mass is weakened by discontinuities even though they do not fully define the failure geometry

Options for assessing rock mass strength include:
- Rock Mass Rating RMR
- GSI (Hoek-Brown)
- Hack SSPC (Hack, 1998)

Figure 6.22 Modes of failure in rock slopes.

of rock, perhaps dipping steeply back into the slope. Large complex failures can involve a number of adverse sets of discontinuities together with some breaking through intact, perhaps weathered rock, allowing the full mechanism to develop.

Deep seated, non-structurally controlled: the rock can be considered an interlocking mass of rock blocks without adverse fabric such as bedding, schistosity or systematic joints.

6.6.1.1 Shallow failures

Steep rock slopes are sources of rockfalls, which can be a major risk, especially where adjacent to a busy road or railway. All rock slope surfaces deteriorate with time (Nicholson *et al.*, 2000). Rock material weathers, vegetation grows and opens up joints and blocks get undermined by erosion (Figure 3.58). Even small blocks can cause accidents. On large lengths of highway through a mountainous region, there will be a need to identify where the risk is greatest so that the risks can be mitigated cost-effectively (Box 6-4). This can be done by using some Rock Mass Rating appraisal system together with software capable of predicting where falling rock might end up, but it is often just a matter of engineering judgement taking account of the history of rockfalls. In such an assessment, it should be remembered that relatively minor rockfalls may be precursors to major rock collapses. Methods of mitigating rockfalls and other potential landslides are discussed below.

Box 6-4   Judging the severity of rockfall hazards and the associated risks

'People – even experts – rarely assess their uncertainty to be as large as it usually turns out to be.'

Baecher & Christian (2003).

The assessment of hazard of rock slope failure is always rather subjective, as illustrated by a visit to the petroglyphs at Anhwa-ri, Goryeong, Korea, in February 2008. The rock exposure shown in Figure B6-4.1, above the rock carvings, appears to be on the brink of failure and one would be tempted to

Figure B6-4.1 Petroglyphs at Anhwa-ri, Goryeong, Korea. Rock slope above petroglyphs (with small protective fence) shows signs of vegetation wedging, with loose blocks resting against trees.

fence off the area, immediately followed by removal of any blocks that cannot be stabilised by dowelling and dentition works. However, the fact that the precarious open-jointed rock is directly above the ancient rock carvings, is evidence that this rock face has not retreated very far over a period of more than 2,000 years. The process of deterioration and collapse is actually quite slow and judgment of the risk as immediate and obvious, requiring urgent action, would therefore err on the conservative side.

Conversely, the slope shown in Figure B6-4.2 is in the Cow and Calf Quarry at Ilkley, Yorkshire, in the UK, and was used to teach MSc engineering geology students to map rock discontinuities for several years. The collapse to the left of the photograph occurred unexpectedly between mapping exercises, despite its repeated examination and systematic logging on scan lines, without the failure mechanism having been identified.

Figure B6-4.2 Unexpected rock failure in Cow and Calf Quarry, Ilkley, West Yorkshire, UK.

These examples illustrate our uncertainty and the difficulties in judging the degree of hazard by examination alone. It is highly likely that even after ground investigation, our ability to judge the severity of the situation is often rather poor. The conclusion must be that consequence should be the priority when assessing the risk of slope failure. If there is a major risk to life, then works should be done. This is the underlying philosophy behind the Landslide Preventive Works (LPM) strategy in Hong Kong where the catalogue of tens of thousands of slopes, prepared in the 70s and 80s, has been compiled and ordered in terms of perceived risk (a function of height, angle and proximity to vulnerable facilities). Each slope is being checked and upgraded in order. Most of these are dealt with using essentially prescriptive engineering works, including soil nails and inclined drains installed to a pattern.

If there is clear danger from the hazard, then it should be dealt with. In the Korean case discussed above, despite the apparently slow retreat of the rock exposure above the petroglyphs, visitors to the site should be protected against the evident rockfall hazards.

Quantitative risk assessment of rockfall to roads

At the site shown in Figure B6-4.3, while it might be intuitively obvious that there is some risk to life from rockfall along the road and some history of such rockfalls, the cost of preventive works may be very expensive. One way to deal with this quandary is to try to quantify the risk and compare this to the cost of reducing the risk.

To do this requires the following data to be measured or estimated:

– Frequency and size of rockfall incidents (per day).
– Number of vehicles per day, average length and velocity.
– Vulnerability of persons in vehicles to rockfall (depends on size of falls).

Figure B6-4.3 Road cut through limestone with very little engineering support or protective measures, Tailuko Gorge, Taiwan.

The annual probability of risk of death can then be calculated and compared to published guidelines on acceptable risk (e.g. Fell *et al.*, 2005). Different sections of road will be shown to have different risk levels, which will allow decisions to be made on where to carry out mitigation works. Quite often such a calculation will show that risks are acceptable even if, judgmentally, the hazard is still intolerable (the situation looks very worrying). It may well be found that relatively simple measures, such as scaling off the most obvious loose rock and providing netting or cheap barriers such as gabions locally, will reduce risk considerably whilst also making the situation feel safer. Further guidance on judging rockfall hazards and the use of rockfall rating systems is given by Bunce *et al.* (1997) and Li *et al.* (2009).

6.6.1.2 Structural

The distinction of failure mechanisms into planar, wedge and toppling, and the discontinuity geometries and conditions responsible for each style of failure, are set out clearly by Hoek & Bray (1974), and this has been updated by Wyllie & Mah (2004). The most common type of failure is sliding on a single discontinuity, and this is simple to analyse. The main difficulties are in assessing shear strength of the rock discontinuities, as set out in Chapter 5, and how to deal with groundwater pressures. Generally, a simple analysis is done in which it is supposed that water pressure at the slope face is zero, increasing back within the slope, to some height below ground surface at the rear of the slope (Figure 6.23). This is often a conservative assumption, in that water pressure will be localised, not acting throughout the whole slope at the same time. Richards & Cowland (1986) discuss a well-investigated site where it would have been unrealistic to design the slope to withstand the maximum water

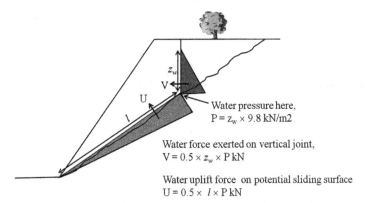

Figure 6.23
Typical model for analysing influence of water pressure on stability of a sliding rock slab.

Water pressure here,
$P = z_w \times 9.8 \text{ kN/m2}$

Water force exerted on vertical joint,
$V = 0.5 \times z_w \times P \text{ kN}$

Water uplift force on potential sliding surface
$U = 0.5 \times l \times P \text{ kN}$

pressures at each location, all acting at the same time, because instruments clearly showed pulses of water pressure travelling through the slope, following a rain storm.

Even small intact rock bridges can provide sufficient true cohesion to stop seemingly hazardous slopes from failing (Figure 5.19). This can be a major dilemma because the rock bridges cannot be seen or identified by any realistic investigation method. Careful geological study has failed to identify a useable link between persistence and any other measurable joint characteristic (Rawnsley, 1990) and, it must be remembered, traces exposed at the Earth's surface may be poor representations of characteristics inside the unexposed mass, because of stress relief and weathering. Because of this uncertainty, designs will typically require the risk of failure to be minimised by incorporating toe buttresses, reinforcement with anchorages of some kind, or some other protection, possibly using an avalanche shelter.

From experience, wedge failures are relatively rare so that even where these are identified as a problem from stereographic analysis, this might not develop in practice. Similarly, most slopes that appear to have a toppling problem do not do so in reality, generally because of impersistence. Care must be taken, therefore, to be realistic in appraising the results of any geometrical analysis that suggests there to be a problem. One factor that must be considered is risk, which is the product of hazard (likelihood of a failure) and consequence (likelihood of injury or damage). One other aspect is that where major failures do occur, it is often found by later inspection that the rock mass was in serious distress long before failing and this might have been discovered by carefully targeted investigation. Key factors to look for are open and infilled joints and distorted trees, though again the situation might be less risky than it immediately appears (Box 6-4). There is no easy answer to this – it is a matter requiring observation, measurement, analysis, experience and judgement, and consideration of consequence.

Monitoring can be conducted in the real time, for example, using total systems that record movements at short intervals automatically or vibrating wire strain gauges with data transferred to the responsible person, as discussed in Chapter 4. Alternatively, periodic examinations using inclinometers, radar or photogrammetry can all be effective.

6.6.1.3 *Deep-seated failure*

Very large rock slope failures often involve some zones where sliding on discontinuities is happening whilst elsewhere the rock mass may be acting as an isotropic fractured mass in a Hoek-Brown way and in other areas intact rock may be failing. Explaining such complex failures is a much easier task than prediction. Many large failures have been studied in detail, and these cases are probably the best place to look for ideas and inspiration when dealing with large slopes (e.g. Bisci *et al.*, 1996; Eberhardt *et al.*, 2004).

6.6.2 *Soil slopes*

For soil slopes, where the ground mass can be regarded as essentially isotropic within each unit or layer (stratum), analysis involves searching through the slope geometry, looking for the potential slip planes with the lowest FoS. This can be done easily using available software such as SLOPE-W and Slide. In weathered rock and indeed many soils, stability might well be controlled by adverse relict joints and other weak discontinuities so that a variety of possible failure modes need to be addressed. An example of hazard models requiring particular analysis within cut slopes in Eocene mudstone at Po Chang in Korea is given in Box 6-5.

Box 6-5 Hazard models for a slope, Po Chang, Korea

The slope shown in Figure B6-5.1 is within a development site near Po Chang, South Korea. The rock comprises weak to strong bedded mudstone containing strong rounded concretions. The slope was excavated several years prior to the photograph and in some areas is deteriorating very rapidly, with large screes of disintegrated mudstone debris.

Apart from bedding, the main discontinuities are orthogonal vertical sets of joints, probably formed during burial. There are also conjugate shear fractures, inclined at steep angles. As Figure B6-5.2 shows, the same rock is exposed in unprotected slopes adjacent to main roads on the outskirts of Po Chang. There are evident recent failure scars in some of the slopes.

This case provides an example of how a single slope or series of slopes may contribute several different hazards, each of which needs to be considered in a different way, as illustrated in Figure B6-5.3.

Figure B6-5.1 View of large cut slope in Eocene mudstone near Po Chang Korea.

Figure B6-5.2 Road side cutting though same sequence of Eocene mudstone on outskirts of Po Chang.

Shallow hazards include boulder fall from the concretions, undermined from the continuing ravelling deterioration of the mudstone. Trees may collapse in a similar way.

There is a risk of structurally controlled failure on the steeply inclined conjugate shear set of joints, with vertical joints providing the release surfaces. There was evidence of such failures in some exposures. Finally, there is a risk of large-scale landslide in these steep slopes, involving a generalised slip surface through the closely fractured rock. The question is how to determine an appropriate set of strength parameters for analysis. It might be reasonable to use the GSI approach (Marinos &

Hoek,2000), but the first step would be to collect more empirical data on the way that the Po Chang mudstone behaves regionally and to identify whether there are any large-scale failures that might be back-analysed.

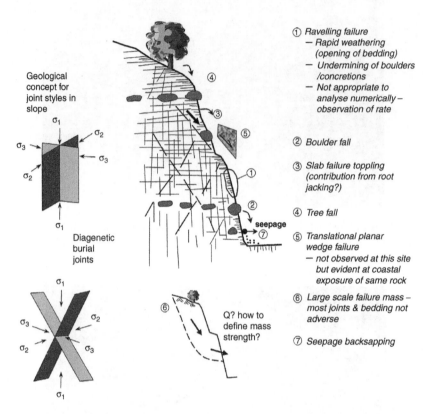

Figure B6-5.3 Ground model for slope. Main discontinuities include bedding which is almost horizontal, orthogonal near-vertical joints and steeply inclined shear joints. Potential failure mechanisms (1 to 7) are identified, each of which need to be considered and assessed individually, both through observation in the field and by numerical analysis where possible.

The simplest type of analysis is undrained in which it is assumed that the soil has a uniform strength, independent of stress level, expressed as cohesion along the potential slip plane. The logic is that any change in normal stress is matched by a change in water pressure so that change in effective stress and frictional resistance is zero. This type of analysis is only appropriate for earthworks in clay, immediately after cutting, and is not considered further here (see also Box 6-3 re pile design).

More generally an effective stress analysis is used where the strength of the soil (or closely fractured rock) is considered to be derived from two components – friction and cohesion, as per the Mohr-Coulomb expression:

$$\tau = (\sigma - u)\tan\phi + c$$

where τ is shear strength; σ is total stress (generally due to weight) normal to the failure plane; u is water pressure reducing σ to an effective stress, σ'; ϕ is angle of friction; and c is cohesion.

Frictional resistance changes with stress conditions, which vary throughout the slope, and to deal with this, a method of slices is used typically to calculate stability. Figure 6.24 shows a slope with the potential failing mass split into four vertical slices. In this diagram, the weights of slices 1 and 3 have been resolved into destabilising shear force, S, parallel to the tangent to the section of slip surface below each slice and a normal force acting normal to the shear surface (N). It is evident that the ratio of S to N varies considerably from one slice to the next. Slices 1 and 2 are being prevented from failing by Slices 3 and 4. The FoS for the slope as a whole is the ratio of the summation of shear resistances beneath each slice to the summation of the shear components. There are many different versions of the method of slices. For some, circular slip planes are assumed, in others, irregular slip surfaces can be analysed (e.g. Morgernstern & Price, 1965). Slice boundaries are generally taken to be vertical and assumptions need to be made regarding the forces at the vertical interfaces between each slice. The method of Sarma (1975) allows non-vertical slices, which gives some flexibility in dealing with more complex geology. Software packages (limit equilibrium) give a range of options regarding the method of analysis and give almost instant answers so the results from the various analytical models can be compared. Sometimes this is done in a probabilistic manner, varying the various strength parameters through their anticipated ranges and distributions (Priest & Brown, 1983). Generally, these analyses are carried out to try to establish that the FoS exceeds some chosen value – typically between 1.2 for a slope with low consequence of failure and 1.4 for a higher risk slope and, empirically, most slopes analysed with such FoS

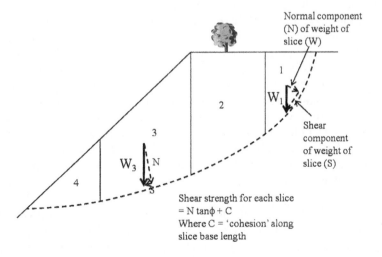

Figure 6.24 The method of slices for slope stability analysis.

Normal component (N) of weight of slice (W)

Shear component of weight of slice (S)

Shear strength for each slice
= N tanϕ + C
Where C = 'cohesion' along slice base length

will stand safely provided that the ground model is correct – probably in part because of inherent conservatism in most assessments of mass strength parameters (see discussion on disturbance in Chapter 4). More sophisticated analyses can be carried out using generalised representations of soils and their properties in both two and three-dimensions (e.g. FLAC SLOPE and FLAC3D – Itasca), and these software packages allow the engineer to see how the failure develops in a time-stepping manner, which is very helpful. The best use of stability analyses is to test the significance of the various assumptions to the outcome. Lumb (1976) addressed some of the problems of the Factor of Safety approach and advocated that engineers think instead in terms of probability: 'forcing the designer to consider the reliability of all his data and to face up to the consequences of his being wrong'. If, for example, water level is shown as critical to stability, then that should lead to a careful assessment of the need to prevent infiltration and to install drainage systems.

In the partial factor approach of Eurocode 7, each part of the analysis – forces and strength parameters – are factored in a prescriptive manner. Commonly used software packages can cope with this. This approach might be regarded as rather limiting and perhaps giving an incorrect impression that everything is understood and that all factors are always the same. For example, the Eurocode partial factor for cohesion is the same as for friction (1.25), whereas it is common experience that friction can generally be measured or estimated with far more confidence than cohesion and changing assumptions on cohesion can have a disproportionate influence on calculated FoS.

All analyses are of course only as valid as the input parameters and especially the geological and hydrogeological models; if the model is wrong, so will be the analysis. In a study of the failures of several engineer-designed slopes, Hencher (1983e) concluded:

'Six of the eight cut slopes that failed had been investigated by drilling in recent years. In five of these cases, important aspects that controlled the failure were missed. In only one case were the true geological conditions recognised, but even then the groundwater levels were underestimated considerably. In all cases, where piezometric data were available and the groundwater level was known by other means, albeit approximately (e.g. observed seepage), the piezometric data did not reflect peak water pressure at the failure surface. This was principally due to failure to observe rapid transient rises and falls in water levels. A further problem was that many of the piezometers were installed at levels where they could not detect the critical perched water tables which developed.'

More recently, Lee & Hencher (2009) document a case study where a slope was subject to numerous ground investigations and analyses (often in response to some relatively minor failure) over many years, before the slope finally collapsed in a disastrous manner. There were fundamental misconceptions about the geological conditions by all of the investigators. The potential for self-delusion that such methods of analysis truly represent actual stability conditions is expressed a little cynically in the song 'Slopey, Slopey, Slopey' in Box 6-6. Lerouiel & Tavernas (1981) used various classic examples of slope failures and their analysis to demonstrate how different assumptions can lead to different results and explanations.

6.6.3   Risk assessment

A decision needs to be made on whether the risk from slope failure is acceptable or not and whether the cost of engineering works can be justified. A modern approach to assessing the need for preventive measures is to use quantified risk assessment, as described by Pine & Roberds (2005). The project described involved remediation and stabilisation of several sections of high cut and natural slopes dominated by potential sheeting joint failures and by the potential for failure of rock blocks and boulders bouncing down exposed sheeting joints to impact the road below. Design of slope cut-backs and stabilisation measures was based on a combination of reliability criteria and conventional FoS design targets aimed at achieving an ALARP (as low as reasonably practicable) risk target, which, in actuarial terms, translated to less than 0.01 fatalities per year per 500 m section of the slopes under remediation. Further examples of quantitative risk calculation are given by Fell *et al.* (2005).

6.6.4   General considerations

Remediation of stability hazards on slopes is often not trivial, especially where the works are to be conducted close to existing infrastructure and implementation of the works can itself increase the risk levels, albeit temporarily. Factors that will influence the decision on which measures to implement include the specific nature of the hazards, topographic and access constraints, locations of the facilities at risk, cost and timing. The risks associated with carrying out works next to active roads, both to road users and to construction workers themselves, need to be addressed (GEO, 2000a). Pre-contract stabilisation works might be needed to allow site access and preparation. Preventive measures such as rock bolting may be carried out at an early stage to assist in the safe working of the site and designed to form part of the permanent works. Options for the use of temporary protective barriers and catch nets to minimise disruption to traffic during the works also need to be addressed, as do contractual controls and alternatives for supervision of the works. Traffic controls may be

Box 6-6   Slopey, Slopey, Slopey (1982)

Figure B6-6.1 Chung Hom Kok, Hong Kong.

– sung (and danced if you wish) to the tune of the Hokey-Cokey.

You put your phi value in
You take your c' value out
You add a bit of suction
And you shake it all about
You do the old Janbu[2] and you turn around
That's what it's all about.

Chorus:
Oh slopey, slopey, slopey
Oh slopey, slopey, slopey
Oh slopey, slopey, slopey
It's so easy
One – point – four[3]!

Written and sung by the GCO Cabaret Stars, 1982

[2] Janbu is the author of a commonly used limit equilibrium method of slices for calculating Factors of Safety of slopes and can be applied to non-circular surfaces. There are two forms: a routine method and a rigorous method (Janbu, 1973). Lumsdaine & Tang (1982) carried out an exercise comparing results of calculations by six Government Offices and 36 others and found a very high proportion of analytical errors and lack of documentation, which of course is over and above any uncertainty in ground model, parameters adopted and assumed groundwater conditions – either positive pore pressure or suction.
[3] A Factor of Safety of 1.4 is generally regarded as an acceptable number to guard against failure in a high-risk slope in Hong Kong (GCO, 1979; 1984).

needed, and in some circumstances it will be necessary to close roads or evacuate areas temporarily, especially where blasting is to be used. The use of a risk register, as piloted for tunnels (Brown, 1999), with clear identification of particular risks and responsible parties, helps to ensure that all hazards and consequences are adequately dealt with during construction. Decision analysis is now widely applied at an early stage to assess whether to mitigate slope hazards (e.g. by rockfall catch nets) or to remediate/resolve the problem by excavation and/or support approaches. If construction of intrusive engineering measures to stabilise hazards might be unduly risky, then passive protection can be adopted instead. A hybrid solution is often the most pragmatic approach for extensive, difficult slopes where some sections might be stabilised by anchors and buttresses, with other sections protected by nets and barriers (Carter *et al.*, 2002; Pine & Roberds, 2005).

6.6.5   *Engineering options*

Some of the options for improving the stability of slopes are illustrated in Figure 6.25 and listed more comprehensively in Hencher *et al.* (2011). These can be split into *passive* options that either deal with the possible failure by controlling surface deterioration at source, or installing preventative reinforcement to increase local factors of safety, or adding walls or buttresses to restrain detached debris before it causes injury or damage, and *active* measures that enhance overall Factors of Safety of larger sections of slope by major engineering works, including cut backs or buttresses or heavy tie-back cable anchors.

6.6.5.1   *Surface treatment*

Many risks can be mitigated cost-effectively through surface treatment to stabilise or remove relatively small blocks of rock. Surface drainage is important, using adequately sized concrete channels with a fall across the slope and channels down the face that may be stepped to reduce velocity of flow. Further guidance is given in GCO (1984a) and in Ho *et al.* (2003).

There is a temptation to use hard slope treatments such as shotcrete to constrain loose blocks at the slope surface but such measures, if not properly designed, can restrict drainage from the slope, hide the geological situation from future investigators and can themselves cause a hazard as the shotcrete deteriorates, allowing large slabs of shotcrete to detach. Furthermore, shotcrete is increasingly an unacceptable solution for aesthetic reasons and there is a push towards landscaping high visual slopes where safety is not compromised (GEO, 2000b).

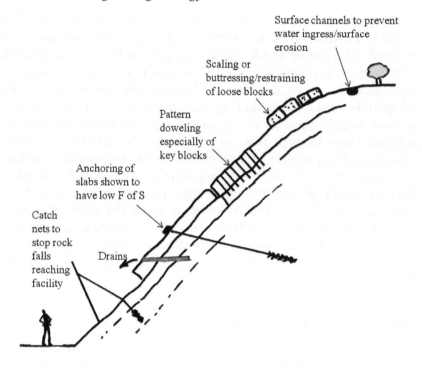

Figure 6.25
Schematic representation of various measures for stabilising rock slopes or protecting public.

Bioengineering is used to generally improve the stability and reduce erosion from natural slopes. Roots bind the soil and vegetation can increase surface runoff. Most bioengineering solutions cannot, however, be relied upon to improve the long-term stability in risky slopes, because vegetation can rot and die or be destroyed by fire. Furthermore, root growth can lead to rock blocks becoming loosened and detached.

6.6.5.2   *Rock and boulder falls*

Where individual rockfall sources are identified, these can be scaled off, reinforced by dowels, bolts, cables or dentition buttresses and/or netted where the rock is in a closely jointed state. Removing large blocks can be difficult because of the inherent risks associated with breakage techniques, including blasting and chemical splitting, which can dislodge blocks unexpectedly. Care must be taken to protect the public and workers during such operations. The most difficult zones to deal with are those with poor access. Implementing passive or active protection needs to start from safe ground and move progressively into the areas of more hazardous stability.

Rockfall trajectory analysis, using widely available software, allows prediction of energy requirements and likely bounce heights and run-out damage zone extent. Where energy considerations allow, toe zone protection measures, catch benches, catch ditches and toe fences

Figure 6.26
Retaining
structures and catch
nets to stop natural
terrain landslides
impacting new
road, Lantau
Island, Hong Kong.

provide the earliest viable mitigation approach, without requiring access to the slope.

Catch nets or fences can be positioned on-slope or in the toe zone of the slope, depending on energy requirements and site restrictions. An example is shown in Figure 6.26. Where energies computed from rockfall analyses are too high for toe zone protection alone to maintain risk levels below prescribed criteria for highway or rail users, on-slope energy protection fences become a necessity to reduce total energy impact at road level. Where the road (or railway) passes under areas prone to continuous rockfall, an avalanche shelter is commonly used (Figure 6.27).

6.6.5.3   Mesh

Wire mesh is commonly used to restrict ravelling-type rock failure and can be fixed at many anchorage points or can simply hang down the face, fixed with anchors at the top and weighted with scaffold bars or similar at the toe. Mesh (varying from chain-link, triple twist, hex-mesh to ring-net, in increasing order of energy capacity) can be placed by a variety of techniques, ranging from climber-controlled unrolling of the mesh to the use of helicopters.

6.6.5.4   Drainage

Deep drainage can be very effective in preventing the development of adverse water pressures, and this is often a combination of surface protection and channelling of water away from the slope and inclined drains drilled into the slope. Regular patterns of long horizontal drain

Figure 6.27 Rock-fall nets and avalanche shelter, near Cape Town, South Africa.

holes can be very effective, but all drains will seldom yield water flows, and the effectiveness of individual drains will probably change with time as sub-surface flow paths migrate. Typically, drains comprise plastic tubes with slotted crests and solid inverts, inserted into pre-drilled holes of tens or even hundreds of metres. Inner geotextile liners might be used that can be withdrawn and replaced if they get clogged up. Drains might need to be flushed out periodically. Attention should be made to detailing the drain outlets properly otherwise the slope face may backsap. If not maintained, vegetation can block outlets reducing their effectiveness.

In rock slopes, there is a need to target sub-surface flow channels, many of which will be shallow and ephemeral. The paths may be tortuous and hard to identify and drainage measures can therefore be rather hit or miss. If the exposed joint is badly weathered, the weak material may backsap and possibly pipe, leading to destabilisation, partially caused by lack of free drainage, and careful detailing will be required to prevent deterioration. No-fines concrete, whilst appearing to be suitable to protect weathered zones, often ends up with lower permeability than designed and should not be relied upon without some additional drainage measures.

As an alternative to deep drains drilled into the slope from the surface, drainage adits and tunnels are sometimes used to lower the water table, generally with drainage holes drilled radially into the rock mass from the tunnel walls. Other solutions include deep caissons constructed at the rear of the slope to intercept through-flow, with inclined drains leading away from the slope at their base (McNicholl *et al.*, 1986). Pumped wells are also occasionally used, pumps being activated when water levels reach critical heights within the slope.

6.6.5.5 *Reinforcement*

Stability can be improved by a variety of reinforcement options. For rough matching joints, provided there has not been previous movement, the interlocking nature provides considerable shear strength. If the joint can be prevented from movement by reinforcing at strategic locations, then full advantage can be taken of the natural shear strength. Depending on configuration, rock may be stabilised by passive dowels, tensioned bolts or cable anchors. Passive dowels allow both mobilisation of a normal force (due to the resistance provided by the fully grouted dowel) plus active shear restraint provided by the steel of the dowels resisting block slide mobilisation (Spang & Egger, 1990).

The Geotechnical Engineering Office in Hong Kong has published some guidelines on prescriptive measures for rock slopes and in particular gives guidance on rock dowelling for rock blocks with volume less than $5\,m^3$ (Yu *et al.*, 2005). In essence, it is advised to use pattern dowels with one dowel per m^3 of rock to be supported, with minimum and maximum lengths of 3 and 6 m respectively, and where the potential sliding plane dips at less than 60 degrees. The dowels are to be installed at right angles to the potential sliding plane, with the key intention to allow the dowels to act in shear, whilst also enhancing the normal restraint due to asperity ride during sliding. In practice, dowels frequently need to be used in more variable orientations. Designs must be checked in the field during installation, to check that the perceived ground model is correct. If not, then the design must be revised.

Sub-horizontal cable anchors can be used if capacities larger than about 20 tonnes per reinforcement member are required. Great care needs to be taken to ensure that such tensioned anchors are adequately protected against corrosion, and regular checking and maintenance will be required. Several cases of anchors that have failed due to corrosion are discussed in Chapter 7. For weaker rock and soil, pattern soil nailing is now commonly used. The nails, which typically comprise 50 mm or so diameter steel bars connected, as necessary, by couplers every 6 m, are usually installed in pre-drilled holes, held centrally by lantern spacers and then pressure grouted over their full length using tubes installed with the nail. Soil nails are usually installed as a passive reinforcement that would only take on load if the slope began to deform prior to failure.

6.6.5.6 *Retaining walls and barriers*

Retaining walls are commonly used to support steep slopes, especially where the slope comprises weak and broken rock and where space is constrained. There are many different types, as illustrated in Figure 6.28.

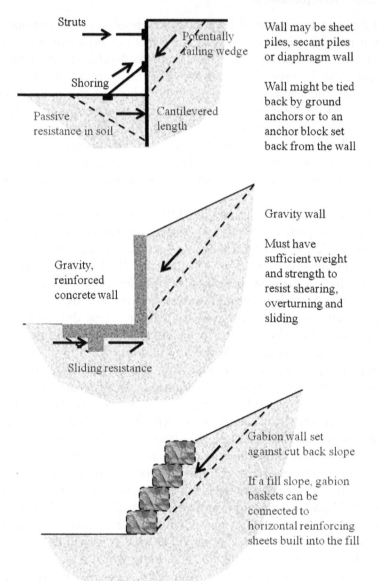

Struts

Potentially failing wedge

Shoring

Passive resistance in soil

Cantilevered length

Wall may be sheet piles, secant piles or diaphragm wall

Wall might be tied back by ground anchors or to an anchor block set back from the wall

Gravity, reinforced concrete wall

Gravity wall

Must have sufficient weight and strength to resist shearing, overturning and sliding

Sliding resistance

Gabion wall set against cut back slope

If a fill slope, gabion baskets can be connected to horizontal reinforcing sheets built into the fill

Figure 6.28 Different types of retaining wall.

Figure 6.29
Concrete retaining
wall under
construction, Hong
Kong.

For temporary works, corrugated steel sheets are generally driven, vibrated or pushed into soil prior to deep excavation, with each sheet linking to its neighbour. As the excavation proceeds, sheets are usually braced by a system of struts and waling beams, although they may also rely on depth of embedment. Diaphragm walls formed by concreting deep trenches excavated under bentonite mud are also used as part of temporary works and then may be incorporated in the permanent structure. Permanent retaining structures are often created using piles. Alternatively, where the ground can be anticipated to stand steeply, temporarily during construction, the full slope is cut back and then a wall of concrete constructed at some short distance in front. The space between the wall and the natural ground is backfilled with granular free-draining material, often with geotextile material at the interface, feeding water down to a drain (Figure 6.29). Drainage is very important if the retaining wall is not going to act as a dam. Gabionstructures are made from galvanised steel or, rarely, plastic baskets, backfilled with rock. The main advantages are that they are free-draining, can be landscaped, and they can be made cheaply on site using locally derived rock to fill locally woven baskets. They are therefore very suitable for forming retaining structures or barriers in remote locations (Fookes *et al.*, 1985). Deflection structures and barriers are commonly used to divert or retain channelised debris flows away from buildings or roads.

6.6.5.7 Maintenance

Whatever the engineering solutions adopted, slopes should be examined periodically for signs of distress and for maintenance such as cleaning out of drainage channels. The requirement for inspection, testing and possible remediation works, should be built into the design of any new slope, with careful consideration for how this is to be achieved. In Hong Kong, the current practice is to prepare maintenance manuals for slopes and to carry out routine engineer inspections at regular intervals. New and newly upgraded slopes are generally constructed with access ladders and often with hand rails provided along berms to allow safe inspection.

6.7   Site formation, excavation and dredging

6.7.1   *Excavatability*

Site excavation is usually carried out by heavy machinery, and the main questions for the engineering geologist are what machinery would be suitable and whether the rock would need to be blasted first.

Where blasting is restricted, then the contractor might need to use some kind of chemical or hydraulic rock splitter, but the noise levels of drilling and rock breaking might still be a problem. Generally, the factors that will control whether or not blasting is needed are intact rock strength and the spacing between joints (MacGregor *et al.*, 1994; Pettifer & Fookes, 1994). As emphasised elsewhere, care must be taken to differentiate between mechanical fractures with low tensile strength and incipient fractures with high strength as this will strongly affect the ability of machines to rip the rock.

6.7.2   *Dredging*

Dredging (underwater excavation) is commonly carried out for port works, to improve navigation on rivers and as part of other land reclamation projects in providing fill material (Bray *et al.*, 1997). There are several types of dredger and these vary in their capacity to deal with soil and weak rock. Where there are few natural fractures in the weak rock, excavations can be difficult, even for the strongest suction-cutter dredgers and then some pre-treatment, normally blasting, will be required. Reviews on dredging practice in various countries, including the USA, UK, Hong Kong and Singapore, are given in Eisma (2006).

6.8   Ground improvement

6.8.1   *Introduction*

At many sites, the ground conditions are too weak or wet to allow construction by the preferred method or even to allow access by heavy construction equipment. Ground improvement might therefore be carried out, often as an alternative to some engineering solution such as piling, and the engineering geologist should be aware of the techniques that might be employed to deal with a particular site condition (Charles, 2002). Ground improvement might be used in temporary works, such as freezing the ground to allow tunnelling through saturated and potentially flowing materials, or the construction of barriers to water flow, or to restrict vibrations during construction. In other situations, ground improvement might provide a permanent solution such as densification or using chemical additives to provide additional strength.

6.8.2   *Dynamic compaction*

One of the simplest methods is dynamic compaction, which involves dropping a large weight, up to about 30 tonnes, from a crane, over a regular pattern and then backfilling the depressions with granular material. Further drops are carried out at closer spacing. The depth of improvement depends upon the weight dropped, size of pounder and the height. Typically, a weight of about 15 tonnes dropped 20 m might be expected to improve ground to about 10 m deep (e.g. Bo *et al.*, 2009). The method is most suitable for improving fills and granular soils generally, but sites underlain by clay have also been improved, although consideration must be given to the pore pressures that might be generated and how these dissipate. Generally, the improvement is measured by tests before and after improvement, using techniques such as the SPT, CPT or the Menard pressuremeter that was developed specifically for this purpose (Menard & Broise, 1975). The technique has been applied successfully for quite prestigious projects involving large-scale reclamation, such as Nice Airport. In Hong Kong, it has been used to densify the upper few metres in old fill slopes in an attempt to improve their stability.

6.8.3   *Static preloading*

If time allows, then an effective way to improve the consolidation characteristics at a site is to preload it, often by placing an embankment of fill material that can be removed again later or re-graded at site, compacted properly in thin layers. The process of consolidation is generally accelerated by introducing a series of vertical drains to increase the mass permeability and allow excess pore pressures to dissipate, monitored using piezometers. The drains can be sand wicks, which are sausages of geotextiles, filled with sand and installed in pre-drilled holes. Other systems include wick drains that are geotextile-covered plastic elements pushed into the ground using a purpose-built machine. At some sites consolidation and strengthening is achieved by a technique termed vacuum preloading; references are given in Charles (2002).

6.8.4   Stone columns

Stone columns can be used to enhance drainage and are installed to depths of 10 m and sometimes more. These are formed by using a vibrating poker, pushed into the soil to form a void and then filling the void with gravel and sand, which is compacted in stages using the same vibrating tool (McCabe *et al.*, 2009). Stone columns have been used to increase mass permeability and prevent liquefaction of loose silty sand during an earthquake, although in such a usage settlement will still occur but in a relatively uniform and non-catastrophic manner. Stone columns are also used generally to improve the bearing conditions at a site, the improvement depending upon the ratio of cross-sectional area of stone columns to untreated ground. Groups and lines of stone columns can be used as weak piles to provide support to structures such as oil tanks.

6.8.5   Soil mixing and jet-grouted columns

Clay soils especially, can be improved by mixing with lime slag and cement, either at the ground surface (to prevent erosion in slopes, for example) or in columns or trenches, using hollow-stem augers and similar equipment. The works will improve the bearing capacity of the ground, although the improvement might be difficult to quantify. Stronger columns can be formed by using jet-grouted columns formed using high-pressure grout jets as a drilling string is rotated and lifted from depth. The resulting column of mixed soil and grout can be used to carry structures or to form cut-off barriers to restrict water flow, for example, beneath dams. Jet grouting is sometimes used to form structural members during temporary works construction of deep excavations (Puller, 2003; also see case study of Nicoll Highway collapse in Chapter 7).

6.8.6   Drainage

For deep excavations and tunnelling, it is commonly necessary to lower the groundwater during construction, although there are many factors that must be considered, not least associated settlement of the ground due to increased effective stress and self-weight compaction and consolidation and drying up of land in adjacent properties (Preene & Brassington, 2003). New and steep flow paths through the soil can lead to seepage piping and lique-faction in the floor of excavations. The cheapest and simplest way to lower water is just to let it happen naturally as the excavation proceeds, to channel water inflow to collection sumps and to then pump this water away, although disposal may be an issue on environmental grounds, and pumping from great depth will require a series of pumps at different levels. Active dewatering is generally conducted using well-point systems or submersible pumps in wells. Details are given by Puller (2003). As noted elsewhere, dewatering is often important to the stability of slopes and semi-permanent solutions include drains, drainage caissons and adits. Emergency pumping systems are sometimes set up to be triggered if piezometric levels become dangerously high.

6.8.7 Geotextiles

Geotextiles are fabric or plastic sheets that have many different uses in ground engineering. A few of these are discussed below.

6.8.7.1 Strengthening the ground

To improve site access, sheets of plastic mesh may be laid on the ground and then a layer of gravel placed and compacted on top. The purpose of the geotextile is to prevent the gravel being pushed into and mixing with the underlying soil that may be wet and soft. In this way, temporary road access can be provided. In other circumstances, more complex solutions might be designed involving elements such as stone columns or piles, together with a geotextile grid draped across and linking the structural elements.

Geotextile mats and strips are also used in the design of reinforced earth structures (as are metal grids and strips), as illustrated in Figure 6.21. Basically, the frictional resistance between the soil and grid or mats, placed horizontally and regularly within a fill structure, enhances the overall strength of the soil mass and prevents it failing. Where facing walls are used or the geotextile is wrapped around at the face to prevent soil erosion, the finished structure can be very steep or even vertical.

Plastic grid boxes, infilled with rock cobbles, have been used to form gabion walls as barriers. Care must be taken that the situation is not one where the finished structure can be destroyed by fire and that the deterioration rate is acceptable given the proposed lifetime of the structure.

6.8.7.2 Drainage and barriers

Geotextile sheets are available that are highly permeable but also designed with a mesh size that restricts soil erosion, in the same way as traditional soil filter systems. Geotextiles are therefore used, for example, as part of the drainage system behind concrete retaining walls. Plastic sheets (geomembranes) are used as barriers to water flow, especially for landfill sites. Great care must be taken to ensure that sheets are welded one to the other and that those welds are tested. Membranes must be resistant to and protected from puncturing. Any leakage may be extremely difficult and expensive to rectify at a later stage. In Chapter 7, an example is given where a combination of permeable geotextiles and impermeable geomembranes were used to reduce leachate loss from a quarry used for landfill.

6.8.8 Grouting

Grouting is generally used to increase strength of a rock or soil mass and to reduce permeability (Warner, 2004). It is routinely used below dams to provide a cut-off curtain to restrict seepage through the foundations. A main consideration is the type of grout – usually cement, but sometimes chemical grouts or resin must be used to penetrate low-permeability ground. The pattern of holes to be used, phases of grouting necessary, and pressures to be adopted are also matters for specialist design. Grouting might jack open existing joints in rock or form new fractures in soil and weak rock (claquage). Grouting is sometimes used to correct settlement or other deformations caused by engineering works such as tunnelling (e.g. Harris *et al.*, 1994), but care must be taken that the grouting does not make matters worse, as per the Heathrow Express Tunnel collapse described in Chapter 7.

6.8.9 Cavities

Cavities that engineering geologists and geotechnical engineers need to contend with include natural cavities such as those often found in limestone areas, more rarely in other rock types, including unlikely candidates such as weathered granite (Hencher *et al.*, 2008). The other main problem is mining. Ground investigation for such voids is a matter of careful desk study (including the mining method that might have been used if that is the hazard of concern), focused investigation, possibly using geophysics such as micro-gravity and resistivity and probing, perhaps using percussive drilling to keep the costs down. If and when voids are found, these can be explored and characterised using cameras, echo sounders and radar. In the case of old mine workings, inspection may be required by suitably equipped and experienced persons following proper safety procedures. Depending on their extent, voids may be backfilled, grouted or structurally reinforced, as appropriate. When extensive mine workings were encountered unexpectedly during tunnelling for the high-speed railway from Seoul to Taejon in South Korea, one proposed solution was to construct a concrete structure through the workings, but this was considered politically unacceptable because the public was already aware of the situation. Instead, the route for the railway had to be moved several kilometres at considerable cost and the completed works were abandoned.

6.9   Surface mining and quarrying

Surface mining and quarrying are industries that have strong demands for geotechnical expertise, including engineering geology. Slope design is often very important and the design practices discussed earlier, used in civil engineering, also apply to quarries and open pits and opencast mines. The main difference is that in such enterprises many of the slopes are always changing in geometry as the works progress. One key to success is establishing a safe layout for operations such as crushing and processing plants and for haul roads, whilst avoiding sterilising valuable resources because of the siting of infrastructure such as site offices and treatment plants. Major haul roads also need to be established in a safe manner to avoid disruption to operations if instability occurs. Other faces may well be temporary and are therefore formed at angles that would be unacceptable as permanent slopes in civil engineering. For large open-pit mine operations, the scale of overall slope formation can be huge, extending hundreds of metres, and predicting stability often requires numerical modelling, tied in to monitoring systems. Excavation of rock usually involves blasting and this is a specialist operation as it is for tunnelling. A good review is given in Wyllie & Mah (2004). Key considerations for all blasting operations are fragmentation, to avoid producing large blocks that cannot be handled easily and need secondary breaking operations, avoiding damage to the remaining rock, avoiding overbreak beyond the design profile, safety and risk from flyrock, gases and vibrations.

Waste from mining needs to be disposed of. In open-pit coal mining, the waste rock is backfilled into the void as part of the ongoing operations and nowadays in the UK at least, the final reinstatement of the area is strictly controlled, with every attempt made to simulate the natural countryside as it was pre-operations. Other wastes are often wet and contaminated and held behind tailings dams that should be designed and analysed with just as much care as a civil engineering structure. Unfortunately, this is often not the case, and there have been many major failures worldwide over the last fifty years which have resulted in severe contamination and many deaths (Rico *et al.*, 2008).

6.10   Earthquakes

There are four major considerations for design:

1. Local ground failure, e.g. because of liquefaction in loose saturated cohesionless sand and silt.
2. Rupture because of fault movement, which can be significant especially for tunnel design.
3. Ground shaking causing inertial forces. Buildings and slopes are especially at risk from horizontal shaking.
4. Remote hazards. These will include landslides from adjacent land where debris run-out could impact the site, and tsunamis.

6.10.1   Ground motion

Most structures need to be designed to withstand dynamic loading. This includes wind loading (to typhoon levels in countries such as Japan, Korea and Hong Kong), earthquakes and blasting/traffic. The main one of these that requires input from the engineering geologist is earthquake loading. The level of hazard is assessed at the site investigation stage (Chapter 4), and there is often a mandatory design code for a particular country. Alternatively, or as a check, the design team will identify some design earthquake or series of such design events with equal probability of occurrence within the lifetime of the structure. For example, statistical analysis of historical earthquake activity might indicate that there is an equal chance of a magnitude 8 (M8) earthquake at 200 km distance, as a magnitude 5.2 (M5.2) earthquake at 10 km. These earthquakes would probably result in very different ground shaking at the project site. From study of recorded data using

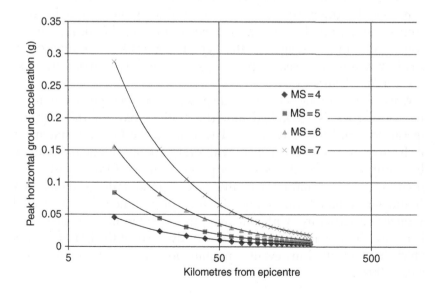

Figure 6.30 Peak acceleration vs. distance for different magnitude earthquakes (European data). Equations given in Ambraseys *et al.* (1996).

strong motion seismographs, attenuation laws have been derived for different parts of the world. Forces are used for engineering design, so acceleration is an important parameter. The equation below has been shown to fit the available European seismic data reasonably well and can be used for prediction (Ambraseys *et al.*, 1996). Data from North America and elsewhere are not very different.

$$\log(a) = -1.48 + 0.266M_s - 0.922\log(r)$$

where a is peak horizontal ground acceleration expressed as a fraction of gravitational acceleration, g, (9.81 m/s^2). M_s is surface wave magnitude and r is essentially the distance between the project site and the earthquake epicentre. Figure 6.30 gives median data and can be refined for degree of confidence and for site characteristics (Ambraseys *et al.*, 1996). Unexpectedly high accelerations do occur, and this is often the result of local ground conditions or topography that amplify the effect, as for the peak accelerations of 1.25g and 1.6g in the abutment of Pacoima Dam, USA, during two separate earthquakes (Bell & Davidson, 1996). The February 2011 earthquake that caused huge damage in Christchurch, New Zealand, involved vertical ground accelerations up to 2.2g and horizontal ground accelerations of up to 1.2g, which are very high for a 6.3M event and can be largely attributed to the very shallow nature of the earthquake (about 5 km, according to the New Zealand Society for Earthquake Engineering).

Peak ground acceleration, although an important starting point, is not enough to give an indication of structural performance. What also matters is the time that the strong shaking continues and the frequency spectrum of the waves carrying the energy. The situation is

Figure 6.31 Predominant periods in rock for different magnitudes of earthquake at different distances (US data).

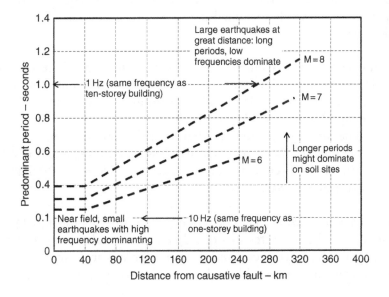

complicated by the way that individual structures respond to repetitive dynamic loading, which is a matter of harmonic resonance. Thus, whilst for the M5.2 design earthquake at 10 km the peak acceleration can be predicted from the equation presented earlier as 0.12g and for the M8 design earthquake at 200 km as 0.04g, other characteristics will be very different. Figure 6.31 shows the predominant period in ground acceleration records for western USA (Seed *et al.*, 1968), which indicates that for a near-field M5.2 quake the predominant period might be less than 0.2 secs, whereas for the distant M8 quake the predominant period could be more than 0.8 secs. Furthermore, the duration of shaking will be significantly longer for the large magnitude earthquake (e.g. Bommer & Martinez-Pereira, 1999). The duration of strong shaking for a M5.2 earthquake might be a few seconds. For the Christchurch Feb 2011 M6.3 earthquake, the strong shaking lasted about 12 seconds. For an earthquake of M8, the duration could be over a minute. With longer duration, the potential for amplification will be much greater and fatigue-type failure can occur.

6.10.2   *Liquefaction*

Liquefaction is a common failure mode in natural soils, fill and sometimes in embankment dams during earthquakes. It occurs in loose saturated cohesionless sand and silt, which, when disturbed, loses its structure and collapses. Because of its low permeability, water cannot escape so natural piping and even general liquefaction occurs as the effective stress and thereby friction reduces to zero. There are many classic examples of whole apartment blocks tilting over and buildings settling. Elsewhere, service pipes float to the surface and sea walls collapse as the retained fill flows into the sea. The potential for liquefaction is readily identified during site investigation. The general rules are:

1. It occurs in un-cemented deposits – fill or geologically recent soil.
2. The most susceptible soils are cohesionless (sands and silts) with a liquid limit less than 35% and water content greater than 0.9 times the liquid limit (Seed & Idriss, 1982).
3. It generally occurs at depths shallower than about 15 m.
4. Generally, SPT N value (corrected) less than 30 (Marcusson *et al.*, 1990) or CPT cone resistance less than 15 MPa (Shibata & Taparasaka, 1988).

Analysis of the hazard might be refined by considering the liquefaction potential vs. the characteristics of a design earthquake, but generally if the area has high seismicity and the granular soil at a site is relatively loose and groundwater table high, then it is probably wise to carry out preventive measures. These might include compaction, grouting or the installation of stone column drains that will help prevent excess pore water pressure development, although they would not prevent settlement. Alternatively, passive mitigation may be the best option – relocate the proposed structures away from the zone of liquefiable soil. If the ground does liquefy, then apart from movement of structures in or on the ground, the settled soil might cause drag down (negative skin friction) on any piles installed through that zone.

6.10.3   *Design of buildings*

For buildings such as one or two-storey houses, there are certain simple rules that, if adopted, can reduce the risk of failure and would limit injuries, especially in developing nations. These include ensuring that walls are tied together, preferably by reinforced ground beams or beams along the tops of walls (Coburn & Spence, 1992).

For larger engineered structures, these need to be designed to withstand the repeated force waves. As outlined in 6.10.1, given a particular design earthquake one can make estimates about the ground motion characteristics that the structure will have to withstand. These include peak acceleration, predominant frequency and duration of shaking for a given return period earthquake. Typically, the return period used is 1 in 500 to 1,000 years but the choice is rather arbitrary and will depend on the nature and sensitivity of the project and the seismic history. These bedrock ground motions may be modified by the local site geology or topography and estimates of the modified shaking characteristics can be made by dynamic analysis using

Figure 6.32 Responses of buildings to earthquake shaking.

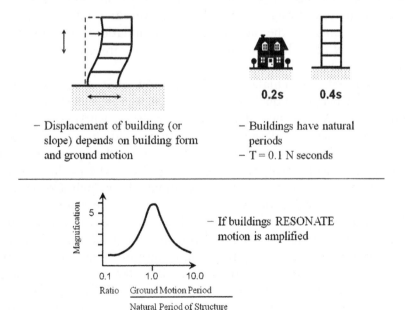

– Displacement of building (or slope) depends on building form and ground motion

– Buildings have natural periods
– T = 0.1 N seconds

0.2s   0.4s

Magnification

5

0.1   1.0   10.0

Ratio   Ground Motion Period

Natural Period of Structure

– If buildings RESONATE motion is amplified

software such as SHAKE or through reference to published ground motion spectra for particular ground profiles. Generally, thick soft soil profiles may lead to relative amplification of longer period waves. The design ground motion then needs to be applied to the structure. Structures have their own dynamic characteristics and if the incoming frequencies match the natural response frequencies of the structure, then movements may be magnified (Figure 6.32). Structural engineers will take the incoming design earthquake characteristics and calculate the response of the structure. For more frequent, smaller earthquakes, the structural engineer will design the structure as far as possible to behave elastically (no permanent displacement). In the event of an extremely large and less probable event, a structure can be designed to be fail-safe. Redundant elements such as additional steel beams can be included that yield under extreme loads but also change the fundamental frequency of the building, damping the response to the shaking. Other options are to put a building on springs of some kind or to include hydraulic actuators or pendulums that again reduce the structural shaking. An example of an innovative aseismic design is the foundations for the Rion–Antirion Bridge constructed in Greece in 2005. The cable-stayed bridge, with five main spans extending 2.25 km across a fault zone, was designed to withstand horizontal accelerations of 0.5g at ground level and up to 2 m offsets between adjacent towers. Underlying each tower is thick soil and the depth of sea is up to 65 m. The towers were founded on 90 m diameter cellular structures placed on a 3.6 m layer of gravel placed on the natural soil, which was reinforced by up to 200×2 m diameter tubular piles to depths of 30 m. The foundation structure is not attached to the piles; the gravel acts as a fuse, limiting the transfer of load to the superstructure. The piles in the underlying soil are there to prevent rotational bearing failure. Details of the design of the foundations are given by Combault *et al.* (2000) and further references are given at the web page for the bridge.

Two recent earthquakes, however, show that even with good design practice, earthquakes can cause damage to a level that is not anticipated. As a result of the February 2011, Christchurch, NZ, earthquake, many small one-and two-storey buildings were destroyed or badly damaged, as one might expect near the epicentre of an earthquake with magnitude exceeding 6.0, where the ground motion might be expected to be dominated by high frequencies. Widespread liquefaction was also a major contributor to the damage of these smaller buildings. However, for this earthquake, because of its shallow nature and possibly other factors that served to concentrate and amplify the ground motion, unexpectedly large accelerations and forces were generated. In the case of the March 2011 earthquake that struck NE Japan (east of Honshu), most engineered buildings on mainland Japan withstood the very strong shaking associated with this 8.9 or even 9.0M earthquake (10,000 times as strong, in terms of overall energy release, as the Christchurch earthquake) and this is testament to the skill and knowledge of the civil engineer designers. The huge damage and large number of deaths caused by the Japan earthquake resulted from a 10m high tsunami wave that came ashore and destroyed whole villages. Regarding engineered structures, several nuclear power stations had been constructed along the shoreline in the impacted region. The structures apparently performed well in terms of withstanding seismic shaking but severe damage did occur because of failure of cooling systems. The initial shaking caused safe shutdown of the reactors, as is the required procedure for nuclear power stations impacted by a major earthquake, but the loss of electrical power stopped the flow of cooling water required to prevent the fuel rods overheating. Backup diesel generators kicked in and provided the necessary power for an hour or so but then they failed because of the tsunami. In hindsight, no doubt the secondary power sources could and should have been designed to survive inundation, as they are for more modern installations, and the risk properly identified using an event tree approach.

6.10.4 Tunnels

Tunnels and mines tend to be safer than surface structures during earthquakes, and this safety increases with increased depth (Power *et al.*, 1998). Except where the tunnel passes through particularly poor ground or intercepts active faults, earthquake resistant design is generally not a high priority. Of course, where the support in a tunnel is inadequate or marginal under static loading conditions, then earthquake shaking might well trigger failure. This is especially true at portals of tunnels; landslides and especially rockfalls are very commonly triggered by earthquakes, as discussed in the next section. Failures in some tunnels, and especially the failure of Daikai subway station during the Kobe earthquake in 1995, have caused a rethink on seismic stability of underground structures. Hashash *et al.* (2001) provide a very useful review and examples of aseismic design. A reinforced concrete lining should have significantly better seismic resistance characteristics than an unreinforced lining. If the tunnel intersects a fault that is suspected of being active, then special measures will be required or, preferably, the fault avoided. Key considerations are the estimated magnitude of the displacement and the width of the zone over which displacement is distributed. If large displacements are concentrated in a narrow zone, then the design strategy may be to enlarge the tunnel across and beyond the displacement zone. The tunnel is made wide enough such that the fault displacement will not close the tunnel and traffic can be resumed after repairs have been made. In some cases, an enlarged tunnel is constructed outside the main tunnel and the annulus backfilled with weak cellular concrete or similar. The backfill has low yield strength to minimise lateral loads on the inner tunnel liner, but with adequate strength to resist normal ground pressures and minor seismic loads. If fault movements are predicted to be small and/or distributed over a relatively wide zone, it is possible that fault displacement may be accommodated by providing articulation of the tunnel liner using ductile joints. This detail allows the tunnel to distort into an S–shape through the fault zone without rupture, and with repairable damage. This may not be feasible for fault displacements more than 75–100 mm. An alternative approach is to accept that damage will occur and to make contingency plans to control traffic and to carry out repairs as quickly as possible in the event of a damaging earthquake.

6.10.5   *Landslides triggered by earthquakes*

Landslides are commonly triggered by earthquake shaking, especially in mountainous areas. The Wenchuan earthquake in Sichuan Province, China, of 12 May 2008, was very large (M8) and quite shallow (14 km) and the active faults ran through populated valleys surrounded by high slopes. Landslides, including rockfall, caused more than 20,000 deaths, with one individual landslide killing more than 1,600 people (Yin *et al.*, 2008). One of the main consequences was the damming of streams, which necessitated emergency engineering works to lower the water levels in the lakes that formed behind the landslide debris before they were overtopped or burst uncontrollably.

6.10.5.1   *Landslide mechanisms*

Slopes affected by strong earthquake shaking can be categorised in three classes, as set out in Table 6.4. These are:

1. Stable slopes: these are defined as situations where the shaking is not strong enough to cause permanent displacement in a slope. This may be because the peak forces are insufficient to overcome the strength of the ground or because different parts of the same slope are out of phase so that whilst some parts are being driven towards failure, other parts are being accelerated in the opposing direction.
2. Permanently displaced slopes: the key aspect of dynamic loading, whether it is from earthquakes or blasting, is its transient nature. The waves pass through the ground and induce inertial forces. In the same way as discussed in Chapter 5 regarding a laboratory experiment (Figure 5.27), at a critical acceleration (k_c) a slope will start to move. The continued positive acceleration above critical will cause the displacement to increase in velocity. However, after

Table 6.4 Performance of slopes under dynamic loading

Class	Condition	Details and examples
1. Stable	Acceleration less than that necessary to cause permanent displacement	
2. Permanent displacement but stable	FoS post-quake > 1.0 Damage may allow deterioration and later collapse	 Tension cracks resulting from Erzincan earthquake, Turkey, 1992
3. Failure	Metastable condition so that acceleration leads to catastrophic failure directly Reducing shear strength so that FoS is less than 1.0 after earthquake shaking has finished. Deteriorating condition following earthquake Rise in water pressure due to collapse of soil structure or regional changes in hydrogeological conditions	 e.g. Tsao ling reactivated landslide, Chi Chi earthquake, Taiwan, 1999 Rockfalls and other landslides may continue for days after an earthquake

a short time (typically a fraction of a second), the exciting acceleration will decrease and then change direction so that the inertial force is back into the slope. This will stop the movement unless the slope is metastable, as discussed later. Sliding friction can be lower than residual (Hencher, 1977, 1981a; Crawford & Curran, 1982; Tika *et al.*, 1990), and by employing pessimistically low shear strength, total displacement can be calculated for a series of acceleration pulses and this used as part of a design decision. Generally, even for a very large earthquake, the permanent displacement in a slope directly attributable to inertial loading will be small, of the order of millimetres or centimetres (Newmark, 1965; Ambraseys & Srbulov, 1995). Nevertheless, small permanent displacements will make the slope prone to accelerated weathering and deterioration if not protected or repaired.

3. Failed slopes: catastrophic landslides during earthquakes can be the result of four different conditions, viz:

   – *Low residual strength*. The inertial displacement during the earthquake reduces shear strength to a residual value so that even after the earthquake shaking, the slope continues to move. Examples of large-scale failures involving sliding on bedding planes with reducing strength are described for the Chi-Chi earthquake (1999) by Chen *et al.* (2003) and Chigira *et al.* (2003), and for the Niigata earthquake (2004) by Chigira *et al.* (2006).
   – *Deteriorated state*. The structure of rock or soil mass is disturbed so that it collapses and a flow can develop.
   – *Geometrically unstable equilibrium*. The initial displacement caused by the earthquake shaking results in unstable equilibrium. A typical example is rockfall from exposed rock cliffs. Once displaced, the rock will fall, sometimes as a progressive failure several days after the earthquake. Rockfalls may become entrained and develop into debris avalanches.
   – *Water-induced failure*. Firstly, loose saturated soil can collapse and liquefy down to depths of about 15 m on slopes inclined at only a few degrees. The collapsed material can spread and flow. As a second mechanism, the general groundwater flow paths can be affected by earthquake loading and this can trigger slope failures.

6.10.5.2   *Empirical relationships*

Keefer (1984, 2002) identifies 14 individual types of earthquake-induced landslide. The three main categories are:

1. Disrupted slides and falls: these include highly disrupted landslides that move down slope by falling, bouncing or rolling, or by translational sliding, or by complex mechanisms involving both sliding and fluid-like flow. They typically originate on steep slopes, travel fast and can transport material far beyond the slope in which they originate. Other than large rock avalanches, failures in this category are thin with initial failure depths less than 3 m.
2. Coherent landslides: these include translational slides and rotational slides. Such failures are typically relatively deep seated (greater than 3 m), slow moving and displace material less than 100 m.
3. Lateral spreads and flows: fluid flow is the dominant mechanism and this mode of failure is typical of liquefied soils.

The most common failures, according to Keefer, are rockfalls, rockslides and disrupted soil slides. This follows from the analysis of Table 6.4 where it can be seen that significant landslides will only occur where there are predisposing factors such as a topographic setting that is in unstable equilibrium or strain softening (due to collapsing structure or low residual strength, for example, through the loss of rock bridge-cohesion during the earthquake shaking). Keefer compiled data from many earthquakes and plotted the area affected by earthquakes vs. magnitude of the earthquake. The upper bound is rather well defined. For a magnitude M5, the affected area might be about 100 km^2, 1,000 km^2 for M6, and 10,000 km^2 for M7. Keefer also presents data on the maximum distance of landslides triggered by earthquakes of given magnitude. He provides separate upper bound curves for disrupted, coherent and flow-type failures. Disrupted landslides such as rockfalls, which are the most common type of earthquake-triggered landslides, are also shown as the most likely to occur at far distances from the epicentre. Rodriguez (2001) has carried out a further review of data, including more recent data from Japan, and his data demonstrate the considerable scatter that can be expected and therefore the difficulties in prediction on a site-specific basis. For example, some M7 earthquakes only cause landslides within an epicentral distance of 10 km whereas others of the same magnitude cause landslides 200 km away. This might be attributed in part to resonance effects associated with ground frequency spectra and duration, as for buildings (Hencher & Acar, 1995).

6.10.6   *Slope design to resist earthquakes*

Traditionally, and in most software packages, there are two main approaches to slope design to withstand dynamic loads (mostly earthquakes). The options for landslide prevention are essentially the same as for the static condition (change geometry, reinforce, reduce water pressure, protect the site below or move the facility at risk).

6.10.6.1 *Pseudo-static load analysis*

One approach is simply to include a horizontal inertial load into the analysis (some authors argue for an inclined force but it really makes little difference considering the inexact nature of the method) and to determine whether or not the FoS reduces below 1.0. The problem with this approach is that if one includes the peak predicted particle acceleration (say from equation 6.1) then very often the slope will be shown to fail, whereas in reality the permanent displacement would be negligible because of the extremely short time that acceleration would be acting. As confirmation, many vertical slopes in quarries are acted on by accelerations approaching or exceeding 1g during production blasting, but landslides due to blasting are very rare. Engineers therefore often choose to use some arbitrarily reduced acceleration, such as a nominal 0.1g, as a pseudo-static force in the stability analysis to check that the slope (or dam) has some degree of resistance to horizontal loading, but this is clearly rather unsatisfactory.

6.10.6.2 *Displacement analysis*

As discussed earlier, given a predicted acceleration against time record, it is straightforward to calculate the likely displacement that might be caused in a slope during an earthquake, and there are options to do so in software such as SLOPE/W. Those displacements will always be small, however, no matter how large the earthquake, and what matters more is the residual state of the slope after the earthquake – is there a situation where the ground is strain softening or is it in unstable equilibrium? These are considerations for the engineer, who must decide whether additional reinforcement might be necessary or other protective measures such as nets and barriers. Other software such as FLAC and UDEC (Itasca) can be used to study the seismic susceptibility of slopes. These being time-stepping software, the mode of failure can be identified, expressed visually and perhaps as a movie. It might, for example, be possible to test the potential failure mechanism of soil nails during an earthquake, each nail modelled specifically. That said, as ever, the models can only be as good as the input data and results will only be indicative.

6.11   Construction vibrations

6.11.1   *Blasting*

Blasting causes noise, ground vibrations, air overpressure and flyrock. All of these can be controlled – generally by using less or different types of explosive and limiting the number of charged drillholes that are detonated at the same time. In particular, using millisecond delays between lines of drillholes will reduce the vibration level considerably. Details are given in Dowding (1985) and many other publications.

Safety is a major issue and an engineering geologist working in a situation where blasting is being conducted may well be involved in blast monitoring, checking fragmentation and reviewing the overall suitability of the blast design, given the changing geological situation as the rock is excavated.

6.11.2 *Piling vibrations*

The other major source of potentially damaging vibrations in civil engineering is from driven piles. Damaging levels are generally limited to about 10m distance, although this depends on the sensitivity and state of repair of the structure. Predictions can be made using empirical formulae into which the main inputs are hammer energy and distance (Head & Jardine, 1992), but these are rarely very accurate.

6.12 Numerical modelling for analysis and design

6.12.1 *General purpose*

There are two main groups of programs commonly used: finite element (FE) and finite difference (FD), time-stepping type software. PLAXIS is a general purpose FE package that allows geotechnical situations – foundations, slopes or tunnels – to be modelled. The model is set up and run to give a quick solution to complex equations – perhaps of deformation or calculation of Factor of Safety of a model that is split into elements – mostly triangular. It can also be used to model fluid flow. As with all sophisticated software, it should only be used by those knowledgeable of the underlying mechanics and the way these are dealt with within the computer program. Following the Nicoll Highway collapse discussed in Chapter 7, it was established that there had been a mistake made in the manner in which the design of the diaphragm walls was carried out using an inappropriate soil model. The same problem would have arisen for any finite element package used in this incorrect manner – it is not unique to PLAXIS. The mistake resulted in excessive deformation of the walls and an under-design of their moment capacity, although these two effects did not have any influence on the final failure. The mistake was in adopting effective stress strength parameters in a Mohr-Coulomb model under undrained conditions and expecting the Mohr-Coulomb model to predict an appropriate undrained strength. For clay, such as the Kallang Formation at the Nicoll Highway site, undrained strength is a function of stress history, in particular overconsolidation ratio, because this determines whether a soil will attempt to contract or dilate as it is sheared in an undrained manner and thus generate positive or negative pore pressures during shearing, which in turn decreases or increases the strength. The Mohr-Coulomb model does not consider dilation or contraction during undrained shearing and, as such, cannot model, for example, soft clays (or dense sands) under undrained conditions. If you use a Mohr-Coulomb model for undrained conditions, then you simply use the undrained strength to control failure conditions and not the effective strength parameters. The Nicoll Highway Committee of Inquiry Report on the collapse includes a well-written section on the problem with the Mohr-Coulomb model (Magnus *et al.*, 2005).

The finite difference program FLAC is probably the second most generally used software for geotechnical design.

Until recently, the program was quite daunting, requiring individual commands to be typed in, but recent versions have a graphic interface, which makes things easier. As for PLAXIS and other sophisticated programs, a great deal of knowledge and understanding is needed if reasonable results are to be achieved. For example, the model must first be set up with proper boundary conditions and brought to equilibrium as natural ground before any engineering works such as excavation are simulated. FLAC progressively calculates and checks solutions. Intermediate stages can be calculated, saved and expressed graphically as a movie which can illustrate how strains are developing with time. FLAC, like its sister program UDEC, can cope with large displacements, more so than typical FE analyses. FLAC is used mainly for soil and rock that can be characterised as continua. UDEC is used for fractured rock and each fracture or set of fractures can be specified individually in terms of geometry and engineering parameters. Both UDEC and FLAC can be used for foundation design, tunnels and slopes.

Other commonly used software include the suite produced by Rocscience, such as Phase^2, and their use is discussed in detail in Hoek *et al.* (1995) and at http://www.rocscience.com/education/hoeks_corner.

Many authors, experienced in the development and use of software, have recommended that sophisticated software should be used in an investigatory way, using many simple models to check sensitivity to assumptions rather than trying to prepare a single complex model in an attempt to simulate all aspects of a situation at the same time (Starfield & Cundall, 1988). Swannell & Hencher (1999) discuss the use of software specifically for cavern design.

6.12.2 *Problem-specific software*

Many suites of software have been developed for particular purposes. SLOPE/W and SLIDE, for example, are commonly used for routine design of slopes. The software calculates stability employing the method of slices, as discussed in section 6.6.2, and gives instant solutions for FoS for a wide range of potential slip surfaces, the broad geometries of which are specified by the operator. Controlling factors such as slope geometry, strength parameters and groundwater conditions can be varied rapidly, allowing sensitivity analysis. Structural elements such as soil nails and rock bolts can be included in the models. There are many similar packages available, all of which are verified and validated against standard mathematical solutions. There is a danger that the ease of use of such software in sensitivity analysis, varying a range of likely parameters, can give a misplaced confidence that all possible conditions have been dealt with. If the ground model is seriously wrong, the results will be meaningless.

Other specialist software packages are used for particular design tasks such as rockfall trajectory analysis, stresses around tunnels, pile design and groundwater and contaminant migration modelling. Details of many of these are reviewed at the web page maintained by Tim Spink: http://www.ggsd.com. Most engineering companies also have in-house spreadsheets (often based on EXCEL) used to solve common analytical problems.

6.13   Role of engineering geologist during construction

6.13.1   Keeping records

Engineering geologists on site should keep careful records as works advance, using daily notebooks. Excavations should be examined, described and photographed as necessary. It is often useful to take photographs and use these as the base for overlays on which to record features such as geological boundaries, strength of materials, discontinuity orientations and style and locations of seepage. Such records will be very helpful in the case of any future disputes over payment or if anything goes wrong. Pairs of photographs taken some distance apart can be used to allow a 3D image to be viewed, and this is particularly useful where access is difficult or hazardous. Where discontinuities are measured, it is important to record the location. In tunnels, description proformas are commonly used as a permanent record, agreed and signed by the contractor and supervisor, as illustrated in Figure 6.33.

6.13.2   Checking ground model and design assumptions

It is fundamentally important that design predictions are checked during construction. Design is usually based on widely spaced boreholes and the interpretation is almost certainly going to be oversimplified. Often this does not have any major consequence but sometimes it does so and the engineering geologist on site should be alert to any indications that the ground model is incorrect or inadequate. Any changed conditions should be flagged up quickly to the designers so that necessary rectifications can be made.

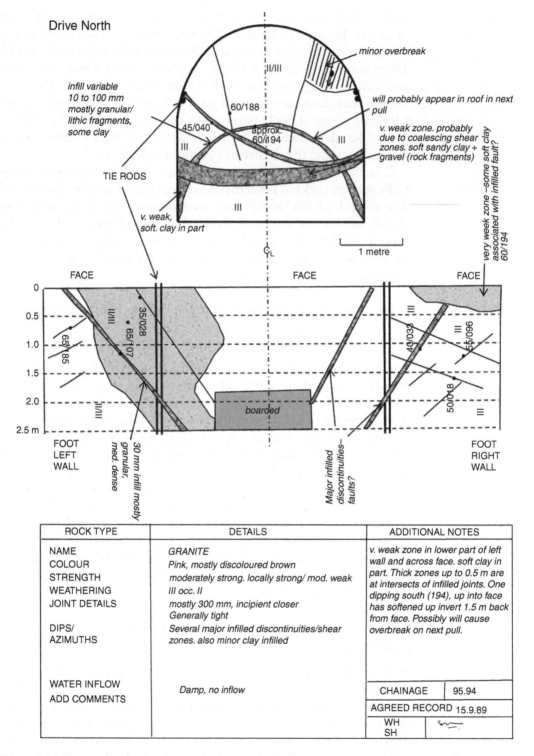

Drive North

minor overbreak

infill variable
10 to 100 mm
mostly granular/
lithic fragments,
some clay

II/III

60/188

45/040

approx.
60/194

III

III

will probably appear in roof in next
pull

v. weak zone. probably
due to coalescing shear
zones. soft sandy clay +
gravel (rock fragments)

TIE RODS

III

v. weak,
soft. clay in part

very weak zone –some soft clay
associated with infilled fault?
60/194

1 metre

FACE

FACE

FACE

0

0.5

1.0

1.5

2.0

2.5 m

II/III

65/85

35/028

65/107

boarded

III

49/033

55/096

50/018

III

II/III

FOOT
LEFT
WALL

30 mm infill mostly
granular,
med. dense

Major infilled
discontinuities–
faults?

FOOT
RIGHT
WALL

ROCK TYPE	DETAILS	ADDITIONAL NOTES
NAME COLOUR STRENGTH WEATHERING JOINT DETAILS DIPS/ AZIMUTHS	GRANITE Pink, mostly discoloured brown moderately strong. locally strong/ mod. weak III occ. II mostly 300 mm, incipient closer Generally tight Several major infilled discontinuities/shear zones. also minor clay infilled	v. weak zone in lower part of left wall and across face. soft clay in part. Thick zones up to 0.5 m are at intersects of infilled joints. One dipping south (194), up into face has softened up invert 1.5 m back from face. Possibly will cause overbreak on next pull.
WATER INFLOW ADD COMMENTS	Damp, no inflow	

CHAINAGE	95.94
AGREED RECORD	15.9.89
WH SH	

Figure 6.33 Example of agreed record of ground conditions in a tunnel. Queens Valley Reservoir diversion tunnel, Jersey, UK.

In rock slope design and construction, the fundamentally important features of discontinuity orientation, lateral persistence, roughness and infill, can only be surmised from typical ground investigation data. Ground models and assumptions need to be checked as the rock is exposed. There are many cases of rock slope failure where sliding occurred on features that were exposed during construction but were either not mapped by the site staff or the significance not recognised. This can be a very difficult situation in that even where rock is well exposed, the presence or otherwise of rock bridges or steps along adversely oriented discontinuities can only be guessed at. Care should be taken to note how easily the rock is being excavated and how stable or otherwise temporary slopes are, as such information will be helpful in judging the risk.

6.13.3   Fraud

As a final note, the engineering geologist should be aware that fraud does occur in civil engineering, as in other walks of life. Whole boreholes in ground investigations have been known to be fictitious, let alone individual test runs. Cases are known where core from one site is placed into core boxes at another site. Tests are sometimes not carried out as specified or results falsified. Data on foundations are sometimes made up – for example, depths, materials used and test results (Hencher *et al.*, 2005). Engineering geologists need to be aware that such practice does occur, albeit rarely, and remain alert in their supervising duties.

7 Unexpected ground conditions and how to avoid them: case examples

'As failure is exactly what engineers do not want it is essential that we learn lessons when it does happen.'

(Blockley, 2011)

7.1 Introduction

When projects go wrong because of ground conditions, it is sometimes because those adverse ground conditions were truly onerous and unpredictable but other times because of poor ground characterisation and modelling. Furthermore, if and when things end up in litigation it is often clear that the problem has been exacerbated by the way the project was set up, managed and contracted (Muir Wood, 2000; Baynes, 2007). Quite often, good practice, which is set out in standards and the literature, is simply not followed because of lack of knowledge, experience or application in the engineering teams or for commercial reasons. Where unexpectedly difficult conditions are encountered during a project, for whatever reason, the consequences can be minimised provided the attitudes of the various parties are to work together to solve the issues. This is often a simple matter of good professional practice on both sides but can be actively encouraged in contracts, as discussed in Chapter 2.

7.2 Ground risks

First, it is worth considering where ground risks arise. Clayton (2001) divides them essentially into three: technical, contractual and managerial. Of the technical risks, these were split down by McMahon (1985) and Trenter (2003) in to:

1. The risk of encountering unknown geological conditions and
2. The risk of using incorrect design criteria.

If one examines failures in projects, however, often the causes are far more complex and it is the interaction of the various predisposing conditions at a site and other construction factors that caused the problems. Very often mismanagement is fundamental to why critical factors are missed, overlooked or not dealt with properly, as discussed for tunnels by Muir Wood (2000). For the following discussion, case examples are grouped according to the predominant cause following the geology plus environment plus construction approach introduced in Chapter 4 (Table 4.1). This is not always easy to do because it is often a combination of factors that led to failure or to the extent of the failure. A fourth category introduced here is systematic failure – where fundamental decisions or concepts seem to have played a major role in what occurred.

7.3   Geology: material-scale factors

Geotechnical hazards occur at a full range of scales from micro (mineralogy, friction) to macro (plate tectonics, typhoons). Material factors are at the scale of hand-held samples or pieces of core. It is the scale of most laboratory and *in situ* tests. Hazards at this scale are associated with the physical and chemical nature and properties of the various geological materials making up the site and used in construction, including their durability.

7.3.1   *Chemical reactions: Carsington Dam, UK*

Carsington Dam failed in July 1984, during construction, when it had almost reached full design height (see Section 7.4.1 for full discussion). One of the lesser-known problems with the Carsington Dam project, however, concerned chemistry of the materials making up the dam, which should have been anticipated. The dam was constructed of locally derived rock fill with a clay core. The rock included black shale with pyrite (FeS) and the riprap was limestone ($CaCO_3$) (Figure 7.1).

Figure 7.1
Chemical reactions between materials making up Carsington Dam.

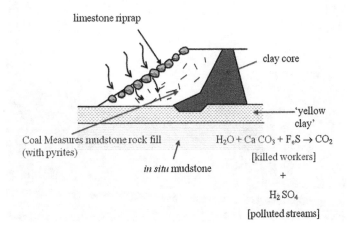

CARSINGTON DAM (1984)

limestone riprap

clay core

'yellow clay'

Coal Measures mudstone rock fill
(with pyrites)

in situ mudstone

$H_2O + Ca\,CO_3 + F_eS \rightarrow CO_2$

[killed workers]

+

$H_2\,SO_4$

[polluted streams]

Sulphuric acid was generated as a result of the geological materials present and this acid polluted local stream courses. Carbon dioxide, because it is heavier than air, collected in underground inspection chambers and four workers died as a consequence. The various chemical reactions and processes are discussed in detail by Pye & Miller (1990). The lesson is that potential chemical reactions should be considered for any project. Other examples are given in Chapter 4.

7.3.2   Strength and abrasivity of flint and chert: gas storage caverns Killingholme, Humberside, UK

Another example of a material-scale factor that should have been anticipated is shown in Figure 7.2, which shows a cavern in chalk under construction for the storage of liquefied gas beneath South Killingholme, Yorkshire. The roadheader pictured had been specified for the excavation of the caverns and had been dismantled and lowered down a narrow shaft for this purpose. Excavation by the road header proved impractical economically, because of wear caused by the presence of bands of extremely strong and abrasive flints and chert in the chalk. Blasting had to be used instead to excavate the caverns (Anon, 1985). Perhaps the presence of bands of flint might have been anticipated in the chalk, even if it had not been logged during site-specific drilling (probably due to inadequate sampling).

7.3.3   Abrasivity: TBM Singapore

Shirlaw *et al.* (2000) provide examples of the high abrasion that can result from high quartz content in soils, weathered and fresh rock during tunnelling. As part of the construction of the North East Line of the Mass Rapid Transit in Singapore, two tunnels were driven from

Figure 7.2 Killingholme LNG caverns under construction. The roadheader is being used to trim blasted caverns rather than excavate them as had been the original intention.

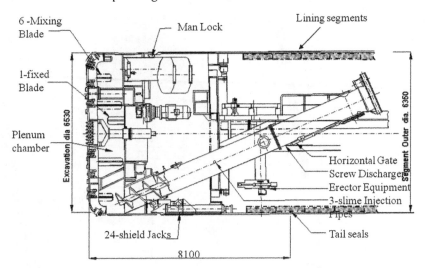

Figure 7.3 EPB Machine in cross section. The pressure at the face is maintained by the rate at which broken down material is removed in the Archimedes screw.

the east bank of the Singapore River to Clarke Quay Station on the west bank. The total length of each drive was only 70 m. A single 6.53 m diameter machine (Figure 7.3) was used to construct the tunnels with the TBM turned around after the first drive. The tunnelling passed from weathered sandstone to marine clay on the southbound tunnel and the reverse on the northbound tunnel. To maintain the face pressure in the weathered sandstone, the EPB shield needed to be operated at almost maximum torque, which resulted in extensive wear to the cutting tools and dangerously high temperatures in the slurry and spoil. Almost all of the discs had to be replaced after 70 m before re-launching the shield for the northbound drive. Furthermore, the sandstone was broken down to abrasive slurry, which, in EPB mode, fills the chamber and the area between the cutting head and the excavated face causing further damage to the machine, including the screw conveyor. In a separate case, for an EPBM tunnelling mainly through Old Alluvium in Singapore, the pressure bulkhead (40 mm thick steel), became so abraded that it failed, resulting in a major loss of ground (Marshall & Flanagan, 2007). The incident caused about 280 mm of settlement and closure of a lane of the central expressway for 12 hours (Shirlaw, personal communication).

Similar problems have been reported and should be anticipated for other soils and rock with high silica content, including granite and soils derived from granite, and can be predicted using Cerchar Abrasivity Index tests. The abrasion can be lessened by appropriate use of additives. In the example above, for the southbound and most of the northbound drives, either water or polyacrylamide were used as a conditioning agent. For the last 10 m of the northbound tunnel, the contractor switched to foam. This had a dramatic effect in reducing the torque required to rotate the cutting head, and the temperature of the spoil.

Figure 7.4 Cracks in face of the Pracana Dam, Portugal, due to alkali-silica reactivity from use of reactive aggregates in concrete (reproduced with kind permission from Ilidio Ferreira, Head of Dam Safety Department at EDP, Portugal).

7.3.4   Concrete aggregate reaction: Pracana Dam, Portugal

Pracana Dam is a 60m high concrete gravity buttress dam built between 1948 and 1951 and located in the centre of Portugal, on the Ocreza River, a tributary of the Tagus River. In 1962, cracks were detected in some buttresses and, in 1964, seepage through the dam increased suddenly (Gomes *et al.*, 2009). Cracking further developed together with progressive upward movement of the crest and downstream movement of the dam. Cores were drilled into the dam and samples taken to analyse deposits in cracks and the aggregate and to conduct expansion tests. It was established that alkali-silica reactivity (ASR) of the locally derived aggregate, including quartzite, milky quartz, feldspar, granite and shale/greywacke, was probably the cause for the swelling phenomenon. A photograph of the crazing resulting from the ASR is given in Figure 7.4. Remedial works included sealing of individual cracks, resin grouting and placing a 2.5mm PVC geomembrane with geotextile on the upstream side of the dam. As illustrated by this case, alkali-silica reactivity in concrete can cause severe deterioration and necessitate expensive remediation and yet is readily avoided if the mineralogy of the aggregate is considered properly and appropriate reactivity and expansion tests conducted when the aggregate is selected for the project (Smith & Collis, 2001).

7.4   Geology: mass-scale factors

7.4.1   *Pre-existing shear surfaces: Carsington Dam failure*

Carsington Dam is infamous because it collapsed over a long length during construction in June 1984 (Figure 7.5). The project was delayed by seven years as a result. A major factor was that the foundation materials were mistakenly considered to comprise undisturbed residual clay derived from *in situ* weathering of the underlying mudstone, whereas they were later identified as periglacial head deposits (Figure 7.6). The head deposits contained pre-existing slip planes along which the shear strength was far lower than had been assumed for the original design (Skempton & Vaughan, 1993). In hindsight, probably the clay above bedrock should have been excavated prior to construction of the dam, as was done for the reconstruction of the dam (Banyard *et al.*, 1992). As noted in Chapter 3, pre-existing shear surfaces are recognised commonly in quite young soils and particularly in the London Clay where they have been associated with landslides.

Figure 7.5 Failure of Carsington Dam. Black area behind water intake structure is tension crack due to failure over 400 m length of the dam crest. White area in front is limestone riprap.

Figure 7.6 Cross section through Carsington Dam, illustrating mode of failure.

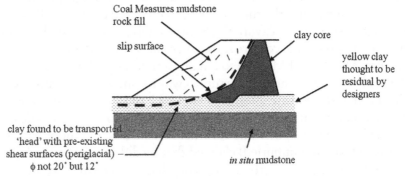

Figure 7.7 Fault in valley at Kornhill, Hong Kong. Poor-quality rock in fault zone resulted in foundations being taken tens of metres deeper than in better rock away from the fault zone.

7.4.2   Faults in foundations: Kornhill development, Hong Kong

Figure 7.7 shows foundations for high-rise structures under construction at Kornhill, Hong Kong. The presence of a major fault (weathered zone in line with the valley) meant that foundations had to be taken locally tens of metres deeper than adjacent foundations. Clearly, the valley was indicative of the potential for poor ground conditions. That said, all valleys are not associated with faults and all major faults are not associated with valleys. At Kornhill, some faults that had been anticipated caused no difficulties whilst other unpredicted faults were discovered during construction (Muir *et al.*, 1986).

7.4.3   Faults: TBM collapse, Halifax, UK

A tunnel was to be constructed in northern England through Carboniferous mudstones, and a tunnel boring machine (TBM) was selected for the construction. The disc cutters of the machine are seen in Figure 7.8. Normally, you should not be able to photograph this view of the machine until the tunnel has been completed and the TBM has entered a reception excavation. Unfortunately, in this case the tunnel had collapsed during construction when the TBM encountered a fault (Figure 7.9). The TBM came to a halt as material collapsed around it. The only way of advancing the tunnel was to sink a shaft in advance of the TBM and excavate a larger tunnel by hand, back to the collapse, and freeing the machine (Figure 7.10). This of course was expensive and led to some dispute. The contractor (constructing the tunnel and whose TBM had got buried) claimed that he did not expect such poor ground. An expert for the client said that he should have done so

Figure 7.8 TBM exposed in fault zone, northern England.

Figure 7.9 Collapse of fault zone around TBM – exacerbated by attempt to move TBM backwards, which necessitated emergency backfill and grouting above the tunnel.

Figure 7.10 Solution – sink shaft in advance of TBM and excavate tunnel by hand back to TBM, allowing it to be protected and freed up.

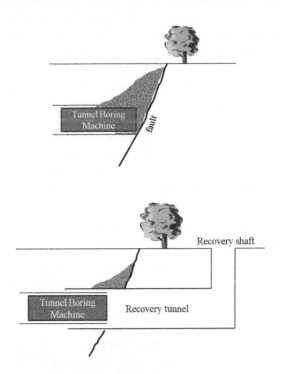

because a fault was clearly indicated by the ground investigation. The contractor agreed that there was a fault shown from the boreholes but argued that the GI did not indicate the degree of disturbance that had led to the collapse. Another expert was called to give evidence. He was asked, 'Was such poor ground an unusual occurrence with such faults?' The expert answered, 'yes, very unusual' – [case going the way of the contractor]. He then went on to say, 'It is so interesting that I bring my students each year to examine it in the quarry near the site where the collapse occurred.' Case dismissed. [The New

Engineering Contract (ICE, 2005) states that in judging the physical conditions, the contractor is assumed to have taken into account:

- The site information
- Publically available information referred to in the site information
- *Information available from a visual inspection of the site*, and
- Other information that an experienced contractor could reasonably be expected to have or to obtain.]

7.4.4   Geological structure: Ping Lin Tunnel, Taiwan

The Ping Lin Tunnel in Taiwan (now called the Hsuehshan Tunnel) was eventually completed after numerous delays, collapses and deaths. The plan was to excavate a pilot tunnel using a 4.8 m diameter rock TBM followed by two 11.74 m diameter TBMs for the main tunnels, but the anticipated tunnelling rates of up to 360 m/month/machine proved hopelessly optimistic because of adverse ground conditions. The geology included a syncline of sandstone with several faults. The pilot tunnel TBM soon ran into trouble as slow seepage from a fault draining the saturated aquifer above rapidly increased and flooded the tunnel (Figure 7.11). Attempts were made to construct bypass tunnels and to advance each of the TBMs using pre-grouting to improve the rock mass, but much of the tunnelling had to be done using drill and blast methods rather than using the purpose-built TBMs. Details of this project are described by Barla & Pelizza (2000).

7.4.5   Deep weathering and cavern infill: Tung Chung, Hong Kong

The new town at Tung Chung is situated close to Hong Kong International Airport and for a large part was built on offshore

Figure 7.11 River of water out of pilot tunnel TBM, Ping Lin, Taiwan.

Figure 7.12 3D geological model, Tung Chung, Hong Kong (courtesy of Dr Chris Fletcher, Fletcher (2004)).

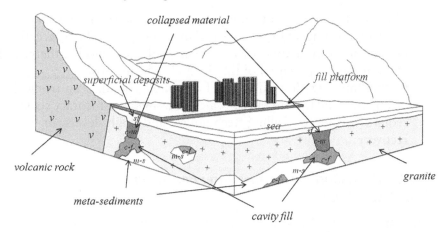

reclamation. Prior to the planning of Tung Chung and the formation of the reclamation, only very limited ground investigations and geophysical surveys were undertaken. As a consequence, the geology of the sub-strate below the seabed was essentially unknown, and extrapolation of the geology from the onshore rock outcrops proved to be misleading.

Initial ground investigations on the newly formed reclamation indicated that the sub-surface bedrock geology comprised rhyolite dykes, marble, metasedimentary rock and skarn. The thickness of weathered bedrock varied greatly across the reclaimed area, from less than 50 m to over 150 m. At the site of a proposed 50-storey residential tower block, a very steep gradient in the rockhead surface was identified. In addition, cavities up to 12 m in thickness and of unknown lateral extent were recorded in the drill logs, deduced from zones of no sample recovery and sudden drops of the drill string. Using this information, a foundation design consisting of 2.5 m diameter bored piles socketed into rock at depths in excess of 120 m was proposed. Ground conditions are illustrated in Figure 7.12. However, the costs and risks involved with this design were considered to be too high, and further ground investigations were undertaken to investigate the geology more fully and to determine whether the cavities were actually present or were filled with soil. The new boreholes used polymer drilling fluids and improved sampling techniques, followed by down-hole electrical cylinder resistivity, gamma density and sonar surveys. No open cavities were identified in the second phase of drilling and a new geological model was proposed. The foundation design was then reassessed, but concern over the mention of open cavities in the original borehole logs still remained and in the event the tower block was never built.

The key points to be learnt from this project are:

– The town planning of Tung Chung should have taken into account the sub-surface geology, so that areas with problematic foundation conditions could have been avoided, if at all possible.

- The use of geophysical techniques, such as microgravity surveys, prior to the formation of the reclamation, would have greatly assisted the formulation of a realistic geological model, thereby maximising the potential of the site and reducing costs.
- Ground investigations in areas of complex ground, deep tropical weathering and the presence of calcareous metasedimentary rocks, require advanced drilling techniques and high levels of soil recovery.

7.4.6   *Predisposed rock structure: Pos Selim landslide, Malaysia*

The Pos Selim landslide in Malaysia is described by Malone *et al.* (2008). The landslide occurred in one of the many large and steep cut slopes along the new 35 km section of the Simpang Pulai–Lojing Highway project, and it is pertinent to ask the question why did it occur there rather than somewhere else?

Failure occurred early on in a small cut slope and affected the natural slope above the cutting. Progressively, the slope was then cut back in response to further failures until the works reached the ridgeline about 250 m above the road (Figure 7.13). The slope then continued to move with huge tension cracks developing near the crest, with vertical drop at the main scarp of more than 20 m in three years.

Clearly, at the site there are some predisposing factors that are causing instability, whereas many other equally steep slopes along the 35 km of new highway show no similar deep-seated failures. The general geology of the site is schist but the main foliation actually dips into the slope at about 10 degrees, so the common mode of failure associated with such metamorphic rocks, of planar sliding on daylighting, adverse schistosity, or on shear zones parallel to the schistosity (Deere, 1971), is not an option to explain this landslide. Following

Figure 7.13 View of Pos Selim landslide, Malaysia.

Figure 7.14 3D model of postulated failure mechanism, Pos Selim.

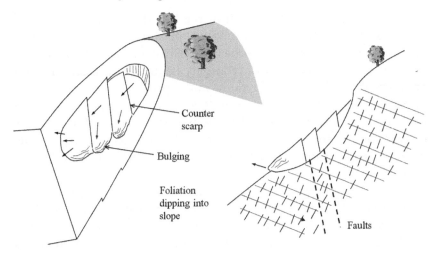

detailed face mapping by geologists, review of displacement data and examination of the various stages of failure, a model was derived that can be used to explain the nature of the failure, the vectors of movement and the fact that it has not yet failed catastrophically but is bulging at one section of the toe (Figures 7.14 & 7.15). Key aspects of the geology are frequent joints that are oriented roughly orthogonal to the schistosity, three persistent faults cutting across the failure and

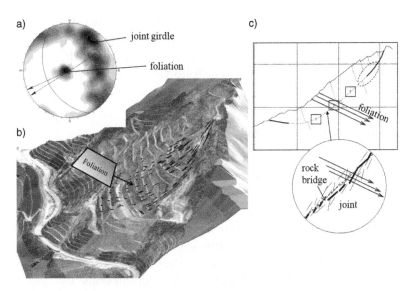

Figure 7.15 (a) Digital model of Pos Selim landslide with movement vectors compiled by comparing an orthophoto with an as-built CAD drawing (courtesy of Dr Andy Hansen). (b) Stereo plot with concentration of poles near centre representing the foliation dipping back into the slope at about 14 degrees. Most joints form a girdle at about 90 degrees to the mean pole of the foliation. (c) Explanation for main body of landslide – exploiting jointing and kicking out at the toe and upwards, along foliation.

another major fault to the north of the landslide area. The derived model is of a mechanism of sliding on the short, impersistent joints that combine with offset sections of schistose fabric to form a shear surface. The shearing forces are largely balanced by sliding friction on joints along the transverse faults and schistosity in one part of the toe where the failure is kicking out. Resistance is also provided by the dilating mass towards the right side toe of the slope (facing). One possible option for remediation that can be derived from this model, therefore involves strengthening the toe area by anchoring or otherwise buttressing.

It is to be noted that this model is not numerical but could certainly be used as the basis for a numerical model that would indeed work for some realistic set of parameters. Without this understanding of geological mechanism, it would be impossible even to begin to design successful remedial measures.

There is some evidence that groundwater is playing a part in the failure (some seepage) and therefore it was recommended that long trial raking drains be installed at points of seepage, in such a way that they also allow water pressures to be monitored within the slope (a cost-effective combination of ground investigation and remedial measure).

7.5   General geological considerations

One of Terzhagi's principles, as reported by Goodman (2002), was to: 'assume the worst configuration of properties and boundary conditions consistent with the data from site investigations', i.e. within the confines of an appropriate ground model. The following two examples illustrate the consequences of failing to do so.

7.5.1   *Tunnel liner failure at Kingston on Hull, UK*

The failure of a tunnel at Kingston on Hull is reported by Grose & Benton (2005). The tunnel was constructed with a TBM through a sequence of saturated Quaternary sediments. During construction, water and then soil migrated through one of the already constructed segmental liner joints and the tunnel had to be abandoned temporarily as the situation deteriorated. A subsequent investigation failed to come up with a definitive answer but it was clearly a matter of soil-structure interaction and possibly construction defects. The discussions by Hartwell (2006) and Shirlaw (2006) are instructive and they come up with various ideas to do with problems with grouting of the liner to explain the failure, which seem feasible. The bottom line seems to be that the tunnel design and construction methodology was not robust enough for the ground conditions. In this case, whatever the geological variability that ultimately brought about the failure, variability was to be anticipated and the design and construction methodology should have coped with this.

7.5.2   *Major temporary works failure: Nicoll Highway collapse, Singapore*

A major failure of temporary works occurred during the construction of part of the Circle Line of the MRT, Singapore, on 20 April 2004. Four persons were killed. The post-failure investigations were presented to a meeting of the British Geotechnical Association (Hight, 2009).

At the time of collapse, excavation was taking place of a 34m deep excavation for a cut and cover tunnel between two diaphragm walls. As the excavation was lowered, the vertical walls were supported by a system of steel struts, waler beams and kingposts. By the time of the failure, there were nine levels of struts and some of these were instrumented with load cells and strain gauges. About six hours before the final failure, some of the struts began to lose load rapidly, whilst others took on more load. This has been interpreted as brittle failure (rapid and unrecoverable loss of strength) of some of the strutting. A detail of the strut beams/waling collection was considered the major factor in the collapse (Magnus *et al.*, 2005).

The excavation was in an area of reclaimed land, but most of the excavation was through very soft and soft to firm predominantly clays of the Kallang Formation, which is an extensive and well-investigated stratum in Singapore (Bird *et al.*, 2003). It was concluded that there had been relative vertical movement between the diaphragm walls and strutting system as part of the failure mechanism. At the meeting reported, it was argued by Hight (2009) that the failure also related to the fact that the temporary works were designed to a FoS of 1.2 yet contained brittle elements, including the steel strutting/waling connection and a concrete strut formed at depth using jet piling. It was suggested that the trigger for the failure was due to ground conditions deviating from the design assumptions, in that the undrained strength profile was lower than assumed and a complex geology, which involved significant variations in stratigraphy between the diaphragm walls. The comment regarding undrained strengths actually does not relate to incorrectly measured or anticipated strengths but an error in the way that strength was dealt with in numerical modelling (Magnus *et al.*, 2005). The comment on complex geology apparently relates to a local deepening of the Kallang Clay but such occurrences are commonplace in Singapore (Bird *et al.*, 2003).

In this case, the fundamental cause of the failure was the structural connection between struts and waling beams that was under-designed. As reported from the BGA meeting, 'The waling detail that yielded and underwent brittle failure had a direct load capacity that was only

marginally lower than the load that was predicted would be applied to it.'
Perhaps if a higher overall FoS (say 1.4) had been adopted for the design
of the walls, it might have compensated for the error in structural design
detail, but if it had done so it would have been essentially by chance.
Blockley (2011), in his paper on engineering safety, discusses a Swiss
cheese model (Reason, 1990) where the components controlling safety
are expressed as a series of layers such as management, design and
ground model, each with defects (holes) in them, including unsafe acts.
These are dynamic – some of the holes move around during construction
or generally with time. Failure occurs where the holes line up. The Nicoll
Highway collapse is an example where there were several such layers
with defects and it just happens that perhaps three out of five or so
lined up at this particular location and these were enough to initiate
failure. The fact that others lined up elsewhere on site (including differing ground conditions and incorrect analysis) contributed to the scale of
the failure. The FoS or partial factor approaches of Eurocode 7 cannot be
expected routinely to cope with structural design faults nor errors in
ground models and analytical mistakes. At the BGA meeting, the
site was described as unforgiving, but this seems outside the sense of
the term as used in Chapter 4 where it is taken to describe particularly
adverse geological conditions that would be very difficult to
anticipate or to investigate using routine approaches which was not the
case here.

7.5.3   *General failings in ground models*

As a general point, when things do go wrong and a detailed examination is made of the ground model assumptions vs. what caused the
failure, it is often found that the ground model was inadequate or
incorrect. Sometimes this can be because the GI was poorly designed or
conducted, but also because practice was poor. Some of the common
errors and poor practices are set out in Table 7.1.

7.6   Environmental factors

Environmental factors include hydrogeological conditions, *in situ*
stresses and earthquake shaking. Such factors should be considered
in preparing the ground model for a site. The environmental factors to
be accounted for depend largely on the nature, sensitivity and design
life of structures and the consequence of failure.

7.6.1   *Incorrect hydrogeological ground model and*
inattention to detail: landfill site in the UK

A quarry in Gloucestershire, UK, was used for disposing domestic
refuse. The quarry had been developed to extract moderately

Table 7.1 Errors in practice that contribute to failure to foresee ground conditions.

	Common failings	Comments
General	Desk study inadequate	These are really tasks for a trained and experienced engineering geologist.
	No site reconnaissance and mapping of exposures.	
	Failure to interpret the landscape – e.g. colour of soil, topography, vegetation, seepage.	
	Lack of anticipation of geological associations, e.g.: – Link between landscape and geological history. – Significance of soil origin. – Metamorphic and igneous associations. – Origins of discontinuity networks. – Lateral and vertical variation, especially in weathered terrain.	
Site-specific	Inadequate or incompetent investigation.	Investigations must be targeted at important unknowns.
	Failure to examine samples (designer).	May miss vital clues.
	Misinterpretation of data.	Over-reliance by designer on interpretation by ground investigation contractor.
	Ignoring significance of lost samples and poor recovery.	Poor recovery often equates with poor ground conditions.
	Designer fails to recognise significance of ground information.	Most soil and rock mechanics textbooks say little about geology and training in civil engineering courses is often superficial.
	Contractor fails to examine samples or site.	Tender restrictions might constrain accessibility but if prevented from so doing contractor should document this.
	Instrumentation inadequate (type, location and monitoring arrangements).	Installation must enhance ground models and allow prediction for project construction and lifetime.

strong to strong Great Oolite limestone with well-defined bedding planes dipping shallowly at about 4 degrees to the northeast. The rock also had many joints orthogonal to bedding, many of which were open, partly as a result of dissolution and partly due to blasting damage. The limestone overlies Fullers Earth, which is a clay-rich formation. The landfill operatives had been advised that the risk of leachate (liquid rich in waste products) migration from the quarry was low, but it turned out that this was based on an incorrect interpretation of local geology. The landfill was then operated using internal clay bunds and a series of drains and lagoons but, crucially, these were constructed overlying a thin layer of limestone that had been left in place in the floor of the quarry as an operating surface for vehicles, both during quarrying and landfill operations. When the quarry was about half full of refuse, leachate emerged at a spring in an adjacent valley about a kilometre away. This polluted a stream, which then impacted a fish hatchery downstream as well as water supply. Reappraisal of the geological model using available published maps indicated that the leachate was probably passing laterally through the lower stratum of jointed limestone that had been left in the base of the quarry and then channelled down a fault to the spring. Tracer tests were commissioned and these confirmed the link between the quarry source and the polluted spring (Smart, 1985). Various options were considered to improve the situation, including grouting, but the preferred remedial measures involved the excavation of landfill along the downstream margin, cutting a trench through the lower limestone and into the underlying Fullers Earth clay and then using geotextile membrane on the inside and an impermeable sheet on the outside wrapped around a drain falling to a sump where leachate could be collected regularly for separate treatment. Upstream of the quarry, a similar membrane and drain system was keyed into the Fullers Earth, collected groundwater throughflow and channelled it around and away from the quarry. As the landfill was completed, it was to be capped to prevent direct rainfall ingress. The works led to significant reductions in migration of leachate and improvement in the quality of water emerging from the spring.

It is to be noted that at the time the quarry was operated a dilute and disperse approach was generally adopted to landfilling in the UK and worldwide, whereby it was assumed that the overall migration of leachate would be insignificant as it mixed with a large volume of groundwater. This case illustrates that where the leachate is transported by local channel flow, rather than volumetric dispersion through a porous mass, dilution cannot be relied upon. Nowadays, in most countries the base and sides of landfills will be sealed using a 0.6 to 1.0m layer of low-permeability soil followed by flexible geomembranes, which again will be covered by a layer of soil to prevent puncturing by traffic or other means. Levels of leachate will also be controlled by internal drainage. While these systems can seldom be completely leak proof, pollution levels should generally be low.

7.6.2   *Corrosive groundwater conditions and failure of ground anchors: Hong Kong and UK*

In the 1970s and 1980s, many ground anchors were installed to stabilise cut slopes in Hong Kong, but this practice largely came to a halt following failure of a number of anchors in the slope shown in Figure 7.16. The problem was recognised following the explosive failure of concrete anchor head covers as the bar anchors broke. Investigation showed that whilst the steel anchor bars were protected from chemical attack along most of their lengths, either by grease or by cement grout, close to a steel coupler there was an air space and inadequate protection, which allowed rusting and failure. Similar deterioration has been observed in the UK on the A685, above the M6 near Tebay, where rock bolts were installed, some with built-in dynamometers to allow loads to be checked periodically (Edwards, 1971). The bolts, with lengths up to 9.15 m, had fixed anchor lengths varying from 0.31 to 0.61 m (epoxy resin). The rest of the length was grouted with either cement or bitumastic, following a nominal loading of 50 kN. Several of the bolts have now lost their anchor plates and therefore are unable to carry their design loads due to deterioration of the rock mass and general rusting of the assemblages (Figure 7.17). Designers and owners need to recognise long-term maintenance requirements, including regular inspection and testing and the likelihood that anchors will need to be replaced periodically. Edwards (1971) identified the need for corrosion protection and long-term maintenance but this appears not to have been done. Even in more recent projects, and despite stringent standards in force for design and construction, including corrosion protection, anchors fail. The design of rock cut slopes at Glyn Bends on the A5 in North Wales required 4,500 m of ground anchors stressed to between 400 kN and 600 kN (Green & Hawkins, 2005). Within 10 years of construction, the new section of road had to be closed after two anchor heads fell off and other anchors failed on testing (North Wales Geology Association, 2006). Anchorages

Figure 7.16
Anchored slope below Clearwater Bay Road, Hong Kong.

Figure 7.17
Deteriorated rock
bolt with proving
ring. Jeffrey's
Mount, above M6
near Tebay, UK.

also fail due to corrosion in other situations, including holding down
anchors for concrete dams (Cederstrom *et al.*, 2005). Similar problems
can arise with corrosion of steel piles or of steel reinforcement within
concrete piles, especially in aggressive environments such as at the coast.
These issues need to be addressed by the designer.

7.6.3   *Explosive gases: Abbeystead, UK*

In 1984, a methane gas explosion destroyed a valve house at
Abbeystead waterworks in Lancashire. Eight people were killed
instantly by the explosion and others died subsequently. The inquiry
into the disaster concluded that the methane had seeped from coal
deposits 1,000 m below ground and had built up in an empty pipeline.
Following a trial and appeal, the designers were held negligent for
failing to exercise reasonable care in assessing the risk of methane in
the finished structure. A review of the various opinions as to which
party (client, contractor and/or engineer) should have foreseen the
accumulation of methane is given by Abrahamson (1992). Other
details are given by Orr *et al.* (1991).

7.6.4   *Resonant damage from earthquakes at great distance: Mexico and Turkey*

One of the interesting aspects of earthquakes is the damage that is
caused at great distances through resonant affects. One of the classic
cases is the earthquake that killed an estimated 10,000 people in
Mexico City in September 1985. The earthquake was large (8.1 magnitude) but the epicentre was more than 350 km away from the city, so

the peak bedrock acceleration was not very high at about 0.04g (as would be anticipated from attenuation equations discussed in Chapter 6). Consequently, in areas of the city built on rock, the earthquake was hardly felt. However, the dominant period of the incoming waves matched the natural period of a basin of lakebed sediments on which part of Mexico City is constructed (about 2.5 seconds) and the ground motion was amplified by the long duration ground motion. That ground motion in turn matched the natural resonant frequencies of some buildings, and almost all damage was caused to structures between 6 and 15 storeys in height.

In a similar manner, whilst most damage from the M6.8 Erzincan earthquake, 1992, occurred close to the epicentre and near to the North Anatolian Fault, anomalous damage occurred at great distances, including complete failure of a six-storey reinforced concrete structure at about 40 km, the very top level of a minaret at 80 km (Figure 7.18) and large landslides at more than 100 km distance. This

Figure 7.18 Damage to top of minaret at about 80 km from epicentre, Sularbasi.

damage almost certainly was caused by long-period wave resonance effects (Hencher & Acar, 1995).

7.7 Construction factors

The third verbal equation of Knill & Price (Knill, 2002) addresses the interaction between the geological and environmental conditions at a site and the construction and operational constraints and interactions. The systematic review and investigation of site geology and environmental factors, discussed in Chapter 4, needs to be conducted with specific reference to the project at hand – how construction is to be achieved and the long-term performance of the structure.

7.7.1 Soil grading and its consequence: piling at Drax Power Station, UK

It is often during construction that things go seriously wrong. An example is provided from the 20,000 piles driven to support the second phase of construction of Drax Power Station in Yorkshire, UK. The ground conditions appeared to be straightforward with about 17 m of firm, varved silt and clay overlying 2–3 m of sand that was typically dense and in turn overlying Triassic sandstone. This profile was proved to be consistent across the site by numerous shell and auger boreholes, with many SPT tests in the sand. Piles were therefore expected to come to a halt during driving at depths of about 18–19 m. In the event, the piles were driven to depths which varied unexpectedly and unpredictably by up to about 4 m. Figure 7.19 shows some of the piles; the holes in the front are where piles disappeared below carpet level without reaching a satisfactory resistance (set) during driving. Where piles were driven to carpet level without

Figure 7.19 Driven piles at Drax Power Station completion. Dark holes in foreground are where pile heads have been driven below carpet level without achieving an adequate set and need re-driving.

achieving the specified resistance, they had to be re-driven contractually. Invariably, it was found that on re-driving, the piles could not be advanced at all. This was not through design or choice and was a costly problem because piles had to be manufactured to cater for longer lengths of penetration to avoid re-drives that were expensive and caused delays to the programme. An investigation was carried out using driven sample tubes into the sand, as well as additional cone penetrometer soundings in groups of piles where they came to an early set and where the depth of penetration was much greater, respectively. In addition, some pneumatic piezometers were installed within the sand horizon and readings of groundwater pressure taken during pile driving as the piling front approached the instrument positions. Data are presented in Figure 7.20. It can be seen that for piles more than 4 m away from the piezometers, there was no influence from the pile driving. However, as the piles approached the piezometer positions, water pressures increased markedly and for the closest piles were off scale at more than twice overburden pressure. Clearly, we were dealing with an effective stress problem. Examination of samples established that an explanation for the different behaviour was in the soil grading. In areas where the piles were driven to refusal at unexpectedly high levels, the sand was uniform and clean. Where they penetrated to deeper levels, the sand had higher contents of silt and clay and clearly had lower permeability. Evidently, the lower permeability allowed higher water pressures to develop, which reduced effective stress and enabled greater penetration. If the driving was stopped, the water pressure dissipated and the piles could not be advanced, however

Figure 7.20
Piezometric
pressure vs.
horizontal distance
between pile and
instrument.

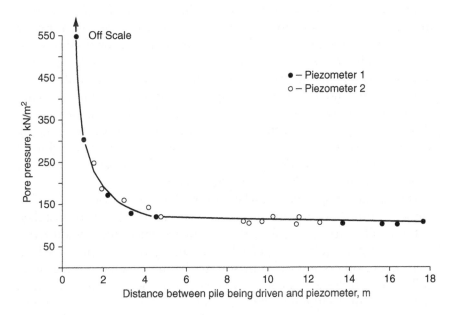

hard they were hit. This was proved by driving some piles to a target depth next to boreholes, 1.5 m into the sand rather than to set or to carpet level. Piles that had not reached set on initial drive were then re-driven (they did not advance) and some were tested without re-driving to 1.5 times working load, which was carried perfectly satisfactorily. In hindsight, the original GI had been too simple; the standard clack valve-type sampling of the sand horizon, typical of UK practice, was not adequate for establishing the true nature of the soil. The variability across the site was due to the outwash fan origin of the sand with local pools of clay and silt-rich soils in the glacial landscape. These difficulties added perhaps 10% to the cost of pile production (Hencher & Mallard, 1989). The problem had not been anticipated and was related entirely to the method of construction. If bored piles had been used instead of driven piles for the Drax foundations, which would have been feasible, then the problem would not have arisen. What was particularly galling was when a senior consultant, who had been involved in the construction of the first phase of the power station, visited the site and said, when informed of what was happening, 'Oh you have found that again, have you?'

7.7.2   Construction of piles in karstic limestone, Wales, UK

Fookes (1997) discusses a number of examples of unexpected ground conditions where he believes that the situation could have been anticipated prior to tender, and includes several dealing with karstic limestone. One that is described is for a river crossing in Wales, and Fookes argues that karstic conditions could readily have been identified from local evidence and boreholes specific to the project. The case is rather more complicated than that. The karst was indeed recognised prior to tender by all parties and, as is good practice (Cole, 1988), a ground improvement contract was let to wash out and grout voids in the limestone around pile locations prior to the piling contract. That being so, the conditions anticipated by the piling contractor were as shown in Figure 7.21. In the event, for various reasons, the grouting was not comprehensive, in particular because some arbitrary rockhead was adopted at some depth below true rockhead (see Chapter 3), which rather negated the point of the grouting.

During the piling contract, where the rock mass was of good quality (either naturally or due to successful grouting), piles were installed without difficulty. Elsewhere, the contractor encountered soil overlying a very irregular rock profile containing voids, some of which were open, others infilled with soil and some grouted. Cased holes collapsed and there was loss of concrete from the holes once formed, with a risk of pollution to the adjacent river. The karstic conditions were not unforeseen, as suggested by Fookes, but the lack of ground

Figure 7.21
Anticipated
conditions for
piling in karstic
limestone,
following ground
improvement
contract.

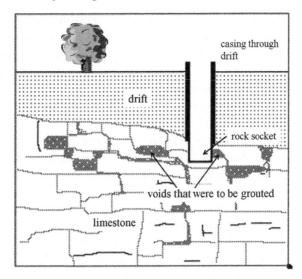

improvements was. Of course, other foundation schemes might have
been conceived that would have placed less reliance on the pre-piling
grouting contract, but as things stood there was a valid claim.

7.8   Systematic failing

Sometimes things go wrong because of some fundamental misunder-
standing of the proper approach to follow or due to some miscommu-
nication or lack of management of the process. Muir Wood (2000)
compares the design of a tunnel to the performance of a symphony.
Whereas in the concert hall, all the players are using the same music
sheets and are watching a conductor, all too often in engineering projects the control is inadequate and all the players have their own agenda.
The result, as Muir Wood puts it, can end up cacophonic. Often when
things go wrong it is argued that ground conditions were unexpected,
but on analysis it is shown that ground characterisation of the site was
fatally incorrect or there were failings in management of the project.

   Regarding the Heathrow Express Tunnel collapse discussed below,
this was described by the Health and Safety Executive, UK (2000), as
an organisational accident with a multiplicity of causes. Other cases
below could be described in similar fashion.

7.8.1   *Heathrow Express Tunnel collapse*

One case from which many lessons can be learned is the well-
documented collapse of a series of tunnels and chambers in London
Clay under construction for the Heathrow Airport Express (HEX) in
October 1994. This case is described in detail, both technically and
regarding management and contractual issues, by Muir Wood (2000),

who was appointed as part of the team investigating the collapse by the Health and Safety Executive (HSE, 2000). The monetary cost of the failure was at least £400 million whereas the original contract value was £60 million. The reputation of civil engineering in the UK was also badly tarnished overseas.

The project was to be conducted by the New Austrian Tunnelling Method (NATM), which was a novel approach for tunnelling in London Clay, although it had been used successfully for trial tunnels at Terminal 4. As described in Chapter 6, the traditional concept of NATM relies on the rock mass locking up as joints and interlocking blocks of rock interact and dilate during convergence towards the excavation. The mass often forms a natural arch within the surrounding rock mass as it relaxes, and no or little support is needed. Optimising support requirements requires prediction of likely convergence rates, making measurements as excavation proceeds and then applying any support necessary to prevent collapse. Generally, shotcrete is used to prevent loss of loose blocks and wedges, which would destabilise the arch and maybe lead to ravelling failure. Further active support options include the use of bolts, lattice girders and steel arches, as appropriate. The principle is that the rock mass carries most of the stress and by waiting until the appropriate time, the engineered works are kept to a minimum. After the rock mass has stabilised, an inner liner can be constructed, with or without waterproofing.

There is some debate over whether the method adopted for HEX really was or should be termed NATM. Muir Wood prefers the term informal support to describe the use of a primary shotcrete liner for the Heathrow Express project; ICE (1996) uses the term sprayed concrete linings (SCL). The common feature between NATM in rock and SCL in weaker ground is the use of shotcrete as a preliminary support, but in weaker ground the shotcrete is used to minimise the deformation and settlement of the ground, which 'effectively reverses the original NATM principle of encouraging controlled ground deformation' (ICE, 1996). It is a subtle but important difference. Even before the Heathrow tunnel project, considerable question marks had been raised over whether the London Clay (an extremely weak, fissured rock mass) would behave as a jointed rock mass, appropriate to the use of a NATM approach. In fact, the design concept was to use sprayed concrete (with steel mesh and lattice girders) to form a load-bearing closed ring (without any bolts or dowels as might be used for rock) to limit settlement – a quite different concept to NATM in rock. One of the features of the failure at HEX was cracking and perhaps repair works of that damage prior to a complete folding in of the shotcrete liner, and this has been blamed as a contributory factor although it was argued, post collapse, by experts, that even with part of the invert completely removed, 'the tunnel, as designed, should still have stood up for up to 80 days' (Wallis, 1999).

Muir Wood summarises the factors he considers as contributing to the collapse of the Heathrow tunnels. He notes that if any one of these had been addressed competently, in all probability the collapse would not have occurred.

Management: an unfamiliar system of project management based on the New Engineering Contract (NEC) (ICE, 2005) with self-certification by the contractor of a design by others. Muir Wood advocates a more partnering approach.

Control of works: it was assumed that the specialist consulting engineers advising the contractor would bring particular knowledge and expertise to the project but commercial factors limited his presence on site and in reality he had limited power to exercise control as should have been done for a design-led system of construction.

Compensation grouting: due to higher than anticipated ground settlement, compensation grouting was carried out beneath one of the buildings adjacent to the works. Muir Wood describes this as, 'in reality grout-jacking, requiring pressures in excess of pre-existing vertical ground stress'. The grouting probably loaded the tunnel and contributed to the collapse.

Lack of reaction to instrument data: there is no evidence that there was any reaction by the contractor and his advisors to the data indicating circumferential movement of the tunnel lining together with depression of the crown. Muir Wood argues that there was no acceptable explanation for this phenomenon other than a weakness of the invert. This should have been clear from distress that had been seen in the invert. The advisors had not established any criteria for acceptable deflections or movement of the lining so there was no quantitative trigger to warn that action needed to be taken. Data were presented in figures and diagrams without any commentary or discussion. Nevertheless, according to Muir Wood, the data caused the client (British Airport Authorities) to question the integrity of the tunnelling, a suggestion that was dismissed by the contractor.

In addition, an important factor was probably the presence of several parallel openings at the time of the failure, which would have allowed failure mechanisms to develop more readily than in the trial tunnels at Terminal 4 (Karakuş & Fowell, 2004).

Little is said in the various papers dealing with the Heathrow collapse about the geology of the site. According to an expert for the prosecution by the HSE, 'The London Clay is a well-documented, largely homogeneous, uniform and extensive body of over consolidated sedimentary clay with very few discontinuities and none identified in the area of the collapse that could have caused a landslip ...'

(Wallis, 1999). This description is somewhat at odds with the observations of Skempton *et al.* (1969) and others of joints and fissures in London Clay, including numerous polished and slickensided ones. The contractor's expert argued that: 'Some unforeseeable and completely unpredictable behaviour or geotechnical mechanism in the clay body is the only explanation for the collapse ...' (Wallis, 1999) linked to the opinions of other experts that the excavated tunnels should have stood up for up to 80 days. It is noted that in subsequent years several authors have highlighted the importance of low shear strength flexural shears in the London Clay (Chandler, 2000; Hutchinson, 2001), which have led to landsliding, including near to Heathrow, but no such surfaces were apparently observed in the post-collapse investigations at HEX (Norbury, personal communication; HSE, 2000).

7.8.2 *Planning for a major tunnelling system under the sea: SSDS Hong Kong*

The Stategic Sewerage Disposal Scheme (SSDS) Stage 1 was an ambitious project to construct a series of shafts and deep tunnels at typical depths of 120–145 m, leading from a number of catchments around Victoria Harbour in Hong Kong to a central treatment works constructed at Stonecutters Island. The layout of the scheme is set out in Figure 7.22 and details are given by McLearie *et al.* (2001).

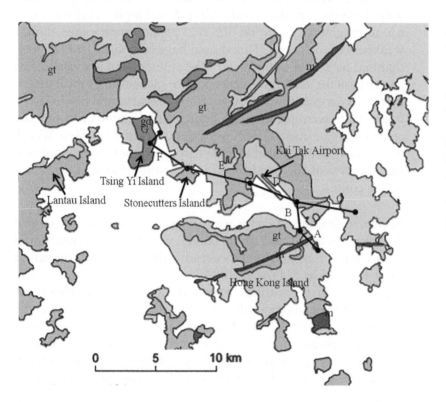

Figure 7.22 Layout of SSDS Stage 1 tunnels A to G, Hong Kong harbour. Darker shades are granitic rock: gt = granite, gd = granodiorite, m = monzonite. Most of the other rocks are mainly of volcanic origin. Base map is from Fletcher (2004).

These tunnels were to be the first bored tunnels under the sea in Hong Kong, and the general thinking amongst the design engineers was that if the tunnels were deep enough, in rock that was grade III or better, then water inflow would be low and ground conditions suitable for excavating with open-face rock TBMs. In the event, most boreholes put down for the project and available at time of tender did not prove ground conditions to the depths of the tunnel alignment that was eventually adopted. Buckingham (2003) summarises as follows:

'When the SSDS project was conceived, site investigation was undertaken along the planned alignment to levels determined by this alignment. This revealed ground conditions at the planned depth to be worse than expected, the tunnels were lowered by several tens of metres below the depth of these boreholes with the idea that they would be below the poor ground. The tender was based upon this assumption that the tunnels would be in better rock with minimal water ingress.'

There was a tender competition and the successful contractors elected to use mostly refurbished open-face rock tunnel boring machines with diameters of between 3.2 and 4.3 m. These TBMs had very limited ability to grout ahead of the tunnels and a low level of shielding for electrical and mechanical devices, which reflected the general assumption by all parties – the HK Government, the design engineers and the bidding contractors – that water inflow would be low. The contractor was required to accept all geotechnical risks such that there was no mechanism for additional payment in the event of unexpected conditions.

In the event, once tunnelling commenced, severe water inflow conditions were encountered in Tunnels F and C especially. In Tunnel A/B there was almost zero water inflow, due to the massive nature of the rock. Fairly onerous conditions had been set for the original contract, including a limit for groundwater inflow of 200 litres/minute/1,000 m length of tunnel. This compares to a final inflow rate of 5,300 l/min for the completed Tunnel F, which equates to 1,400 l/min/1,000 m (McLearie *et al.*, 2001), despite extensive ground treatment and re-letting of the contract, as discussed below.

The original contractor had great difficulty in grouting ahead of the tunnels because the machines and ancillary drilling rigs were not designed for such poor conditions. The original specification was for 120 degree drilling capability rather than 360 degrees, which is more appropriate for stability improvement than preventing water ingress. In Tunnel F, measured water pressures – to be overcome by grouting – was up to 14 bars. Post-excavation grouting behind the TBM proved ineffective in meeting the requirements of the contract. When water was stopped at one location, it reappeared somewhere else. Figure 7.23 shows water conditions in Tunnel F at one stage.

Figure 7.23
View along
Tunnel F (1995).
Note flow weir in
invert.

The original contractor halted work in June 1996 and took advice on the likely time to complete the tunnels from specialist consultants brought in to assess the situation. Following negotiations between the contractor and the government, the contractor was removed from the SSDS project in December 1996 and this was followed by arbitration.

In 1997, the project was re-let as three separate contracts, which were completed by November 2000. The contractual arrangements are unknown in terms of acceptance of ground risks but what is known is that considerable difficulties were encountered, including tunnelling through major fault zones (as had been predicted by the specialists advising the original contractor at the time that he stopped work). In Tunnel F, the Tolo Channel Fault Zone comprised 268 m of mostly poor-quality rock, which required up to 20 grout holes of typically 54 m length around the full perimeter, with grout quantities averaging 5,000 kg/m of tunnel through most of the fault zone (McLearie, *et al.*, 2001). Some details of the additional works and additional costs resulting are given in the Legislative Council Panel on Environmental Affairs report (2000).

Other problems associated with the re-let tunnelling contracts included settlement in an area of recent reclamation about 1 km away from Tunnel C. Inflows into Tunnel C from a discrete highly fractured fault zone peaked at 10,400 l/min (compared to a permitted limit of about 1,000 l/min for the full length of the 5.3 km Tunnel C) and this led to considerable drawdown, settlement and damage to several housing developments (Maunsell, 2000; Kwong, 2005).

The final breakthrough of the tunnels was four years late and US $200 million over budget (Wallis, 2000).

Buckingham (2003) summarises the project:

> 'Inadequate thought and planning during the site investigation stage lead to poor equipment and method selection. Also the fact that the contractor was open to all ground condition risks, which eventually led him to pull of the job resulting in lengthy and expensive arbitration, created additional problems.'

It was argued at the time of the failed Stage 1 works, in defence of the government's case, that the investigation for the SSDS tunnels was the most extensive seen in Hong Kong (albeit that most boreholes did not reach eventual tunnel depth, failed to sample or intersect the major faults along the route and provided almost no data on permeability conditions). That argument resonates in some ways with a quote regarding nuclear works (Nirex application, see below), highlighted by Green & Western (1994):

> 'If a problem is too difficult to solve, one cannot claim that it is solved by pointing to all the efforts made to solve it.'
>
> (Bulletin of Atomic Scientists, 1976)

7.8.3   Inadequate investigations and mismanagement: the application for a rock research laboratory, Sellafield, UK

An application by Nirex to construct an underground rock research laboratory at Sellafield to investigate the site's suitability for disposal of nuclear waste was rejected following a public inquiry. This essentially brought a halt to investigations in a dramatic way. Considering the considerable cost of investigations up to that time and the consequences for the nuclear industry and Britain's energy policies, this can be considered a major failed project. Moreover, the failure was basically a matter of ground modelling and interpretation.

Nirex in the 1990s were given the task of developing a site for the disposal of waste in the UK. The government set strict safety guidelines that would need to be met for a final application for a repository. Following a high-quality ground investigation that was 'the most expensive "single" geological investigation carried out in Britain other than the North Sea oil projects' (Oldroyd, 2002), Nirex decided that part of their studies should include an underground rock research laboratory. Other parties were not convinced that Nirex were ready for this stage because of doubts regarding:

1. the general suitability of the site at Sellafield
2. the capability of the Nirex team to control the necessary science

3. an opinion that the baseline hydrogeological conditions had not yet been established and that these would be disturbed irrevocably by the construction of a laboratory. The baseline conditions would be a crucial part of establishing a safety case in the future.

It was argued that there was a major risk that construction of the underground laboratory would itself damage the site irretrievably. A research laboratory should be constructed elsewhere to investigate and refine the science, even if the final disposal site was to be at Sellafield. Furthermore, various parties were suspicious that the rock research laboratory would be a 'Trojan Horse'. Once considerable money had been invested in a rock research laboratory, it would be very difficult for the UK Government to argue that Sellafield was fundamentally unsuitable as a repository site. Many of the arguments put forward at the public inquiry as expert evidence are published by Haszeldine & Smythe (1996).

Some authors have interpreted the failure at the public inquiry to represent a fundamental ruling on the unsuitability of the Sellafield site. Others are less convinced that that point was established, simply that Nirex were not ready at that time to make the case to proceed with an underground laboratory at the site where waste might be disposed of. Even at the time of the public inquiry, various people were of the opinion that the fundamental problem was the government's insistence on disposal (such that the site would not need to be monitored or any provision made for retrieval), rather than stored underground whilst a safety case was established and tested, possibly over many years. Warehousing waste underground would reduce many of the risks of storage at the ground surface, although it would not be the final solution desired by the politicians. There are arguments for disposal, if it can be achieved, not least to remove a burden from future generations, but perhaps that is simply unrealistic at the moment. It seems highly likely that radioactive waste might be regarded as less dangerous in the future due to advances in medical science, and there may be ways of modifying or even using the waste in the future as an energy source. There certainly will be improved methods for investigating and modelling the geology and hydrogeological conditions at a site.

A year before the public inquiry – which brought the whole process to a halt – Green & Western (1994) wrote on behalf of Friends of the Earth that the government should ensure (amongst other things) that:

1. Radioactive wastes are held in interim, retrievable and monitorable storage until scientific knowledge has advanced to enable permanent solutions to be adopted.

2. An ongoing and comprehensive research programme is initiated for all waste streams into waste conditioning, retrievability and long-term management and the long-term behaviour of radioactive wastes in the environment.

These recommendations still seem eminently sensible, and if that advice had been accepted then the investigations at Sellafield would have proceeded but with a different focus and programme. Oldroyd (2002) comments, 'in retrospect, the whole Nirex enterprise appears to have been in too much of a hurry'. As a result, rather than an ongoing, well-planned and managed approach to deal with the radioactive waste disposal problem, the UK efforts seem to have come to a halt due to fundamental mismanagement.

7.8.4 Landslide near Busan, Korea

The failure of slopes and the subsequent costs of remedial works are often the result of insufficient geological investigation and inadequate interpretation of ground conditions prior to design. This is compounded by poor investigations into the causes of failures and systemic problems associated with poorly defined responsibilities for the stability of cut slopes. This was illustrated by the repeated failure of a large slope in Korea (Lee & Hencher, 2009). The original ground investigation and design were deficient, particularly considering the predictable complexity of the geological conditions. Subsequent investigations were similarly deficient. As a consequence, the slope failed six times despite nine reassessments by various professional engineers and the implementation of several different remedial schemes over a period of seven years up to a disastrous failure in 2002. During the history of design, failure and reassessment, the height of the cut slope increased from 45m to 155m and the cost increased from 3.3 million to 26 million US dollars.

The investigation, design and management of excavation of this cut slope can all be strongly criticised. There were many warnings that the slope was not safe and yet opportunities over a four-year period, to prevent the final failure, were not taken. Investigations, instrumentation, monitoring and design fell far short of international good practice and failed to meet the then current Korean (rather poorly specified) guidelines. It appears that all (or nearly all) engineers involved in investigations and review thought that failure would be restricted to soil. That being so it was only considered necessary to drill and prove rock for a metre beneath the soil to establish a ground model for analysis. There was neither appreciation nor understanding by the engineers that they were dealing with a weathered rock slope with structural control of mass shear strength and hydrogeology. The slope was modelled and analysed as if it comprised layers of cheese, albeit soil mechanics cheese, overlying stainless steel rock which could never fail. This certainly indicates a lack of engineering geological thinking and may reflect some fundamental problems with training and the sometimes-unhealthy compartmentalisation of geotechnics into rock mechanics and soil mechanics. This is not just a Korean problem. As noted in Chapter 4, in many situations sites are forgiving in that an incorrect model does not inevitably lead to disaster. The geotechnical engineer proceeds in normal fashion, ignorant of the real geological conditions, but gets away with it. Unfortunately, this was one of those unforgiving sites in a seriously adverse geological situation and where a superficial and rather poor approach to ground investigation and modelling proved inadequate.

7.8.5   *A series of delayed landslides on Ching Cheung Road, Hong Kong*

One section of Ching Cheung Road in Hong Kong has been the focus for a series of large and unusual landslides since it was first built in 1963 to 1967. Figure 4.6 shows the section of road and was used earlier to illustrate the usefulness of aerial photographs at desk study stage.

In 1972, two major landslides occurred on Ching Cheung Road. The HK Government engaged consultants to study the two land-slides, partly because they were unusual in that both occurred on dry days. The smaller of the two landslides occurred four days after a major rainstorm, the larger ten days after the rainstorm. The investigation concentrated on a theoretical consideration of the infiltration and storage characteristics of the ground, but no explanation could be given for the delayed nature of the failures. The consultant designed the slopes to be cut back, but during the remedial works it was noted that one of the slopes was issuing water and that movements were occurring. A pragmatic solution was recommended to install a series of raking drains, and this seems to have been effective in stopping the movements. The two slopes subsequently survived effectively for several years without any major event. In 1982, another major delayed failure occurred on this section of the road, away from the area that had been cut back with raking drains in the 1970s. As part of the investigation for that failure, the early photographs were found and interpreted showing the pre-disposing poor condition of the hillside. The delayed nature of the failure was explained conceptually through a delayed rise in the main groundwater table (Hencher, 1983c; Hudson & Hencher, 1984).

Following the 1983 failures, the section of Ching Cheung Road was selected for investigation and upgrading, as necessary under the Landslide Preventive Measures (LPM) programme. Several boreholes were put down, analysis conducted and a programme of cutting back and vegetating of the slopes instigated. In so doing, all the previous channels leading from the series of raking drains, installed after the 1972 failures, were stripped away as part of the change from hard covering to a vegetated finish. In essence, the drains and their function had been forgotten about.

In 1997, another large failure occurred essentially at the same location as one shown in the 1940s photographs and one of the 1972 failures. Photographs of the failure show raking drains hanging out of the slope (HAP, 1998). Post-failure investigations showed evidence of extensive natural pipes that were almost certainly associated with previous movements in the slope. Details are given in Hencher (2006).

There are two systematic failings illustrated by this example. Firstly, all of the subsequent failures might have been avoided by proper site investigation prior to construction of the road. The photographs illustrating the predisposing factors were obviously available in government but were not consulted, perhaps because API was not routinely used for engineering projects in the 1960s. The second failing was a management error in disregarding the previously successful engineering works when designing the upgrading works, which probably made things worse rather than better and may have contributed to the 1997 landslide.

Appendix A: Training, institutions and societies

A.1 Training

Thanks are due to professional colleagues worldwide who kindly found the time to advise on current situations re training and professional matters. Apologies for gaps in coverage – information on practice elsewhere can probably be gained from contacting the international society secretariats (IAEG, ISRM, ISSMGE), details of which are given later.

A.1.1 United Kingdom

Most engineering geologists from the UK begin their careers with an undergraduate degree in geology or some closely related subject, and there is no real shortcut for gaining an adequate understanding of geological processes. Portsmouth University has for many years offered a practical undergraduate course in engineering geology, which provides graduates with a suitable entry level of training for the profession but inevitably some of the basic geology teaching has to be omitted in such a course. Furthermore, as indicated in Table 1.1, there is considerably more fundamental geology to learn, even for graduates whose undergraduate courses offer a much higher modular content in geology.

Generally, the best way to gain the next level of knowledge is through formal training via an MSc course but even after that it will take some years of experience before the engineering geologist becomes a person that can contribute fully to a geotechnical team. Geotechnical engineers would normally follow a similar career path, starting with a degree in civil engineering and then taking a specialist MSc course. Many MSc engineering geology courses will fulfil the need for specialist training for both geologists and civil engineers – but the career route for the MSc graduate will really follow from his initial training as a geologist or engineer. In the case of a geologist graduating with an MSc in engineering geology, he will still know that he is out of his depth if asked in his first year in employment to check the structural design of a strutted excavation. Civil engineers who have proceeded to take an MSc in engineering geology are similarly unlikely to have gained enough knowledge of geological processes and relationships to identify realistic geological models, other than for simple situations. It is important that even an MSc-qualified individual follows a period of training, and this is often arranged in a formal and structured manner by large consulting or contracting companies or by government departments. In the UK and several other countries such as Hong Kong, the aim of the individual is to become either a chartered geologist (CGeol) through accreditation with the Geological Society of London or to become a member of one of the institutions of engineering and a chartered engineer (CEng). A chartered status indicates that the individual has gained adequate experience in various facets of the profession. As stated earlier, many engineering geologists have careers in civil engineering and become skilled in geotechnical engineering (including design of structures) and should then aspire to becoming members of engineering institutions and to become chartered engineers. That can only be achieved through further study, possibly formal exams in engineering subjects and/or extensive proven experience. Further and continuing study (self taught, reading technical and scientific journals and attending lectures and seminars) is a formal requirement of membership of most institutions. Some of the career routes are set out in Chapter 1, Table 1.3, and details of institutions and learned societies, what they do and offer, and routes for membership are presented later.

A UK Register of Ground Engineers (the term has been specifically chosen in preference to geotechnical engineers) has recently been established and includes engineering geologists. The scheme was drawn up by the ICE, the Geological Society (Engineering Geology Group) and IOM[3] and administered by the ICE. The scheme is open to chartered members from the three professional bodies. Applicants are required to demonstrate their competence on six specific topics (innovation, technical solutions, integration, risk management, sustainability and management) and there are three levels of Registrant: Professional, Specialist, and Adviser. As at 2011, the scheme does not have any particular legal status. The emphasis is for applicants to demonstrate competence in the various areas.

A.1.2 Mainland Europe

In continental Europe, higher education is moving towards the UK model. Through the Bologna Declaration, most European countries (including the UK) agreed that university systems and degrees should have the same standards within the so-called European Higher Education Area. The background of the declaration is that it facilitates the employment and study of

European citizens in any European country, which is one of the main goals of the European Union.

Not every country has implemented the Declaration to the same level, however. In the Netherlands, the first degree for an engineering geologist will now generally be a BSc in civil and geo-engineering, geology, geography or any other subject that has some relevance to geotechnical engineering or engineering geology. Depending on any deficiencies in the first degree, additional subjects may be required to be studied before or during an MSc degree.

The German university system has changed to a system comparable to the Netherlands and UK with the introduction of a split system of first and second degrees, so the situation for engineering geology education is now broadly the same as in the UK and the Netherlands. Other EU countries are at various stages of implementing the Bologna Declaration.

The European Federation of Geologists has adopted a system of multilateral recognition between affiliated national associations, which is incorporated in the professional title european geologist (EurGeol). As with CGeol status in the UK, the title EurGeol is open to all geologists, whether they work in government, academia or industry, and therefore gives no indication of competence in engineering geology.

A.1.3   *United States of America*

As in the UK, engineering geologists from the United States generally start off with an undergraduate degree in geology or geological engineering. Typically, most engineering geologists are initially trained and educated in geology, primarily obtaining undergraduate degrees in geology. A number of universities also offer Bachelor of Science (BS) degrees in geological engineering. These programme provide the student with a general background in fundamental geology and geophysics, and geological engineering design, including such subjects as soil and rock engineering, geological and geophysical exploration, geological hazard evaluation, groundwater hydrology, geographic information systems (GIS), hazardous waste management and environmental science. The undergraduate student choosing this field of study will learn to apply geologic principles to engineering solutions related to design of geotechnical/civil infrastructure such as tunnels, dams, bridges, excavations and waste disposal sites but as with similar undergraduate degree courses in the UK, this is inevitably at some expense regarding the depth of geology taught. The same applies to others aspiring to become engineering geologists whose initial degree contains a relatively small amount of geological training (e.g. physical geographers and even many earth scientists whose geological modules may make up perhaps only 30% of the course). Such individuals obviously have other skills and learning that will help them in their careers but they will find that there is still a lot of geological topping-up to do as well as all the engineering during their early years in employment (Box 1-1).

In the USA, a minimum of a Master of Science (MS) is now normally required for persons seeking employment in geological or geotechnical engineering fields. For most engineering geologists, the MS degree programme includes studies in rock and soil mechanics, geotechnics, groundwater, hydrology, strengths and permeability of soil and rock, and civil engineering design. Once in employment, geological engineers are encouraged to obtain their professional engineering licence or registration, particularly if working for a smaller consulting firm or are working for a state agency. As at 2011, 31 states have Geologist Licensing Boards (California, Florida, Oregon and Washington to name a few). Typically, education requirements consist of graduation from an accredited college or university with a degree in geology, hydrogeology or engineering geology. To be licensed as a geologist, typically one must have at least five years (Oregon requires seven years under direct supervision of a registered geologist) of documented and verifiable professional geological practice or, if applying for a specialty such as engineering geologist, three to five years of specialty practice that is acceptable to the review board. In some states, an undergraduate degree and/or each year of graduate study may count as one year or more of experience. Geological research or teaching at college or university level may be credited year for year if, in the opinion of the board, it is comparable to experience from practice of geology or a specialty. In most states, applicants must also pass a geologist and/or specialty geologist examination.

In the states of Oregon and California, the licensure title is Certified Engineering Geologist, and in Washington State the title is Licensed Engineering Geologist. There are some states that have reciprocity (i.e. California Board for Geologists and Geophysicists and the Washington State Geologist Licensing Board agree to reciprocity). Applicants requesting license through reciprocity must, however, have certified proof from the state where they are licensed.

A.1.4 *Canada*

In Canada, engineering programmes must be accredited by the Canadian Council of Professional Engineers. Accreditation is normally evaluated every six years by a visiting team, some of whom evaluate the general programme and university environment. Students graduate with the subject matter that is required for professional registration. Graduates may register as Engineers in Training, but require four years of experience, supervised by a Professional Engineer, before they can become Professional Engineers in their own right. They have to pass a

Professional Practice Exam, which covers engineering law and engineering ethics, before the P.Eng is conferred. The accreditation review is completed by a national organisation (the Canadian Council of Professional Engineers), but the Law and Ethics exams, as well as the evaluation of the applicant's file are administered by the provincial organisation (in Ontario, for example, the organisation is called Professional Engineers Ontario).

Some Universities in Canada, which offer earth sciences, have put together a package of courses that is expected to meet the course requirements for registration as Professional Geologists. While Geoscientists Canada publishes national guidelines as to the courses required, these are not binding, and programmes are not directly accredited. Rather, the experience and course review by each provincial organisation is completed for each individual applicant. Three available electives are considered within the Professional Geologist designation: geology, geophysics and environmental geoscience. Each applicant must pass a Professional Practice and Ethics exam.

In both cases (P.Eng and P.Geol), because registration is completed on a provincial level, the geoscientist or geological engineer must become registered in all of the provinces where they plan to work. Furthermore, in both cases, where an applicant is missing core courses, they must complete subject area exams to demonstrate their technical proficiency in the subject.

In summary, engineering geologists in Canada have a legal requirement to register as professional geoscientists in each province where they wish to work but there is no specific professional qualification for engineering geologists. P.Eng applies specifically to engineers but engineering geologists with long experience in civil engineering can register – i.e. hold both P.Geol and P.Eng qualifications.

A.1.5   *China*

China's universities provide courses in both geology and engineering geology as first degrees. Some level of professional status as a geologist or engineering geologist is provided by membership of the Geological Society of China (GSC). To become a member requires five years experience after graduation or two years experience after completing a postgraduate MSc or MEng. Engineering geologists can become members of the Engineering Geology Committee (EGC) (the China National Group of IAEG), which is a professional committee within the GSC. Engineering geologists' status in the workplace is identified by positional title, a ranking system – junior to senior. This professional ranking is reviewed and entitled by a committee called the positional title audit panel, organised by local government, or within some big institutes/companies authorised by the government. Salaries and allowances are largely determined by this positional system.

Currently in China, there are professional qualifications of Registered Civil Engineer, Registered Geotechnical Engineer, Registered Structural Engineer, Registered Mining Engineer, and so on, and it is likely that the role of Registered Geologist will be established and recognised in China soon. Currently, some engineering geologists do achieve the status of Registered Geotechnical Engineer by taking professional examinations.

A.1.6   *Hong Kong*

Hong Kong, which is a Special Administrative Region (SAR) of China, has its own Institution of Engineers with an equivalent status to the Institution of Civil Engineers in the UK. Engineering geologists, qualified as chartered engineers through one of the UK institutions can become members of HKIE through mutual recognition of societies, or may become members through normal routes requiring graduation from a recognised course with adequate engineering input, or by taking additional professional examinations. This initial step must be followed by a period of additional training and professional experience over a period of typically four or five years. After at least a year's relevant professional practice in Hong Kong, members of HKIE can apply to become Registered Professional Geotechnical Engineers RPE(G). Many other geologists and engineering geologists in Hong Kong achieve chartered geologist (CGeol) status through the Geological Society of London, and this requires the individual to follow a prescribed course of training and experience, as detailed below. As noted earlier, however, the qualification CGeol can be achieved by all geologists who are members of the Geological Society of London and does not in itself indicate knowledge or experience of engineering practice.

A.2   Institutions

A.2.1   *Introduction*

There are a number of professional institutions in the UK that govern professional practice, to some degree, by setting out training routes and requirements for their members, thereby setting standards. Membership is conferred and recognised by letters that can be appended to the member's name, as in MICE (Member of the Institution of Civil Engineers). Most institutions also act as learned societies, publishing journals, books and offering their support to conferences and meetings. Similar bodies exist in different countries around the world and offer reciprocal recognition of qualifications, allowing a member from one country to practice professionally in another. All the UK institutions are open to foreign members, and there are well-established regional groups in different countries (and throughout the UK).

For geologists, in the UK, direct membership is possible to the Institution of Geologists, which has its home within the Geological Society of London. Through the Institution of Geologists, one can become a chartered geologist (CGeol). Other institutions that the engineering geologist might wish to join because of their importance to the industry are the Institution of Civil Engineers and the Institution of Mining, Metallurgy and Minerals (which is obviously more attuned to mining and the extractive industries). Through both of these institutions, a suitably trained and experienced engineering geologist can become a chartered engineer (CEng) – a title conferred by the Engineering Council and which applies to many types of engineers – e.g. mechanical, aeronautical or structural.

A.2.2   *The Institution of Geologists (IG)*

The IG provides a process for the professional validation of geologists, mostly but not exclusively in the UK, a system of regional groups, external relations with government, industry and other professional bodies. It is a founder member of the European Federation of Geologists. It publishes *The Geologist's Directory*, which lists members.

The title chartered geologist (CGeol) is awarded to suitable geological graduates with a period of training and experience on a par with that required of engineers to become chartered engineers (CEng). It is open to all geologists.

Candidates must, via a professional report, supporting documentation and through a professional interview, prove their competence against each of the following criteria:

1. Understanding of the complexities of geology and of geological processes in space and time in relation to their speciality.
2. Critical evaluation of geoscience information to generate predictive models.
3. Effective communication in writing and orally.
4. Competence in the management of Health and Safety and Environmental issues, and in the observance of all other statutory obligations applicable to their discipline or area of work.
5. Clear understanding of the meaning and needs of professionalism including a clear understanding of the Code of Conduct and commitment to its implementation.
6. Commitment to continuing professional development throughout their professional career.
7. Competence in their area of expertise.

Usually, the training period prior to successful application is several years. Large companies and government organisations might well run in-house training schemes under a nominated supervisor, which encourages the junior geologist to get the range of experience he needs. Further details can be found from the Geological Society web page http://www.geolsoc.org.uk.

A.2.3 *The Institution of Civil Engineers (ICE)*

The Institution of Civil Engineers is primarily the governing body for practising graduate civil engineers in the UK but other disciplines, including engineering geologists, can join if they can demonstrate sufficient engineering in their education, take additional professional exams and/or have a proven track record of experience. The usual route is first degree followed by a period of training to meet a range of achievements, usually to by signed off by an engineering supervisor within the employing company. After the period of training, the candidate will apply and need to demonstrate his competence in a professional interview. There are two main grades that an engineering geologist might aim for:

A.2.3.1 *Member*

Membership of the Institution of Civil Engineers (MICE) can be awarded to a wide range of engineers practising in the broad area of civil engineering

A.2.3.2 *Fellow*

Fellow is the highest grade of membership of the Institution and may be awarded to those engaged in a position of responsibility in the promotion, planning, design, construction, maintenance or management of important engineering work.

Chartered engineer (CEng) status can be awarded by the ICE or other engineering institutions. Applicants must be able to demonstrate their professional engineering experience and managerial skills. They must have practical knowledge and understanding of the engineering principles relevant to the disciplines of the Institute. They must also demonstrate the use of such knowledge to contribute to the design, manufacture, maintenance, testing and safety of components, devices and structures, or the control of process plant. Details of the Engineering Council that administers the scheme can be found at: http://www.engc.org.uk/

Full details of the ICE, benefits and membership requirements are given at: http://www.ice.org.uk/Membership.

A.2.4   Institution of Materials, Minerals and Mining (IOM³)

As for ICE, there are two main grades that would interest engineering geologists: Professional Member and Fellow. As for ICE, Members and Fellows can become chartered engineers (CEng) if suitably qualified. Other grades are for those with appropriate technical qualifications and experience.

The IOM³ can be contacted at http://www.iom3.org.

A.2.5   Other countries

Many other countries have their own institutions that govern practice, maintain standards and act as learned societies in the same way as UK institutions; many, such as the American Society of Civil Engineers (ASCE) are open to foreign membership.

A.3   Learned societies

A.3.1   Introduction

Membership of professional institutions provides some certification that the member is competent to practice in a particular field and that he follows some code of conduct in his professional activities. Engineering geologists might also find benefit from joining various societies, most of which produce journals, newsletters and organise meetings for their members. The societies sometimes take it upon themselves to provide advice on practice, although generally this has no legal status and, historically, different societies have offered different advice on the same subject, which can be somewhat confusing.

The following is a list of those societies that might be of particular interest to an engineering geologist. Several are really UK-focused but similar groups can be found worldwide and often are very active and good fun, organising social events as well as dealing with local practice. Over recent years, the IAEG, ISRM and ISSMGE, which are international in nature, have forged better links between themselves with joint commissions looking at different aspects of good practice and research, which may go some way to avoiding the overlaps (and blinkered approaches) between the concepts of engineering soil and rock, as addressed in Chapter 1.

A.3.2   Geological Society of London

The Geological Society of London (also known as The Geological Society or Geol. Soc.) is a learned society, based in the UK, with the declared aim of 'investigating the mineral structure of the Earth'. It is the oldest national geological society in the world and the largest in Europe, with over 9,000 Fellows entitled to the title FGS (Fellow of the Geological Society) – over 2,000 of whom are also chartered geologists (CGeol). Membership is open to any geology graduates (greater than 25% earth science subjects) but also to other graduates who are particularly interested in geology or where they work in a profession where geology is a core subject. FGS is therefore open to most geotechnical engineers who are particularly interested in geology. The Geol. Soc. produces the *Quarterly Journal of Engineering Geology and Hydrogeology*.

A.3.3   International Association for Engineering Geology and the Environment

The International Association for Engineering Geology and the Environment (IAEG) was founded in 1964.

According to http://www.iaeg, the aims of the International Association for Engineering Geology and the Environment are:

- to promote and encourage the advancement of engineering geology through technological activities and research
- to improve teaching and training in engineering geology, and
- to collect, evaluate and disseminate the results of engineering geological activities on a worldwide basis.

A journal, *The Bulletin of Engineering Geology and the Environment*, is produced, and the IAEG runs regular conferences and organises commissions with the aim of improving practice.

A.3.4   British Geotechnical Association (BGA)

According to its website: http://bga.city.ac.uk the British Geotechnical Association (BGA) is the principal association for geotechnical engineers in the United Kingdom. It performs the role of the ICE Ground Board, as well as being the UK member of the International Society for Soil Mechanics & Geotechnical Engineering (ISSMGE) and the International Society for Rock Mechanics (ISRM).

The BGA organises renowned events like the annual Rankine Lecture, as well as supporting young engineers and professional and technical initiatives throughout the field of geotechnics. Membership includes a copy of the monthly magazine *Ground Engineering*, which generally has a practical bias with articles on topical subjects and case studies.

A.3.5   Association of Geotechnical and Geoenvironmental Specialists

The Association of Geotechnical and Geoenvironmental Specialists (AGS) is a non-profit-making trade association established to improve the profile and quality of geotechnical and geoenvironmental engineering. Information can be found at http://www.ags.org.uk. The membership comprises UK organisations and individuals having a common interest in the business of site investigation, geotechnics, geoenvironmental engineering, engineering geology, geochemistry, hydrogeology and other related disciplines. The AGS is also active in Hong Kong.

The AGS produces guidelines on what it considers good practice and organises meetings. It has played a particularly important role in defining ways for transfer of geotechnical data electronically.

A.3.6 *International Society for Rock Mechanics*

As noted above, the ISRM and sister society ISSFGE in the UK are taken under the wing of the British Geotechnical Association, which is, in turn, provided support by the ICE.

The following is taken from the ISRM webpage: http://www.isrm

The International Society for Rock Mechanics (ISRM) is a non-profit scientific association supported by the fees of the members and grants that do not impair its free action. The Society has 5,000 members and 46 national groups. The field of rock mechanics is taken to include all studies relative to the physical and mechanical behaviour of rocks and rock masses and the applications of this knowledge for the better understanding of geological processes and in the fields of engineering (ISRM Statutes).

The main objectives and purposes of the Society are:

- to encourage international collaboration and exchange of ideas and information between rock mechanics practitioners;
- to encourage teaching, research and advancement of knowledge in rock mechanics;
- to promote high standards of professional practice among rock engineers so that civil, mining and petroleum engineering works might be safer, more economic and less disruptive to the environment.

The ISRM holds rock mechanics congresses every four years and also sponsors and supports other conferences internationally.

A.3.7 *International Society for Soil Mechanics and Geotechnical Engineering*

The aim of the ISSMGE is the promotion of international co-operation amongst engineers and scientists for the advancement and dissemination of knowledge in the field of geotechnics, and its engineering and environmental applications (which of course overlaps with other societies).

Like the ISRM and IAEG, it produces a regular newsletter and organises congresses and supports other conferences.

Its webpage is at: http://www.issmge.org

Appendix B: Conversion factors (to 2 decimal places) and some definitions

To convert from	To	Multiply by
LENGTH		
inches	millimetres	25.4
feet	metres	0.30
metres	feet	3.28
kilometres	miles	0.62
miles	kilometres	1.61
AREA		
square metres	square feet	10.76
square feet	square metres	9.29×10^{-2}
VOLUME		
cubic metres	cubic feet	35.31
cubic feet	cubic metres	0.02
litres	cubic metres	$0.001 m^3$
cubic metres	litres	1,000
gallons (UK)	litres	4.55
litres	gallons (UK)	0.22
cubic metres	gallons (US)	264.17
cubic metres	gallons (UK)	219.97

Key volumetric parameters in soil mechanics are porosity, void ratio and specific volume.

The three key components in soil of a particular volume (V) are the volume of solids (V_s) (i.e. mineral grains), volume of voids between the solid particles (V_v) and the volume of water (V_w) within the voids. $V = V_s + V_v$.

POROSITY (n %) = V_v/V, i.e. the volume of voids as a percentage of the total volume of soil.

Some workers prefer to use the closely related VOID RATIO (e) = V_v/V_s. Conceptually, they are essentially the same – as the soil is compressed the void volume decreases and the porosity and void ratio also decrease.

In critical state soil mechanics, workers sometimes use the term SPECIFIC VOLUME, which is defined as 1 + e.

Degree of saturation is the ratio of water volume to total voids. So SATURATION (S_r) = V_w/V_v expressed as a percentage. If all the voids are full of water, S_r is 100%.

WATER CONTENT is ratio of mass of water to mass of solid particles (M_w/M_s) so will vary according to the specific gravity of the mineral grains, all other things being the same. These and other basic definitions are clearly set out in most soil mechanics textbooks (e.g. Craig, 1992).

To convert from	To	Multiply by
FORCE		
pounds	kilograms	0.45
	tons (metric)	4.53×10^{-4}
	newtons	4.44
tons (metric)	kilograms	1,000
	pounds	2204.62
	kilonewtons	9.80
kilonewtons	pounds	224.81
	tons (metric)	0.10
	kilograms	101.97
STRESS or PRESSURE		
pounds/square foot	kilonewtons/square metre	0.04
pounds/square inch	kilonewtons/square metre	6.89
kilograms/square centimetre	pounds/square inch	14.22
	kilonewtons/square metre	98.06
tons (metric)/square metre	kilograms/square centimetre	0.10
	kilonewtons/square metre	9.80
kilonewtons/square metre	pounds/square foot	20.88
	kilograms/square centimetre	0.01
	tons (metric)/square metre	0.10
	1 bar = 100 kPa = 0.1 MPa	

The fundamental unit of FORCE in the metric system is the newton (N), which is roughly the weight (mass × gravitational acceleration) of an apple. A kilogram would comprise about 10 apples and weigh 10 N (actually 9.81 N but usually rounded to 10, which is within the accuracy of most geotechnical analyses). In the USA, Imperial units (feet, inches) are generally used for civil engineering works and the unit of weight is the pound.

STRESS: A 10m high column of granite of one square metre in plan would weigh 270kN and exert a vertical pressure or TOTAL STRESS on its base of 270kN/m^2. If the column of granite was submerged in water, its weight would be buoyant (Archimedes principle) and the EFFECTIVE STRESS at the base of the column of granite would be $(270-100) \text{kN/m}^2$ (the stress at 10m depth of water), i.e. 170kN/m^2. One kN/m^2 is often referred to as 1 kilopascal (1kPa) and $1,000 \text{kN/m}^2$ expressed as 1MN/m^2 or 1 megapascal (MPa). Disregarding buoyancy, 1 MPa is the anticipated vertical stress at a depth of about 40m in rock.

In tunnelling, the term BAR is often used for water pressure or in compressed air environments (used to support the tunnel face and stop water inflow). One bar is 100kPa, i.e. the water pressure at a depth of about 10m.

To convert from	To	Multiply by
UNIT WEIGHT		
tons (metric)/cubic metre	grams/cubic centimetre	1.00
	megagrams/cubic metre	1.00
	pounds/cubic foot	62.42
pounds/cubic foot	tons(metric)/cubic metre	0.01
	kilonewtons/cubic metre	1.57×10^{-1}
kilonewtons/cubic metre	tons (metric)/cubic metre	0.10
	pounds/cubic foot	6.36

A cubic metre of water weighs about 10,000 N = 10 kilonewtons = 10 kN. By comparison, a cubic metre of granite weighs about 27 kN because the SPECIFIC GRAVITY of quartz, one of the main components of granite, is 2.69, i.e. it weighs 2.69 times the weight of the same volume of water. The UNIT WEIGHT is WEIGHT/UNIT VOLUME. The UNIT WEIGHT, γ_R, of granite is 27 kN/m^3 and for water γ_W = 10 kN/m^3. The unit weight of soil γ_S is lower than that of granite and typically between 12 and 18 kN/m^3. This is because, whilst the minerals in soil usually have similar specific gravities to the minerals in granite, soil has many voids filled either with air or water.

Instead of 'weight' we often use the term DENSITY, which is MASS/VOLUME and usually expressed in Mg/m^3. DRY DENSITY – the mass of the dry sample divided by volume – is an easy index test to measure in the laboratory and often correlates quite well with degree of weathering or consolidation of a rock or soil. BULK DENSITY is the mass of natural soil (with contained moisture) per unit volume. RELATIVE DENSITY is a term used to describe the relative compaction of soil compared to its densest condition achievable naturally or by compaction. The terms used are:

Relative density (%)	Term
0–15	Very loose
15–35	Loose
35–65	Medium
65–85	Dense
85–100	Very dense

For natural granular materials (mostly sands), the same terms are used, correlated to SPT 'N' values, as described in Chapter 4.

Appendix C: Soil and rock terminology for description and classification for engineering purposes

C.1 Warning

It would be very nice to set out here a guide to soil and rock description that could be applied worldwide. Unfortunately, that is not possible – different countries and different international societies have their own terminologies and things are getting worse – not better. In this appendix, I try to set out what is common practice or at least widely used. Where this is not possible, I set out some of the disagreements and in some instances argue the case.

Engineering geologists and geotechnical engineers, working in any country, need to comply with local regulations, and this will often apply to standards for soil and rock description. This can be difficult and many will feel uncomfortable at using strange and perhaps poorly defined terms. One option, sometimes, is to avoid the use of classifications and to give actual data instead. We might disagree internationally over what 'strong' or 'highly weathered' rock means but UCS = 45 MPa is clear and a metre is a metre everywhere. That said, there are different testing methods and different capabilities re testing so even 'absolute' numbers must be treated with some caution.

C.2 Introduction and history

Standardisation of soil and rock description has been an aspiration of many individuals and working groups over the last 50 years. A brief review is given below.

Much of the soil description and classification terminology has been fairly standardised since the days of Terzhagi and Casagrande (1948). Deere (1968) made various proposals re rock strength classification and discontinuity spacing and introduced the Rock Quality Designation (RQD), and much of the terminology he recommended continues to be used in some parts of the world, not least the USA (Hunt, 2005). One of the earliest attempts to provide fairly comprehensive guidance for logging rotary core through rock was the Geological Society Engineering Group Working Party Report (Anon, 1970). That report and another on maps and plans (Anon, 1972) set many standards that were then followed, particularly in the

UK. Unfortunately, conflicting use of terms was already a feature, especially regarding rock strength, as illustrated later. Other bodies (ISRM and IAEG in particular) were meantime setting up their own working groups and coming up with sets of terms to describe rock features that were in conflict with those suggested by others. The ISRM publication on Suggested Methods for the Quantitative Description of Discontinuities (ISRM, 1978) is a particularly well-illustrated and useful guide, but some of the terminology is its own.

This was followed in the UK by the preparation of fuller guidance on the description and classification of both soil and rock in the BS 5930: 1981 Code of Practice for Site Investigations. Following its publication, a conference was held to review the BS and papers and discussion were published as Geological Society Engineering Geology Special Publication No. 2 (Hawkins, 1986). The Geotechnical Control Office (1988) published Geoguide 3 on Soil and Rock Descriptions, which largely followed British practice but with some distinct differences, especially regarding the description and classification of weathered rock, which is of particular importance to Hong Kong.

BS 5930 was revised and republished in 1999 and is a better document. Most recently, amendments have been made as part of the introduction of the Eurocode, and some of the changes are not necessarily improvements, as discussed by Hencher (2008). Meanwhile, other countries have adopted their own schemes (e.g. Australia, New Zealand and China) and whilst there are common aspects, often the same terms (and certainly the same properties) are redefined in different ways, which is confusing to say the least. In China, there are currently separate standards for site investigation for different industries: transportation, railway, houses and buildings, oil and gas, mining, hydro-electricity, geological survey and for exploration. US practice is illustrated by Hunt (2005) and CALTRANS (2010).

C.3   Systematic description

Systematic description is generally applicable to ground investigation – logging boreholes and exposures. Descriptions should be thorough and unambiguous so that the end user, perhaps in a design office, will know what has been observed on site. The scope and style of routine description of soil and rock for engineering purposes in logging are well established. Examples are presented in Appendix D and serve to illustrate the differences in practice internationally.

C.3.1   *Order of description*

No preferred order is given in EN ISO 14688 (BSI, 2002) or 14689 (BSI, 2003) so, for the UK, the BS 5939 recommendations should be adopted in logging. This is largely to encourage the logger to consider all aspects.

C.3.1.1 *Soil*

BS 5930:1999 states that soils should be described in the following order:

A) Mass characteristics

    1. density/compactness/field strength
    2. discontinuities
    3. bedding

B) Material characteristics

    1. colour
    2. composite soil types: particle grading and composition; shape and size
    3. principal soil type (name in capitals, e.g. SAND), based on grading and plasticity
    4. shape

C) Stratum name: geological formation, age and type of deposit; classification (optional)

Examples presented in BS 5930 are:

> *Firm, closely-fissured, yellowish-brown CLAY (LONDON CLAY FORMATION).*
> *Loose, brown, sub-angular, fine and medium flint GRAVEL (TERRACE GRAVELS).*

It is advised that materials in interstratified beds may be described as follows:

> *Thinly interbedded dense yellow fine SAND and soft grey CLAY (ALLUVIUM).*

and that any additional information or minor details should be placed at the end of the main description after a full stop, in order to keep the standard main description concise.

C.3.1.2 *Rock*

For rock, the BS recommended order is as follows

A) Material characteristics

    1. strength
    2. structure
    3. colour
    4. texture

5. grain size
6. rock name (in capitals, e.g. GRANITE)

B) General information

1. additional information and minor constituents
2. geological formation

C) Mass characteristics

1. state of weathering
2. discontinuities

BS 5930 (BSI, 1999) should be consulted for all terms and definitions as used in British practice. A detailed commentary is given by Norbury (2010). Other countries have their own terms and practice guidance, so the engineering geologist needs to be aware of local usage wherever he is working.

C.4   Soil description

Tables C1 and C2 set out common terminology to be adopted for soil description and Table C3 presents common terminology for describing the strength and compactness of detrital sediments. Such terms and index tests may not be appropriate or adequate for describing and classifying soil-like materials derived by *in situ* weathering.

C.5   Rock description and classification

For rock, key issues are intact rock strength, nature of discontinuities, weathering and rock mass classification.

C.5.1   *Strength*

Intact strength of rock material is very different from mass strength, as addressed in Chapter 5, and usually to a greater degree than for soils, albeit that soil also contains joints, shear planes and general fissures and fractures that influence mass behaviour.

Strength can be estimated quite readily in absolute terms (MPa) by simply hitting it with a hammer or trying to break it by hand and quite often no further testing will be required (see Box 5-1). Despite this, the geotechnical community has been inconsistent in its terminology for 50 years. Some of the definitions used worldwide

Table C1 Basic soil types – based on GCO (1988) but generally compatible with much international practice.

SOIL TYPE	PARTICLE SIZE (mm)		IDENTIFICATION
BOULDERS	–	>200	Only seen complete in pits or exposures.
COBBLES	–	60–200	Often difficult to recover by drilling.
GRAVEL	Coarse	20–60	Visible to naked eye; little or no cohesion but where cemented should state so; particle shape and grading can be described.
	Medium	6–20	
	Fine	2–6	
SAND	Coarse	0.6–2	Well-graded means wide range of grain sizes. Poorly graded is the opposite. Can use the terms uniform or gap-graded as appropriate.
	Medium	0.2–0.6	
	Fine	0.06–0.2	
SILT	Coarse	0.02–0.06	Coarse silt is barely visible to naked eye – easily seen with hand lens; exhibits little plasticity (can't roll into a cohesive sausage) and marked dilatancy (when wet and squeezed, it increases in volume so that water will disappear); slightly granular or silky to the touch. Disintegrates in water; lumps dry quickly; may possess cohesion but can be powdered between fingers. Silt is often detrital quartz (not clay minerals).
	Medium	0.006–0.2	
	Fine	0.002–0.006	
CLAY	–	<0.002	Dry lumps can be broken by hand but not powdered between the fingers. Disintegrates in water more slowly than silt; smooth to the touch; exhibits plasticity but no dilatancy; sticks to the fingers and dries slowly; shrinks appreciably on drying, usually showing cracks. These properties more noticeable with increasing plasticity. Quartz of clay size (often glacial rock flour) has very different properties than true clay minerals.
ORGANIC CLAY, SILT OR SAND	–	varies	Contains much organic vegetable matter; often has a noticeable smell and changes to different colours because of oxidation (red/yellow) or reducing environment (green/blue/black).
PEAT	–	varies	Predominantly plant remains; usually dark brown or black in colour; often with distinctive smell; low bulk density.

are presented in Tables C4 and C5 and the potential for confusion is obvious. The engineering geologist must be cautious, especially where using an empirical guideline, say for allowable bearing pressure or rippability of rock based on intact rock strength (e.g. Table 6.1). Refer to the actual UCS in MPa and not the descriptive term that could mean different things, depending on who has logged the sample and in which country.

Table C2 Dealing with composite soil types – based on GCO (1988) but generally compatible with much international practice.

PRINCIPAL SOIL TYPE	TERMINOLOGY SEQUENCE	TERM FOR SECONDARY CONSTITUENT	% OF SECONDARY CONSTITUENT
Very coarse (BOULDERS & COBBLES) (>50% of soil >60 mm)	Secondary constituents (finer material) after principal[1]	With a little	<5
		With some	5–20
		With much	20–50
Coarse (GRAVEL & SAND) (>65 % gravel & sand sizes)	Secondary constituents before principal (excluding cobbles & boulders)	Slightly (silty, clayey or silty/ clayey)[2]	<5
		Silty, clayey or silty/clayey	5–15
		Very (silty, clayey or silty/clayey)	15–35
		AND/OR	
		Slightly (gravelly or sandy)	<5
		Gravelly or sandy	5–20
		Very (gravelly or sandy)	20–50
Fine (SILTS & CLAYS) (>35% silt & clay sizes)	Secondary constituents before principal (excluding cobbles & boulders)[3]	Slightly (gravelly or sandy or both)	<35
		(gravelly or sandy)	35–65

Examples: slightly silty/clayey, sandy GRAVEL. Slightly gravelly, sandy SILT. Very gravelly SAND. Sandy GRAVEL with occasional boulders. BOULDERS with much finer material (silty/clayey, very sandy gravel).

For fine soils, plasticity terms should also be described where possible, viz: 'non-plastic' (generally silts), 'intermediate plasticity' (lean clays), 'high plasticity' (fat clays).

Notes:

[1] Full name of finer materials should be given (see examples).
[2] Secondary soil type as appropriate; use 'silty/clayey' when a distinction cannot be made between the two.
[3] If cobbles or boulders are also present in a coarse or fine soil, this can be indicated by using one of the following terms relating to the very coarse fraction after the principal: 'with occasional' (<5), 'with some' (5–20), 'with many' (20–50), where figures in brackets are % very coarse material expressed as a fraction of the whole soil (see examples).

Table C3 Guide to soil strength terminology – based on GCO (1988) but generally compatible with much international practice.

SOIL TYPE	TERM	IDENTIFICATION
Very Coarse (COBBLES & BOULDERS)	Loose / Dense	By inspection of voids and particles packing in the field
Coarse (SAND & GRAVEL)	Very loose	SPT 'N' value 0–4.
	Loose	'N' value 4–10; can be excavated with spade, 50 mm peg easily driven.
	Medium dense	SPT 10–30.
	Dense	'N' value 30–50, requires pick for excavation; 50 mm peg hard to drive.
	Very dense	'N' value >50.
Fine (CLAY & SILT)	Very soft	Undrained shear strength (S_u) <20 kPa; exudes between fingers when squeezed in hand.
	Soft	S_u 20–40 kPa; moulded by light finger pressure.
	Firm	S_u 40–75 kPa; can be moulded by strong finger pressure.
	Stiff	S_u 75–150 kPa; cannot be moulded by fingers; can be indented by thumb.
	Very stiff or hard	S_u >150 kPa; can be indented by thumbnail.
Organic (ORGANIC CLAY, SILT SAND & PEAT)	Compact / Spongy / Plastic	Fibres already compressed together. Very compressible and open structure. Can be moulded in hand and smears fingers.

C.5.2   *Joints and discontinuities*

Joints, faults and fractures are of major importance to rock engineering and yet are rather poorly defined.

Norbury (2010) states that:

Discontinuities are synonymous with fractures – mechanical breaks which intersect the soil or rock.

Joints are breaks in the continuity of a body of rock along which there has been no visible displacement.

Fissures are exactly the same as joints but the term is reserved for soil.

Incipient fractures are natural fractures which retain some tensile strength and so may not be readily apparent on visual inspection.

This largely tallies with ISRM (1978).

Table C4 Intact rock strength.

The left-hand column classification is used in several textbooks and standards for foundation design (e.g., BS 8004, 1986; Tomlinson, 2001). The changes in the amended BS 5930 (right-hand column) are inconsistent; note in particular the very different definitions of the term weak. Care must be taken when interpreting boreholes to check the governing standard at the time of the works. Other terms are used in different standards in different countries (Table C5).

BS 5930: 1999; GCO (1988)		BS EN ISO 14689-1:2003; ISRM (1981); amended BS 5930: 1999 (2010)	
Term & UCS (MPa)	Identification	Identification	Term & UCS (MPa)
Extremely weak <0.5	Easily crumbled by hand; indented deeply by thumbnail.	Indented by thumbnail.	Extremely weak <1.0
Very weak 0.5–1.25	Crumbled with difficulty; scratched easily by thumbnail; peeled easily by pocket knife.		
Weak 1.25–5	Broken into pieces by hand; scratched by thumbnail; peeled by pocket knife; deep indentations (to 5 mm) by point of geological pick; hand-held specimen easily broken by single light hammer blow.	Crumbles under firm blows with point of geological hammer, can be peeled with pocket knife.	Very weak 1–5
Moderately weak 5–12.5	Broken with difficulty in two hands; scratched with difficulty by thumbnail; difficult to peel but easily scratched by pocket knife; shallow indentations easily made by point of pick; hand-held specimen usually broken by single light hammer blow. Point load strength (PLS) 0.2–0.5 MPa.	Can be peeled by a pocket knife with difficulty, shallow indentations made by a firm blow of geological hammer.	Weak 5–25
Moderately strong 12.5–50	Scratched by pocket knife; shallow indentations made by firm blow with point of pick; hand-held specimen usually broken by single firm hammer blow. Point load strength (PLS) 0.5–2 MPa.	Cannot be scraped or peeled with a pocket knife, specimen can be fractured with single firm blow of geological hammer.	Medium strong 25–50
Strong 50–100	Firm blows with point of pick cause only superficial surface damage; hand-held specimen requires more than one firm hammer blow to break. PLS 2–4 MPa.	Specimen requires more than one blow of geological hammer to fracture it.	Strong 50–100
Very strong 100–200	Many hammer blows required to break specimen. PLS 4–8 MPa.	Specimen requires many blows of geological hammer to fracture it.	Very strong 100–250
Extremely strong >200	Specimen only chipped by hammer blows. PLS > 8 MPa.	Specimen can only be chipped with geological hammer.	Extremely strong >250

Table C5 Examples of other standard terminology in use for intact rock strength.

New Zealand Geotechnical Society (2005)

Strength	UCS MPa	Is (50) (MPa)
Extremely weak	<1.0	<1
Very weak	1–5	
Weak	5–20	
Medium strong	20–50	1–2
Strong	50–100	2–5
Very strong	100–250	5–10
Extremely strong	>250	>10

Australian Standard AS1726

Strength	Is (50) (MPa)
Extremely low	Generally N/A
Very low	<0.1
Low	0.1–0.3
Medium	0.3–1
High	1–3
Very high	3–10
Extremely high	>10

China National Standard (2009)

Strength	UCS (MPa)
Extremely soft	<5
Soft	5–15
Moderately[4] soft	15–30
Moderately hard	30–60
Hard	>60

[4] Term used in Chinese code (original translation) is 'secondly' rather than 'moderately'.

BS EN ISO 14689-1:2003 does not say what a joint is but defines discontinuity:

Discontinuity: surface which breaks the rock material continuity within the rock mass and that is open or may become open under the stress applied by the engineering work; the tensile or shear strength across or along the surface is lower than that of the intact rock material.

In other words, whereas a discontinuity, as defined by Norbury and lSRM, is a mechanical fracture, the European standard broadens this to include incipient fabric (such as cleavage or bedding planes) that might open up under stress. The implication is that a feature might be described as a discontinuity in some logs but not in others.

C.5.3   *Discussion*

The distinctions might appear trivial but where it matters is when characterising the rock mass. Rock mass classifications discussed later mostly incorporate discontinuity or joint spacing as a fundamental parameter. Rock Quality Designation (RQD) is a measure in boreholes of the proportion of pieces of 'sound' core more than 100 mm in length between discontinuities in the sense of mechanical fractures (natural and not drill-induced). If one were to include incipient fabric such as intact bedding and schistosity, then the whole measure of RQD would be changed and this has knock-on effects for rock mass classifications systems such as Q and RMR.

It is best to follow the Norbury/ISRM definition that a discontinuity is a mechanical break (and this should be used in RQD). Other fabrics that retain tensile strength (including bedding and cleavage) should be described as incipient and not counted in RQD. Incipient discontinuities should be logged and characterised as high tensile strength (close to that of intact rock), intermediate or low (readily split). Geologically, however, incipient fabric such as cleavage, where it has little effect on intact strength, would be called faint, as in faintly defined cleavage.

Geometrical description and measurement of discontinuities is covered in the ISRM guidelines.

C.5.4   *Weathering*

As discussed in Chapter 3, many rocks are weathered to great depths, especially in tropical and sub-tropical areas of the world. Weathering effects should be described and recorded and may be interpreted directly from changes in colour, discontinuity spacing, infill on joints and intact strength. In some circumstances, weathering classifications are useful to characterise rock at the scale of an intact, uniform sample or at the mass scale. The classification provides a shorthand description, which is often treated synonymously with mass strength.

Unfortunately, there are many different classifications, some of which use the same terminology to describe different conditions and profiles; others describe the same phenomena and profiles using different terms, all of which is very confusing (Martin & Hencher, 1986; Hencher 2008).

C.5.4.1. *Material weathering classifications*

Having used weathering classifications a great deal in practice, the author is convinced that material classifications such as that used in Hong Kong and presented in Table C6 are the most useful for dealing with rock that weathers from a strong condition progressively to a soil so that thick profiles of saprolite are sometimes found. It has been used for logging many tropically weathered igneous and sedimentary rocks, even in temperate climates, and is essentially a strength rating.

The Australian standard weathering scheme is presented in Table C7 and is essentially the same as that in Table C6 except that different terminology is used and distinctly weathered (DW) includes a very wide range of strengths (including grades III and IV from Table C6). In

Table C6 Material classification: Geoguide 3 (GCO, 1988).

This classification from GCO (1988) is applicable to uniform samples of weathered igneous and volcanic rocks and other rocks of equivalent strength in fresh state. It is broadly compatible with recommendations of Anon (1995), BS5930: 1999 and used in other standards such as CP4 (2003) in Singapore. It has stood the test of time (from Moye, 1955) as a useful tool in logging core and describing thick weathered profiles. The classification should be supplemented by other descriptive terms, and it is often helpful to qualify with index tests such as Schmidt hammer readings on exposures (not applicable for core logging).

DECOMPOSITION TERM	GRADE SYMBOL	TYPICAL CHARACTERISTICS
RESIDUAL SOIL	VI	Original rock texture completely destroyed; can be crumbled by hand and finger pressure into constituent grains.
COMPLETELY DECOMPOSED	V	Original rock texture preserved; can be crumbled by hand and finger pressure into constituent grains; easily indented by point of geological pick; slakes[1] in water; completely discoloured compared with fresh rock.
HIGHLY DECOMPOSED	IV	Can be broken by hand into smaller pieces; makes a dull sound when struck by hammer; not easily indented by point of pick; does not slake in water; completely discoloured compared with fresh rock.
MODERATELY DECOMPOSED	III	Cannot usually be broken by hand; easily broken by hammer; makes a dull or slight ringing sound when struck by hammer; completely stained throughout.
SLIGHTLY DECOMPOSED	II	Not broken easily by hammer; makes a ringing sound when struck by hammer; fresh rock colours generally retained but stained near joint surfaces.
FRESH ROCK	I	Not broken easily by hammer; makes a ringing sound when struck by hammer; no visible signs of decomposition (i.e. no discolouration).

Notes:

[1]A sample placed in a container of water will lose strength and become a slurry. The rapidity and ease of slaking can be used as a qualifying index test at site.

Table C7 From Australian Standard AS 1726 (1993).

In the Australian Standard, a material weathering classification is prescribed (but no mass classification). The various grades are essentially the same definition as in Table C4 (Geoguide 3) but using some different terms. DW includes material ranging from the strength of fresh rock to material that falls apart in water.

TERM	SYMBOL	DESCRIPTION	COMMENTS
RESIDUAL SOIL	RS	Soil developed on extremely weathered rock; the mass structure and substance fabric are no longer evident; there is a large change in volume but the soil has not been significantly transported.	Same as Geoguide 3.
EXTREMELY WEATHERED ROCK	XW	Rock is weathered to such an extent that it has soil properties, i.e. it either disintegrates or it can be remoulded in water. However, it retains rock structure.	Same definition as completely weathered rock (Moye, 1955) and Geoguide 3.
DISTINCTLY WEATHERED ROCK	DW*	Rock strength usually changed by weathering. The rock may be highly discoloured, usually by iron staining. Porosity may be increased by leaching, or may be decreased due to deposition of weathering products in pores.	
SLIGHTLY WEATHERED ROCK	SW	Rock is slightly discoloured but shows little or no change of strength from fresh rock.	
FRESH ROCK	FR	Rock shows no sign of decomposition or staining	

*Covers highly weathered and moderately weathered classes commonly used internationally. HW and MW classes may be used if noted in explanatory notes.

practice, the boundary between grades III and IV is often taken as distinguishing between rock-like and soil-like behaviour for slope analysis, is readily identifiable, and is often used in specifying target depths for piling, so the advantage of grouping them together is not obvious. The Australian and Hong Kong example borehole logs presented in Appendix D illustrate the use of material weathering grades in practice. Larger-scale mass exposures, such as those with corestone development, can be described with reference to the distribution of the various material classification grades as illustrated in Figure 3.61.

C.5.4.2.   *Mass weathering classifications*

Formal mass schemes are commonly prescribed in different standards because they are readily drawn on paper (percentages of soil and rock) but in practice are often difficult to apply, inflexible and not particularly useful. The scheme used in Hong Kong is presented in Table C8. This was prepared following the work of Martin & Hencher (1986) and was largely a response to the perceived failings of the scheme used in the then BS 5930:1981 and similar schemes adopted by the ISRM. The zones have some sense to them and it seems to work reasonably well in practice, although conditions are found where the application is simply impossible and/or unhelpful. Describing profiles according to material grade distributions is the preferred method of the author. Elsewhere, the ISRM recommendation is still to use a mass scheme by preference – the old scheme of BS 5930:1981 – despite it being impossible to use this in logging (though some people try). The

Table C8 Mass weathering zones (GCO, 1988).

Comment: largely similar to scheme recommended in Anon (1995) and BS5930: 1999. This classification was developed for classifying thick, heterogeneous weathered profiles. The zone boundaries make good sense in terms of engineering significance. Principles are often useful starting points for differentiating engineering units in a ground model.

TERM	ZONE SYMBOL	TYPICAL CHARACTERISTICS
RESIDUAL SOIL	RS	Residual soil derived from *in situ* weathering; mass structure and material texture/fabric completely destroyed: 100% soil.
PARTIALLY WEATHERED ROCK MASS (PW)	PW 0/30	Less than 30% rock. Soil retains original mass structure and material texture/fabric (i.e. saprolite). Rock content does not affect shear behaviour of mass, but relict discontinuities in soil may do so. Rock content may be significant for investigation and construction.
	PW 30/50	30% to 50% rock. Both rock content and relict discontinuities may affect shear behaviour of mass.
	PW 50/90	50% to 90% rock. Interlocked structure.
	PW 90/100	Greater than 90% rock. Small amount of the material converted to soil along discontinuities.
UNWEATHERED ROCK	UW	100% rock. May show slight discolouration along discontinuities.

Table C9 International Society for Rock Mechanics (1978) vs. Eurocode 7 schemes. See Hencher (2008) for further discussion.

The Eurocode 7 scheme is the 1978 ISRM scheme but with different grade numbers. This type of scheme can be applied as descriptive shorthand in areas where weathering is a relatively minor consideration, as illustrated for the greywacke and schists of southern Portugal (Pinho *et al.*, 2006) and can also be used for more deeply weathered heterogeneous zones. It is difficult, however, to log core using this scheme and this causes difficulties for many workers. Deere & Deere (1988) redefine 'sound and hard' rock for RQD as fresh or slightly weathered (ISRM); if the rock core is 'moderately weathered', they say that RQD should be annotated RQD*. This overlooks the dilemma for a logger that grade III (ISRM, 1978) can include up to 50% 'soil' by definition.

	ISRM (1978)		EUROCODE 7	
TERM		DESCRIPTION		DESCRIPTION
FRESH (100% ROCK)	I	No visible sign of rock material weathering; perhaps slight discolouration on major discontinuity surfaces.	0	No visible sign of rock material weathering; perhaps slight discolouration on major discontinuity surfaces.
SLIGHTLY WEATHERED (100% ROCK)	II	Discolouration indicates weathering of rock material and discontinuity surfaces. All the rock material may be discoloured by weathering and may be somewhat weaker than its fresh condition.	1	Discolouration indicates weathering of rock material and discontinuity surfaces.
MODERATELY WEATHERED(>50% ROCK)	III	Less than half of the rock material is decomposed and/or disintegrated to a soil. Fresh or discoloured rock is present either as a discontinuous framework or as corestones.	2	Less than half of the rock material is decomposed or disintegrated. Fresh or discoloured rock is present, either as a continuous framework or as corestones.
HIGHLY WEATHERED (<50% ROCK)	IV	More than a half of the rock material is decomposed and/or disintegrated to a soil. Fresh or discoloured rock is present either as a discontinuous framework or as corestones.	3	More than half of the rock material is decomposed or disintegrated. Fresh or discoloured rock is present, either as a discontinuous framework or as corestones.
COMPLETELY WEATHERED(100% SOIL)	V	All rock material is decomposed and/or disintegrated to soil. The original mass structure is still largely intact.	4	All rock material is decomposed and/or disintegrated to soil. The original mass structure is still largely intact.
RESIDUAL SOIL (100% SOIL)	VI	All rock material is converted to soil. The mass structure and material fabric are destroyed. There is a large change in volume, but the soil has not been significantly transported.	5	All rock material is converted to soil. The mass structure and material fabric are destroyed. There is a large change in volume, but the soil has not been significantly transported.

latest move in this sad story is that Eurocode 7 has reverted to a mass scheme for classification, which is essentially the same as the ISRM but with different zone numbers and even poorer definitions (Table C9). For most countries, this is relatively unimportant, but where the engineering geologist finds himself dealing with severely weathered rock, he will probably find it useful to adopt the approaches used in Hong Kong and in Singapore (refer to GCO (1988), Anon (1995), BS 5930:1999 and CP3 (2004)).

Other schemes are used in different countries; many of these have been prepared from agricultural or soil science perspectives and some are discussed by Selby (1993). In Japan and Korea, distinctions are made between hard rock, soft rock and weathered rock. In China, a pragmatic classification is used based on the ability to cut material with a shovel (Table C10).

As discussed at length in Anon (1995) and adopted by BS5930: 1999 and Eurocode 7, weathering in limestone needs special consideration, as do rocks that weather in a relatively uniform way, such as chalk and some mudstones. Norbury (2010) discusses these schemes comprehensively.

Table C10 Chinese Standard GB50021-2001 (2009 Edition).

Classification is essentially a practical mass scheme, which does not deal with weathered rock masses that develop as corestone-rich profiles. Not applicable for core logging in boreholes.

TERM	DESCRIPTION (SIMPLIFIED FROM ORIGINAL)	COMMENT
RESIDUAL SOIL	Structure fully destroyed. Easily dug with shovel.	Actually, many residual soils are quite strong, cemented with iron especially, so can be quite difficult to dig.
FULLY WEATHERED	Structure basically destroyed but still recognisable. Residual structural strength existing. May be dug with shovel.	
HIGHLY WEATHERED	Majority of structural planes destroyed. May be dug with shovel.	From experience in Hong Kong, highly decomposed (weathered) rock often shows the highest degree of fracturing – they become healed by the completely weathered and residual soil stages (Hencher & Martin, 1982).
MODERATELY WEATHERED	Structure partially destroyed; secondary minerals along joints; weathered fissures developed. Hard to cut with shovel.	
NON-WEATHERED	Structure basically unchanged.	

C.6   Rock mass classifications

This review is not comprehensive but considers the most commonly used rock mass classification systems. A particularly useful review, mostly with respect to tunnelling and underground structures, is given by Professor Hoek at: http://www.rocscience.com/hoek/corner/3_Rock_mass_classification.pdf

C.6.1   RQD

Rock Quality Designation (RQD) was proposed by Deere (1968) as an index of rock quality by which a modified core recovery percentage is measured by counting only pieces of 'sound and hard' core 4 inches (100 mm) or greater in length as a percentage of core run. This measure is usually recorded on core logs and has stood the test of time, partly because it is simple (like the SPT test for soils and weak rock). Many authors have pointed out inconsistencies and suggested modifications, for example, over core lengths to be considered and have questioned the definition of sound and hard, but most of these have not been widely adopted. Basically, one adds the lengths of intact core greater than 100 mm in length along the centre line of core and expresses this as a percentage of core run. If there is 100% recovery and all sticks between natural discontinuities are >100 mm in length, the rock has an RQD of 100%. If all sticks of core were 95 mm, the RQD would be 0%. It is a crude measure but used in practice and is a key parameter in other rock mass classifications, RMR and Q, as discussed below. Details of measurement are given in BS 5930:1999 and most other standards, including Geoguide 3 (GCO, 1988). Various authors have used RQD directly to correlate with rock mass parameters such as Young's modulus or to estimate bearing capacity for foundations (Peck *et al.*, 1974). It should be borne in mind that rock with 100 mm spacing of open joints (100% RQD) is in fact severely fractured rock that would be described as closely spaced (60 mm to 200 mm). According to BS 8004 (BSI, 1986), any rock with joint spacing less than 100 mm would require 'tests on rock', as discussed in Chapter 6, to assess allowable bearing pressure.

Palmström (1982) suggests that RQD can be estimated from the number of discontinuities per unit volume based on visible discontinuity traces surface exposures, using the following relationship:

$$RQD = 115 - 3.3 J_v$$

where J_v is the sum of the number of joints per unit length for all discontinuity sets and known as the volumetric joint count. There is a practical difficulty here in that traces of discontinuities do not necessarily equate with natural breaks in core (the incipient joint problem, as discussed in Chapter 3). The engineering geologist must beware of counting all visible traces in defining RQD, otherwise the ground will be assigned a much lower quality rating than is really justified. This can have major consequences in assessing the potential of tunnel boring machines or roadheaders to make progress when cutting rock.

C.6.2   More sophisticated rock mass classification schemes

Various rock mass classifications have been developed, largely as methods of estimating support requirements for underground excavations linked to case histories, although GSI (covered below) is aimed instead at predicting engineering parameters rather than engineering performance and support requirements. As discussed in Chapter 6, these classifications really come into their own whilst tunnelling and where a decision has to be made quickly, on mucking out, as to the level of support that is required to stabilise the rock mass. In a drill and blast operation, there is no time for the engineering geologist to take sophisticated measurements and carefully weigh up the potential modes of failure and rock loads whilst the drilling crew (no doubt on a bonus linked to advance rate) wait patiently ...

However, not all tunnel engineers are great fans of rock mass classifications schemes. Sir Alan Muir Wood (2000) comments, 'Currently, much time and effort tends to be wasted in assembling prescribed data, often painstakingly acquired at the tunnel face, to enable calculation of a RMC algorithm which is then filed in a geological log book but not applied to serve any further purpose.' And, 'For weak rocks, the contribution from RMC is more limited ... based on material ... as well as discontinuities. Attempts to base support needs for weak rocks on RMC have been notably unsuccessful.' He continues, 'RMC are inadequate to provide reliable information on failure modes', and 'the way ahead must be to identify important parameters for a particular situation and to present these in a multidimensional way'.

C.6.2.1   RMR

The Rock Mass Rating (RMR) of Bienawski (1976, 1989) has also stood the test of time as a useful classification system, despite many question marks over definitions. The five parameters for the RMR system and their ranges of assigned points are as follows:

1. Uniaxial compressive strength of rock material (0–15)
2. Rock Quality Designation (3–20)
3. Spacing of discontinuities (5–20)

4. Condition of discontinuities (0–30)
5. Groundwater conditions (0–15)

Points are assigned for each parameter and then summed to rate the rock as very good to very poor. It is to be noted that the degree of fracturing and condition of those fractures covers 70% of the Rock Mass Rating and that there is a high level of double counting between factors 2 and 3. There is also somewhat of a conundrum in that rock with RQD of 90–100% is allocated the full 20 points but rock with joint spacing of 60–200 mm is only allocated 8 out of 20 points.

Orientation of discontinuities is used to adjust the summed rating according to whether discontinuities are adverse relative to the engineering project.

RMR is used (as is RQD by itself) to correlate with rock mass parameters, including rock mass strength and deformability.

C.6.2.2 Q SYSTEM

The Q system of Barton *et al.* (1974) is commonly used to classify the quality of rock mass – as a predictive tool in estimating tunnel support requirements for a planned tunnel or cavern, for judging whether ground conditions during tunnelling were as expected or not (contractual issues) and for making decisions on temporary and permanent support requirements during tunnelling (Chapter 6). Barton (2000, 2005) also discusses ways of using the Q system in predicting TBM performance. Q has a value range from 0.001 to 1,000.

$$Q(\text{quality}) = \left(\frac{\text{RQD}}{J_n}\right)\left(\frac{J_r}{J_a}\right)\left(\frac{J_w}{\text{SRF}}\right)$$

where:
RQD is Rock Quality Designation; J_n is joint set number; J_r is joint roughness number; J_a is joint alteration number; J_w is joint water reduction factor; and SRF is a Stress Reduction Factor (relating to loosening of rock, rock stress in competent rock and squeezing conditions in incompetent rock).

Ranges and descriptions for each parameter are given in the original publications by Barton (2000, 2005) but also in Hoek & Brown (1980) and at Hoek's corner, as referred to at C6.

C.6.2.3 GSI

The Geological Strength Index (GSI) of Hoek (1999) provides a means of estimating rock mass strength and deformability through broad classification based on rock type and quality of rock mass, as illustrated in Table C11. Ways in which the GSI can be used to estimate rock mass parameters are dealt with in Chapter 5.

Table C11 GSI.

GEOLOGICAL STRENGTH INDEX FOR JOINTED ROCKS (Hoek and Marinos, 2000)	SURFACE CONDITIONS	VERY GOOD — Very rough, fresh unweathered surfaces	GOOD — Rough, slightly weathered, iron stained surfaces	FAIR — Smooth, moderately weathered and altered surfaces	POOR — Slickensided, highly weathered surfaces with compact coatings or fillings or angular fragments	VERY POOR — Slickensided, highly weathered surfaces with soft clay coatings or fillings
From the lithology, structure and surface conditions of the discontinuities, estimate the average value of GSI. Do not try to be too precise. Quoting a range from 33 to 37 is more realistic than stating that GSI = 35. Note that the table does not apply to structurally controlled failures. Where weak planar structural planes are present in an unfavourable orientation with respect to the excavation face, these will dominate the rock mass behaviour. The shear strength of surfaces in rocks that are prone to deterioration as a result of changes in moisture content will be reduced if water is present. When working with rocks in the fair to very poor categories, a shift to the right may be made for wet conditions. Water pressure is dealt with by effective stress analysis.						
STRUCTURE		DECREASING SURFACE QUALITY ⟹				
INTACT OR MASSIVE - intact rock specimens or massive in situ rock with few widely spaced discontinuities	DECREASING INTERLOCKING OF ROCK PIECES	90 / 80			N/A	N/A
BLOCKY - well interlocked un-disturbed rock mass consisting of cubical blocks formed by three intersecting discontinuity sets			70 / 60			
VERY BLOCKY - interlocked, partially disturbed mass with multi-faceted angular blocks formed by 4 or more joint sets				50		
BLOCKY/DISTURBED/SEAMY - folded with angular blocks formed by many intersecting discontinuity sets. Persistence of bedding planes or schistosity				40	30	
DISINTEGRATED - poorly inter-locked, heavily broken rock mass with mixture of angular and rounded rock pieces					20	
LAMINATED/SHEARED - Lack of blockiness due to close spacing of weak schistosity or shear planes		N/A	N/A			10

C.6.3   *Slope classifications*

Various authors have devised soil and rock mass classifications as a means of judging slope stability (Romana, 1991; Selby, 1993; Hack, 1998). In the experience of the author, these are less commonly applied than are rock mass classifications for tunnels, perhaps because there is not usually the same degree of urgency to make decisions and need to link observations to empirical experience. When assessing a slope, one can usually take some time to examine and investigate the geological conditions in some detail and carry out rational analysis and design – as of course should also be done wherever time allows for underground openings, as discussed in Chapter 6. Slope stability assessment classifications can be useful where, for example, carrying out a rapid comparative hazard and risk survey along a highway.

Appendix D: Examples of borehole and trial pit logs

In this appendix, examples are presented of borehole (or drillhole) and trial pit logs from the UK, Hong Kong and Australia. Locations have been omitted in some examples. They contain a great deal of information, not only as records of the ground conditions encountered but also on the way in which the investigation was conducted, the machinery used, the tests conducted and their results and the groundwater conditions. These serve to illustrate the variety of use of descriptive terms and classifications in description as well as typical techniques used in ground investigation.

D.1 Contractor's borehole logs

Contractor's logs are carried out following standards and codes in the country where the work is carried out. As discussed in Chapter 4, they tend to provide somewhat simplified descriptions of ground conditions. The engineer who, unlike the contractor, has designed the GI should have carried out a thorough desk study and be aware of the factors that will be crucial to the success of a project; he may need to examine samples himself to ensure that key features at a site have been correctly identified, described and highlighted.

D.1.1 UK example

The borehole log D1 is courtesy of Geotechnics Ltd. and for a hole taken to 20m depth. Descriptions of soil and rock encountered are to BS 5930:1999.

All of the following detail (and much more) can be read directly from the log without any report or further explanation.

The first 1.2m was excavated as an inspection pit. This is usual practice, just in case there might be services such as electric cables or pipes not identified on plans or using detecting equipment. Water was encountered at 0.4m.

A cable percussion rig was then used to advance a hole of 0.15m diameter. The hole was advanced essentially following the strategy

BOREHOLE RECORD - *Cable Percussion and Rotary*

Project **SAMPLE PROJECT**	Engineer **GEOTECHNICS LIMITED**	Borehole **CP5/RC5** Project No **PC060000**
Client **GEOTECHNICS LIMITED**	Local Grid Coordinates **9826.53 E** / **1424.53 N**	Ground Level **30.48** m OD

Sampling / Properties / Strata — Scale 1:50

Depth	Sample Type	Depth Cased & (to Water)	Strength kPa	w %	SPT (Fl)	Description	Depth	Legend	Level m OD
0.00- 1.00	B				38	Very soft dark grey brown gravelly sandy clay with occasional rootlets. Gravel is angular to subrounded fine to coarse brick, mudstone and slag. [MADE GROUND]	G.L.		30.48
0.50	E								
1.00	D								
1.00	E								
1.20- 1.70									
1.20- 1.65	D	1.00 (0.40)			S5	Very soft brown and dark brown slightly sandy gravelly clay. Gravel is subangular brick and slag. [MADE GROUND]	1.40		29.08
2.00	D				38		2.10		28.38
2.00	E								
2.20- 2.70	B					Very soft grey slightly sandy organic CLAY. [ALLUVIUM]			
2.20- 2.65	D	2.00 (DRY)			S8				
3.00	D				45	Below 3.20m, with occasional plant remains and occasional black speckling.			
3.20- 3.70	B								
3.20- 3.65	D	3.00 (DRY)			S6				
4.00	D				63	Very soft grey slightly sandy slightly gravelly CLAY. Gravel is subrounded fine to coarse sandstone and mudstone. [ALLUVIUM]	3.90		26.58
4.20- 4.70	B						4.20		26.28
4.20- 4.65		4.00 (3.70)			S16	Medium dense brown slightly clayey sandy GRAVEL with a low cobble content of sandstone. Gravel is subrounded fine to coarse sandstone. [RIVER TERRACE DEPOSITS] At 5.00m, recovered as very soft slightly sandy very gravelly clay. Below 5.20m, becoming very dense.			
5.00	D								
5.20- 5.70	B								
5.20- 5.65		5.20 (4.65)			C53				
6.00	D								
6.20- 6.70	B								
6.20- 6.39		6.20 (4.70)			C50/70				
7.00	D					Very stiff grey mottled brown slightly sandy CLAY with many fine to medium gravel sized mudstone lithorelics. [COAL MEASURES] Between 7.00-7.20m, recovered as soft.	7.00		23.48
7.00	W								
7.20- 7.70	B								
7.20- 7.65	D	7.20 (5.70)			S35				
8.00- 8.30	D	7.20 (5.70)			S50/ 146		8.30		22.18

Core Run Core Dia	Depth Cased	TCR/SCR %	Length Max/Min	RQD %	SPT (Fl)	Continued by Rotary techniques — General / Detail	Depth	Legend	Level m OD
8.30- 9.00	8.30	90 / 28	0.05 / 0.01	0		Extremely weak light grey thinly laminated MUDSTONE. Extremely closely to very closely spaced subplanar slightly polished discontinuities with slight clay smear dip 60-70 degrees to core axis. [COAL MEASURES]	Between 8.30-8.45m and 10.06-10.22m, stepped subvertical discontinuity with some orange staining penetrating up to 1mm. Between 9.00-9.75m, assumed zone of core loss.			
9.00-10.50	8.30	50 / 22	0.05 / 0.01	0	(>20)				

Boring

Depth	Dia	Technique	Crew
1.20		Inspection Pit	JS/CS
8.30	0.15	Cable Percussion	JS/CS
20.00	0.12	Rotary Core	GC

Progress

Depth of Hole	Depth Cased	Depth to Water	Date	Time
G.L.			03/12/09	08:00
8.30	7.20	5.70	03/12/09	18:00
8.00	7.20	5.50	04/12/09	08:00
16.50	8.00	7.40	04/12/09	18:00
16.50	8.00	1.40	07/12/09	08:00
20.00	8.00	11.90	07/12/09	18:00

Groundwater

Depth Struck	Depth Cased	Rose to	in Mins	Depth Sealed	Remarks on Groundwater
0.40		0.40	20	1.50	
4.20	4.00	3.70	20		Slow.
11.10	8.00	6.40	20		Fast.

Remarks

Inspection pit hand excavated to 1.20m depth.
A 50mm standpipe was installed to 11.00m with a slotted section from 1.00m to 11.00m with flush lockable protective cover. Detail as follows from base of hole: bentonite seal up to 12.00m, gravel filter up to 2.00m, bentonite seal up to 0.20m, concrete up to ground level.
Chiselling: 6.30-6.60m for 60 minutes and 7.50-8.00m for 60 minutes.
Flush: 8.00-20.00m, Water, 80% returns.

Symbols and abbreviations are explained on the accompanying key sheet.

All dimensions are in metres. Logged in accordance with BS5930:1999

Logged by	NG/LJ
Checked by	TNH
Figure	1 of 2
	17/12/2010

geotechnics

Figure D1 UK drillhole.

BOREHOLE RECORD - *Cable Percussion and Rotary*

Project	SAMPLE PROJECT	Engineer	GEOTECHNICS LIMITED	Borehole	CP5/RC5
				Project No	PC060000

Client	GEOTECHNICS LIMITED	Local Grid Coordinates	9826.53 E / 1424.53 N	Ground Level	30.48 m OD

Drilling		Properties/Sampling					Strata		Scale		1:50

Core Run/Depth (Core Dia)	Depth Cased & (to Water)	Type TCR/SCR%	Length Max/Min	RQD %	SPT (FI)	Description General	Description Detail	Depth	Legend	Level m OD
10.50–12.00	8.30	53 9	0.05 0.04	0	(NR)		Between 10.20-10.38m, recovered as angular gravel size fragments with much soft to firm clay. Between 10.58-10.74m, weak to moderately strong siltstone horizons. Between 11.08-11.18m, extremely weak black vitreous coal horizon. Between 11.18-11.30m grey gravelly clay. Between 11.40-12.00m, assumed zone of core loss.	12.00		18.48
12.00–13.50	8.30	98 47	0.23 0.01	15	(NI) (15)	Extremely weak grading to weak grey thinly laminated SILTSTONE. Extremely closely to closely spaced undulating dull discontinuities with a dusting of silt, dip 50-60 degrees to core axis. [COAL MEASURES]	Between 12.00-12.35m, recovered as angular/tabular gravel sized fragments with some brown stained surfaces.	13.25		17.23
13.50–15.00	8.30	100 80	0.23 0.06	60	(5)	Medium strong, occasionally strong, light grey fine grained SANDSTONE with local light brown staining. Closely to medium spaced discontinuities with light brown staining penetrating up to 3mm, dip 50-60 degrees to core axis. [COAL MEASURES]				
15.00–16.50	8.30	83 24	0.20 0.01	13	(15) (NR)	Extremely weak to weak grey occasionally thinly laminated MUDSTONE with occasional to rare dark grey gravel size plant fossil fragments. Extremely closely to closely spaced undulating discontinuities dull with frequent clay smear dip 40 - 60 degrees to core axis. [COAL MEASURES]	Between 15.40-16.00m, undulating subvertical discontinuity with increasing light brown staining penetrating up to 3mm. Between 15.53-16.00m and 17.59-17.64m, recovered as angular to subangular tabular gravel sized fragments with some clay.	15.25		15.23
16.50–18.00	8.30	97 31	0.09 0.02	0	(>20)					
18.00–20.00	8.30	96 55	0.32 0.01	36	(11) (9)	Weak occasionally medium strong grey occasionally thinly laminated SILTSTONE. Extremely closely to medium spaced discontinuity undulating, dull occasionally slightly micaceous dip 60-70 degrees to core axis. [COAL MEASURES]	Between 18.50-18.75m, rough orangish brown and occasionally dark brown stained discontinuity dip 10 degrees to core axis. Between 18.73-18.75m and 19.48-19.51m, weak, recovered as angular gravel size fragments.	18.20 20.00		12.28 10.48

End of Borehole

Drilling					Progress						Groundwater					
Depth	Hole Dia	Technique		Crew	Depth of Hole	Depth Cased	Depth to Water	Date	Time		Depth Struck	Depth Cased	Rose to	in Mins	Depth Sealed	Remarks on Groundwater

Remarks

Symbols and abbreviations are explained on the accompanying key sheet.

All dimensions are in metres. Logged in accordance with BS5930:1999

Logged by	NG/LJ
Checked by	TNH
Figure	2 of 2
	17/12/2010

geotechnics

Figure D1 (continued).

Project SAMPLE PROJECT   **Project No**   PC060000

Client   GEOTECHNICS LIMITED

Hole	Depth m bgl	Level m OD	Type	SWP (mm)	Seating Drive		Test Drive				SPT 'N' Value	Uncorrected SPT 'N'				
					0-75 (mm)	75-150 (mm)	0-75 (mm)	75-150 (mm)	150-225 (mm)	225-300 (mm)		10	20	30	40	50
CP5/RC5	1.20	29.28	S	0	1	-	1	1	2	1	5	*				
CP5/RC5	2.20	28.28	S	0	2	2	2	2	2	2	8	*				
CP5/RC5	3.20	27.28	S	0	1	2	1	2	2	1	6	*				
CP5/RC5	4.20	26.28	S	0	3	4	4	3	5	4	16		*			
CP5/RC5	5.20	25.28	C	0	8	9	14	12	13	14	53					*
CP5/RC5	6.20	24.28	C	0	9	16/41	50/70				50/70					>
CP5/RC5	7.20	23.28	S	0	8	8	9	9	9	8	35				*	
CP5/RC5	8.00	22.48	S	0	11	14	24	26/71			50/146					>

			Remarks
Driller	John Smith		Equipment checked and calibration carried out in accordance with BS EN ISO 22476-3: 2005
Hammer No.	SD1		
Energy Ratio, Er (%)	60.77		
Calibration Date	01/07/2010		

-/-   Blows/penetration (mm) after seating   S - Standard Penetration Test (SPT)
-*/-   Total blows/penetration (mm)   C - SPT with cone
SWP Penetration under own weight (mm)   L - Split Spoon with liner used

Figure D1 (continued).

outlined in Figure 4.23. Sample types are B (bulk or bag sample of the arisings), E (environmental for contamination testing), W (water sample) and D (small disturbed tub or jar sample, generally taken by a split tube sampler in an SPT test). SPT results (N-values) are given in a separate column (S5 and so on) and more details of the SPT tests are given in a separate sheet. In this example no undisturbed samples were taken. If they had been, they would have been reported as U (open 102 mm tube) or UT (thin-walled open drive sample) together with a record of the number of blows taken to drive the sample (for general information only, as the driving force from the hammer is not usually standardised). The cable percussion driving was continued to a depth of 8.30 m, by which time the investigation had encountered weathered Coal Measures mudstone (described as very stiff slightly sandy CLAY with lithorelicts of the parent rock). An SPT was attempted in the weathered rock but only penetrated 146 mm (rather than 300 mm) for 50 blows of the hammer when the test was terminated, which is standard practice in the UK. In the cable percussion section of hole, water was encountered near the surface but then as casing was installed the hole proceeded in the dry until the base of alluvium at 4.20 m depth. Water rose in the borehole to 3.90 m.

From 8.30 m, the hole was advanced using rotary drilling using water flush with a diameter of 0.12 m. Advance rates are given at the bottom of page 1. The first drilling run was only 0.5 m, but after that 1.5 m runs were adopted up to 18 m when a 2 m run was carried out. Total core recovery (TCR) was not bad, generally over 80%, but from 10.50 to 12.00 m 47% of core was lost (TCR = 53%). Solid Core Recovery (SCR) is defined as percentage of core with full circumference. RQD is percentage of core (on a drilling run basis – NOT rock type) in full sticks of sound rock >100 mm in length. Fracture index (no. per metre) is also recorded. Note that core run lengths are not consistent with geological changes – no reason why they should be unless the driller noted some sudden change in advance rate or loss of flushing fluid perhaps, which might cause him to stop and extract the drill string to investigate the cause. In the contractor's own style of log, he has chosen to split the description into general and detail columns, which is helpful. Water was encountered in the rock at a depth of 11.10 m and rose to 6.40 m. When the hole was complete, a standpipe piezometer was installed – details of the installation are on the log. Also noted is chiselling time – this is useful information for the designer (perhaps if he is considering using driven piles) but is also a record for payment purposes. As a comment, note that there is no attempt to describe weathering state (the strength consequence is clearly recorded so there is no need) – this contrasts with the HK and Australian examples presented later. Initials of the crew who carried out the boring, who logged the materials and prepared the log and who checked the whole report are given.

D.1.2   Hong Kong example

The drillhole record D2 is provided courtesy of Gammon Construction Ltd, HK, for a hole taken to 14.78m depth. Descriptions of soil and rock are to GCO (1988).

As for the UK example, the hole commenced with an inspection pit to 1.50m to check for services. After that, the hole was advanced by rotary drilling using water flush. PX casing with an outside diameter (OD) of 140mm was installed to 3.50m and then HX (OD 114mm) to 6.50m at the top of rockhead, from which depth the hole was uncased. To a depth of 8.50m, an intermittent sampling strategy was adopted, with Mazier samples and SPT tests. To 5.70m, the rock is all described as grade V, completely decomposed granite. At 5.70m there is a 0.8m thick basalt stratum, completely decomposed to very stiff, slightly sandy SILT (also grade V). The basalt was sampled in an SPT test, which gave an N-value of 91 (in the UK the test would normally have been terminated at 50 blows).

From 6.50m, the drilling was continued using a Craelius T2-101 double-tube core barrel with an outside diameter of 101mm and core diameter of about 84mm. The rock recovered is all described as moderately strong or strong, moderately (grade III) or slightly decomposed (grade II) granite. Recovery in the rock is generally good and RQD high. There is some information about discontinuities, their nature, closeness and orientations, but no real detail. If this drillhole was for a slope stability assessment, then more information would be required, especially regarding the dip direction of the joints. If it was for a foundation design, then such information would probably not be necessary. Water levels have been recorded each morning and evening but these would have been affected by the drilling process and water flush so little can be interpreted from these data. A piezometer was installed at a depth of 6.30m (at the basalt horizon). No details are given on this log but probably the piezometer would have been installed in a sand/gravel pocket extending above the basalt into the sandy decomposed granite.

D.2   Consultant's borehole log, Australia

The drillhole record D3 is provided courtesy of Golder Associates (GA)/Sinclair Knight Merz, Brisbane, and with acknowledgements to the Department of Transport and Main Roads, Queensland, for permission. Only the first three of seven sheets are presented. Soil and rocks are described according to Australian Standards (1993). Shorthand descriptors and other terms are given in sheet D3 (terms).

GA are consulting engineers responsible for investigating the site where this borehole has been put down. It is the site of a complex landslide so cores have been logged not only by the GI contractor (not

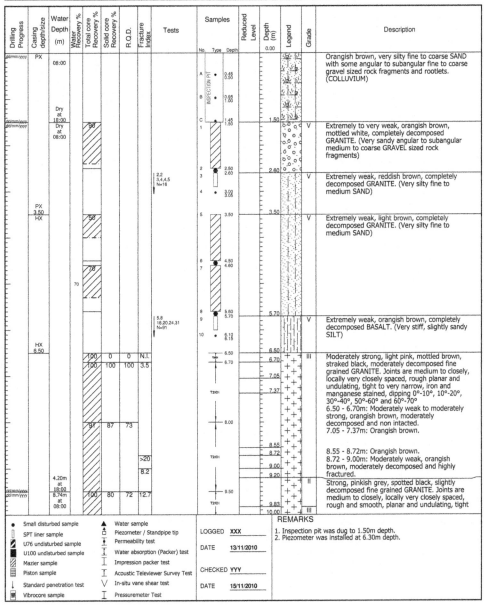

Gammon Construction Limited
Ground Engineering Department
DRILLHOLE RECORD

HOLE No.	ID
SHEET 1 of 2	
CONTRACT NO. 1234	

Project Title

METHOD Rotary	CO-ORDINATES	PROJECT No. 1234
MACHINE & No. 123	E N	DATE from DD/MM/YYYY to DD/MM/YYYY
FLUSHING MEDIUM WATER	ORIENTATION Vertical	GROUND LEVEL mPD

Drilling Progress	Casing depth/size	Water Depth (m)	Water Recovery %	Total core Recovery %	Solid core Recovery %	R.Q.D.	Fracture Index	Tests	Samples No. Type Depth	Reduced Level	Depth (m)	Legend	Grade	Description
dd/mm/yyyy	PX	08:00							A 0.45/0.50		0.00			Orangish brown, very silty fine to coarse SAND with some angular to subangular fine to coarse gravel sized rock fragments and rootlets. (COLLUVIUM)
		Dry at 18:00							B 0.95/1.00					
dd/mm/yyyy		Dry at 08:00	90						C 1.45/1.50 1		1.50		V	Extremely to very weak, orangish brown, mottled white, completely decomposed GRANITE. (Very sandy angular to subangular medium to coarse GRAVEL sized rock fragments)
								2.2 3,4,4,5 N=16	2 2.50/2.60 3 4 3.00/3.05		2.60		V	Extremely weak, reddish brown, completely decomposed GRANITE. (Very silty fine to medium SAND)
	PX 3.50 HX		50						5 3.50		3.50		V	Extremely weak, light brown, completely decomposed GRANITE. (Very silty fine to medium SAND)
			78						6 4.50/4.60 7					
			70					5.8 18,20,24,31 N=91	8 5.60/5.70 9 10 6.10/6.15		5.70		V	Extremely weak, orangish brown, completely decomposed BASALT. (Very stiff, slightly sandy SILT)
	HX 6.50		100	0	0	N.I.			6.50 6.70	6.50 6.70			III	Moderately strong, light pink, mottled brown, straked black, moderately decomposed fine grained GRANITE. Joints are medium to closely, locally very closely spaced, rough planar and undulating, tight to very narrow, iron and manganese stained, dipping 0°-10°, 10°-20°, 30°-40°, 50°-60° and 60°-70°
			100	100	100	3.5			T2101	7.05 7.37				6.50 - 6.70m: Moderately weak to moderately strong, orangish brown, moderately decomposed and non intacted. 7.05 - 7.37m: Orangish brown.
			91	87	73				8.00		8.00			8.55 - 8.72m: Orangish brown. 8.72 - 9.00m: Moderately weak, orangish brown, moderately decomposed and highly fractured.
						>20			T2101	8.55 8.72 9.00				
	4.20m at 18:00					8.2				9.20			II	Strong, pinkish grey, spotted black, slightly decomposed fine grained GRANITE. Joints are medium to closely, locally very closely spaced,
dd/mm/yyyy dd/mm/yyyy	8.74m at 08:00		100	80	72	12.7			T2101 9.50	9.83			III	rough and smooth, planar and undulating, tight
										10.00				

REMARKS

Legend	
• Small disturbed sample	▲ Water sample
SPT liner sample	Piezometer / Standpipe tip
U76 undisturbed sample	Permeability test
U100 undisturbed sample	Water absorption (Packer) test
Mazier sample	Impression packer test
Piston sample	Acoustic Televiewer Survey Test
↓ Standard penetration test	V In-situ vane shear test
Vibrocore sample	Pressuremeter Test

LOGGED **XXX**

DATE **13/11/2010**

CHECKED **YYY**

DATE **15/11/2010**

1. Inspection pit was dug to 1.50m depth.
2. Piezometer was installed at 6.30m depth.

Figure D2 HK drillhole.

Gammon Construction Limited
Ground Engineering Department
DRILLHOLE RECORD

HOLE No.		
ID		
SHEET	2	of	2
CONTRACT NO. 1234		

Project Title

METHOD Rotary	CO-ORDINATES	PROJECT No. 1234
MACHINE & No. 123	E N	DATE from DD/MM/YYYY to DD/MM/YYYY
FLUSHING MEDIUM WATER	ORIENTATION Vertical	GROUND LEVEL mPD

Drilling Progress	Casing depth/size	Water Depth (m)	Water Recovery %	Total core Recovery %	Solid core Recovery %	R.Q.D.	Fracture Index	Tests	Samples			Reduced Level	Depth (m)	Legend	Grade	Description
									No.	Type	Depth		10.00			
			100	80	72					T2101			10.08		II	to extremely narrow, iron and manganese
											10.40		10.30			stained, chlorite coated, dipping 10°-20°,
			97	82	70								10.45		III	20°-30°, 40°-50°, 60°-70° and subvertical.
										T2101					II	9.83 - 10.08m: Moderately strong, orangish
											11.00					brown, moderately decomposed.
			100	86	79											10.30 - 10.45m: Moderately strong, orangish
						3.4										brown, moderately decomposed.
										T2101						
						15.6										
		70	100	100	100	4.5										
		5.13m at 18:00								T2101	12.35					
		9.94m at 08:00	98	98	82						12.87					
26/mm/yyyy 03/mm/yyyy										T2101						
						>20 4.7										
			100	100	96						13.78					
		4.50m at 18:00				11.4				T2101			14.31			14.31 - 14.78m: Black, spotted light grey,
12/11/2010											14.78		14.78			altered GRANITE.
																End of hole at 14.78m depth.
													20.00			

• Small disturbed sample	▲ Water sample
SPT liner sample	Piezometer / Standpipe tip
U76 undisturbed sample	Permeability test
U100 undisturbed sample	Water absorption (Packer) test
Mazier sample	Impression packer test
Piston sample	Acoustic Televiewer Survey Test
Standard penetration test	V In-situ vane shear test
Vibrocore sample	Pressuremeter Test

REMARKS

LOGGED **XXX**

DATE **13/11/2010**

CHECKED **YYY**

DATE **15/11/2010**

Figure D2 (continued).

REPORT OF BOREHOLE: BH380

SHEET: 1 OF 7

CLIENT: SKM	COORDS: 25881.5 m E 139728.85 m N LOCAL
PROJECT: SWTC Package B	SURFACE RL: 97.90 m DATUM: AHD
LOCATION: Cut 4 South, Bench 2	INCLINATION: -90°
JOB NO: 04632111	HOLE DIA: 96/76 mm HOLE DEPTH: 59.60 m

DRILL RIG: FD500
DRILLER: Foundril Pty Ltd
LOGGED: MGM/NAC DATE: 21/4/09
CHECKED: LRC DATE: 3/8/09

Drilling	Field Material Description		Defect Information
METHOD / WATER / TCR % / RQD % / DEPTH (meters) / DEPTH RL / GRAPHIC LOG	ROCK / SOIL MATERIAL DESCRIPTION	WEATHERING / INFERRED STRENGTH Is$_{(50)}$ MPa	DEFECT DESCRIPTION & Additional Observations / FRACTURE FREQUENCY (Defects per unit metre length)

Drilling method: RT (0–6.15), HQ

Water: 0820hrs, 28/04/09

0 — 97.90
Rock roller bit used from 0-4m - No core or samples recovered.

4.00 — 93.90 TCR 100, RQD 0
TRACHYTE
Fine grained, massive, grey to greenish grey, orange staining along joints
Weathering: SW

Defects:
4.00-5.37m: J Set, 50-60°, typical spacing 40-50mm, Pl-Un, Sm, orange sandy silt to 2mm thick, closed with iron oxide staining
4.40m: J, 80°, Pl, Sm, In, orange silty sand

5.00m: J, 80-85°, Pl, Sm, In, orange silty sand

TCR 100, RQD 22

5.47m: J, vertical, St, Ro, Sn, FeO
5.61m: J, vertical, St, Ro, Sn, FeO
5.61-5.82m: J, 90°, St, Ro, FeO
5.86m: J, 45°, Un, Sm, Sn, FeO

6.15 — 91.75 TCR 100, RQD 0
. pronounced increase in number of joints, increased weathering and infill along joints
Weathering: DW-SW

6.15-7.35m: J Set, 50mm spacing, Pl, Sm, some with FeO stained veneer, others open with up to 40mm infill of silty sand with gravel, all FeO stained

7.35
7.50 — CORE LOSS
7.50 — 90.40 TCR 82, RQD 0
TRACHYTE
Fine grained, massive, grey to greenish grey, orange staining along joints
Weathering: DW-SW

7.50m: CZ

8.20
89.70 — 8.42 — CORE LOSS
89.48 — 8.72 TCR 58, RQD 0
TRACHYTE
Fine grained, massive, grey to greenish grey, orange staining along joints
Weathering: DW-SW

8.42-12.20m: Core broken, recovered mostly angular gravel sized trachyte fragments

89.10 — CORE LOSS
TRACHYTE
Fine grained, massive, grey to greenish grey, orange staining along joints
Weathering: MW-SW
TCR 92, RQD 0

8.80-9.80m: Crosscutting J Set, 45-80°, Un, Ro, Sn, FeO, In, clay, Silty Sand in filling

9.80
88.10 — CORE LOSS
10

This report of borehole must be read in conjunction with accompanying notes and abbreviations. It has been prepared for geotechnical purposes only, without attempt to assess possible contamination. Any references to potential contamination are for information only and do not necessarily indicate the presence or absence of soil or groundwater contamination.

GAP gINT FN. F02b
RL2

Figure D3 (1) BH380 Borehole report.

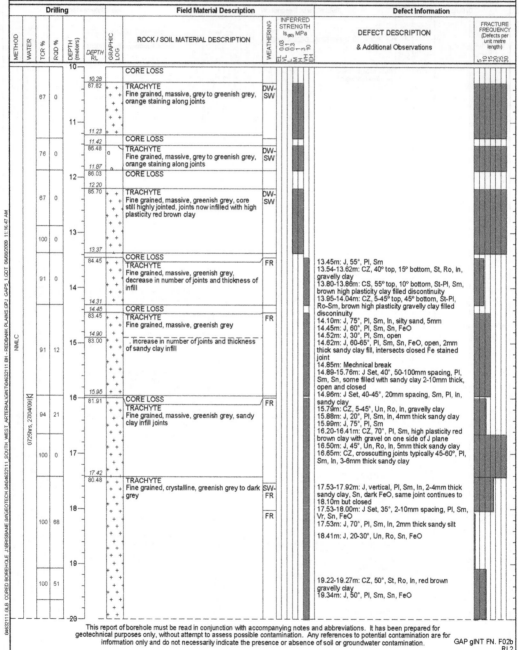

Figure D3 (continued).

CLIENT:	SKM	COORDS: 25881.5 m E 139728.85 m N LOCAL	DRILL RIG: FD500
PROJECT:	SWTC Package B	SURFACE RL: 97.90 m DATUM: AHD	DRILLER: Foundril Pty Ltd
LOCATION:	Cut 4 South, Bench 2	INCLINATION: -90°	LOGGED: MGM/NAC DATE: 21/4/09
JOB NO:	04632111	HOLE DIA: 96/76 mm HOLE DEPTH: 59.60 m	CHECKED: LRC DATE: 3/8/09

Drilling				Field Material Description			Defect Information		
METHOD	WATER	TCR % / RQD %	DEPTH (meters) / DEPTH RL	GRAPHIC LOG	ROCK / SOIL MATERIAL DESCRIPTION	WEATHERING	INFERRED STRENGTH Is(50) MPa	DEFECT DESCRIPTION & Additional Observations	FRACTURE FREQUENCY (Defects per unit metre length)

TRACHYTE
Fine grained, crystalline, greenish grey to dark grey — FR
- 100 / 79

TRACHYTE BRECCIA (21.85 / 76.05)
Fine grained, grey to greenish grey, locally orange along joints, angular fine to coarse gravel sized trachyte in fine matrix — HW
- 100 / 33 (22.25 / 75.65)

SANDSTONE
Fine to medium grained, massive to locally bedded, greyish brown to locally orange, localised fine sand and soft clay filled joints, .04m thick interbed of baked sandstone/trachyte mix at 22.25-22.29m, black, comprised of 2-4mm angular sandstone fragments
- 91 / 20

Carbonaceous CLAYSTONE (24.15 / 24.25 / 73.65)
Fine grained, black, coarse sand to fine gravel sized claystone fragments in medium plasticity sandy clay matrix
CORE LOSS
- 20 / 0 (25.05)

BASALT (25.25 / 72.59)
Fine to medium grained, porphyritic, grey, locally brecciated — HW / HW
CORE LOSS
BASALT
Fine to medium grained, porphyritic, grey, locally brown
- 96 / 20 (26.32 / 71.58)
. localised hydrothermal alteration (26.75 / 26.85 / 71.05)

CORE LOSS
BASALT BRECCIA — HW
Fine to coarse grained, angular basalt porphyry fragments in silty gravelly sand matrix, dark grey to greenish grey
- 88 / 0 (27.56 / 70.34)
CORE LOSS (27.84 / 70.06)

BASALT BRECCIA — HW
Gravel fragments of basalt breccia and sandy matrix
- 64 / 0 (28.24 / 69.66)
CORE LOSS (28.71 / 69.19)

BASALT BRECCIA — HW
Gravel fragments of basalt breccia and sandy matrix, becoming vesicular, purple (29.02 / 68.88) — RS-EW
CLAYSTONE
Fine grained, green to dark green, crushed angular, gravel to trace cobble sized claystone fragments in clay matrix, possible shear zone
- 68 / 0 (30.00)

Defect Description:
- 19.98m: J, 55°, Pl, Sm, Sn, FeO
- 20.17-20.59m: J Set, 70°, 200mm spacing, Pl, Sm, Sn, FeO, closed
- 20.70m: J, 30°, Un, Ro, Sn, FeO
- 20.70m: J, 55°, Pl, Sm, Sn, FeO
- 20.76-20.90m: J Set, 15-20°, 10-70mm spacing, Pl, Sm, Vr to locally 3mm sandy clay filled, all Fe stained
- 20.90m: J, 35°, Un, Ro, In, orange brown sandy silt
- 21.16m: J, 35°, Un, Ro, In, orange brown sandy silt
- 21.18-21.29m: J Set, Un, Sm, In, orange brown sandy silt
- 21.29m: IS, 80°, Pl, Sm, In, 3-7mm orange brown sandy silt
- 21.78m: J, 80°, Pl, Sm, In, 1-2mm orange brown sandy silt
- 22.25-22.29m: C, 5°, Un, Ro, Black, baked contact along base of trachyte
- 22.29m: C, 5°, Un, Ro
- 22.33m: DS, 20°, Pl, Sm

- 26.63m: J, 40°, Pl, Sm

- 26.95m: CZ, 50° top, Pl, Sm, 2mm thick green clay along contact with brecciated zone below

- 29.00-29.58m: SZ, 80° bottom, dark green hydrothermal altered at bottom, crushed claystone throughout

This report of borehole must be read in conjunction with accompanying notes and abbreviations. It has been prepared for geotechnical purposes only, without attempt to assess possible contamination. Any references to potential contamination are for information only and do not necessarily indicate the presence or absence of soil or groundwater contamination.

GAP gINT FN. F02b RL2

Figure D3 (continued).

	TERMS FOR ROCK MATERIAL STRENGTH & WEATHERING AND ABBREVIATIONS FOR DEFECT DESCRIPTIONS
Golder Associates	

STRENGTH

Symbol	Term	Point Load Index, Is$_{(50)}$ (MPa)	Field Guide
EL	Extremely Low	< 0.03	Easily remoulded by hand to a material with soil properties.
VL	Very Low	0.03 to 0.1	Material crumbles under firm blows with sharp end of pick; can be peeled with knife; too hard to cut a triaxial sample by hand. Pieces up to 30 mm can be broken by finger pressure.
L	Low	0.1 to 0.3	Easily scored with a knife; indentations 1 mm to 3 mm show in the specimen with firm blows of pick point; has dull sound under hammer. A piece of core 150 mm long by 50 mm diameter may be broken by hand. Sharp edges of core may be friable and break during handling.
M	Medium	0.3 to 1	Readily scored with a knife; a piece of core 150 mm long by 50 mm diameter can be broken by hand with difficulty.
H	High	1 to 3	A piece of core 150 mm long by 50 mm diameter cannot be broken by hand but can be broken with pick with a single firm blow; rock rings under hammer.
VH	Very High	3 to 10	Hand specimen breaks with pick after more than one blow; rock rings under hammer.
EH	Extremely High	>10	Specimen requires many blows with geological pick to break through intact material; rock rings under hammer.

ROCK STRENGTH TEST RESULTS

▼   Point Load Strength Index, I$_s$(50), Axial test (MPa)

◀   Point Load Strength Index, I$_s$(50), Diametral test (MPa)

Relationship between I$_s$(50) and UCS (unconfined compressive strength) will vary with rock type and strength, and should be determined on a site-specific basis. UCS is typically 10 to 30 x I$_s$(50), but can be as low as 5.

ROCK MATERIAL WEATHERING

Symbol		Term	Field Guide
RS		Residual Soil	Soil developed on extremely weathered rock; the mass structure and substance fabric are no longer evident; there is a large change in volume but the soil has not been significantly transported.
EW		Extremely Weathered	Rock is weathered to such an extent that it has soil properties - i.e. it either disintegrates or can be remoulded, in water.
DW	HW	Distinctly Weathered	Rock strength usually changed by weathering. The rock may be highly discoloured, usually by iron staining. Porosity may be increased by leaching, or may be decreased due to deposition of weathering products in pores. In some environments it is convenient to subdivide into Highly Weathered and Moderately Weathered, with the degree of alteration typically less for MW.
	MW		
SW		Slightly Weathered	Rock is slightly discoloured but shows little or no change of strength relative to fresh rock.
FR		Fresh	Rock shows no sign of decomposition or staining.

ABBREVIATIONS FOR DEFECT TYPES AND DESCRIPTIONS

Defect Type		Coating or Infilling		Roughness	
B	Bedding parting	Cn	Clean	Sl	Slickensided
X	Foliation	Sn	Stain	Sm	Smooth
C	Contact	Vr	Veneer	Ro	Rough
L	Cleavage	Ct	Coating or Infill		
J	Joint	**Planarity**			
SS/SZ	Sheared seam/zone (Fault)	Pl	Planar	**Vertical Boreholes** – The dip (inclination from horizontal) of the defect is given.	
CS/CZ	Crushed seam/zone (Fault)	Un	Undulating		
DS/DZ	Decomposed seam/zone	St	Stepped		
IS/IZ	Infilled seam/zone			**Inclined Boreholes** – The inclination is measured as the acute angle to the core axis.	
S	Schistocity				
V	Vein				

GAP Form No. 7
RL6

Figure D3 (continued).

presented here) but also by GA. In the original report, the logs are accompanied by high-quality photographs. The borehole diameter is 96 mm and core diameter 76 mm. The log only provides information on the rock recovered; for information on how the borehole was

Project	SAMPLE PROJECT	Engineer	GEOTECHNICS LIMITED	Trial Pit	TP8
				Project No	PC060000

Client   GEOTECHNICS LIMITED   Ground Level   31.10   m OD

Samples and Tests				Strata			Scale 1:50	
Depth	Type	Stratum No	Results	Description	Depth	Legend	Level m OD	

Samples and Tests:

Depth	Type	Stratum No	Results
0.50	D		mc=17%
1.00	B		mc=38%
1.50	B		
2.00	D		

Strata Description:

Soft dark brown slightly sandy silty clay with frequent roots and rootlets.
[TOPSOIL]

Soft brown slightly gravelly sandy CLAY with frequent rootlets and a low cobble content of sandstone and quartzite. Gravel is subrounded to rounded fine to medium quartzite.
[ALLUVIUM]

Brown slightly clayey sandy GRAVEL with a high cobble and boulder content of sandstone and quartzite. Gravel is angular to subrounded fine to coarse shale, mudstone, sandstone and quartzite.
[GLACIAL DEPOSITS]

End of Excavation

Depth / Legend / Level m OD:

Depth	Level m OD
G.L.	31.10
0.20	30.90
1.30	29.80
2.30	28.80

Excavation

Plant	JCB 3CX	Width (B)	0.80
Date	20/01/2010	Length (C)	2.20
Shoring	None.		

Date Backfilled   20/01/2010

Stability   Unstable during excavation.

Groundwater

Depth Observed	Depth of Pit	Details
		None encountered during excavation.

Remarks   All sides of pit collapsing below 1.80m, unable to excavated below 2.30m depth.

Symbols and abbreviations are explained on the accompanying key sheet.

All dimensions are in metres.   Logged in accordance with BS5930:1999

Logged by	DF
Checked by	DRB
Figure	1 of 1
	17/12/2010

geotechnics

Figure D4   Trial pit UK.

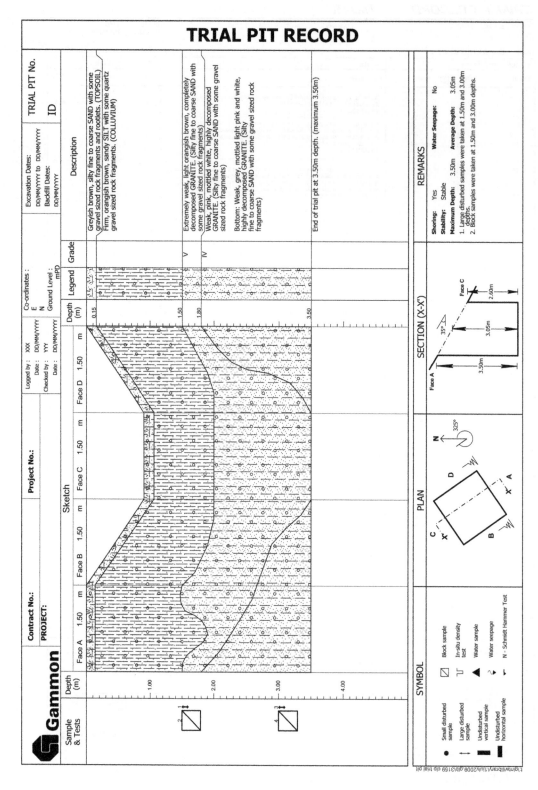

Figure D5 HK trial pit.

conducted, how long it took, casing used, information on flush and groundwater encountered, etc. one would need to examine the GI contractor's log for the same hole. The factor that distinguishes this log from the HK and UK examples is the attention to detail in describing each discontinuity, with information on roughness, infill and mineral coatings, as applicable. That said, this log is really only a preliminary step. After considering other details regarding the site, not least results from instruments monitoring ground movements and water pressures, then sections of core where movement is suspected were examined and logged in even greater detail, with samples taken for strength testing, microscopic examination and chemical analysis.

D.3   Contractor's trial pit logs

Examples from the UK and from Hong Kong are presented in D4 and D5 respectively. Descriptions and classification used reflect local standards and codes. The UK example is essentially the same as a borehole log (one-dimensional). The HK example is more graphic and shows all four sides of the pit. As is common practice in HK, large block samples were taken at depths of 1.5m and 3.0m. These are cut by hand, covered in foil, waxed and placed in a wooden box for transportation to the laboratory, where they are opened, described and prepared for testing, as illustrated in Figure 4.12.

Appendix E: Tunnelling risk

Appendix E-1   Example of tunnelling risk assessment at project option stage for Young Dong Mountain Loop Tunnel, South Korea

The Young Dong Railroad Relocation Project for the Korean National Railways (KNR) included a single-track railway tunnel, in rock, approximately 16.3 km long with a span of approximately 8 m. As at 2011 it is the longest tunnel in Korea. The tunnel had to be constructed as a large radius loop to limit the gradient of the track, as illustrated in Figure AE-1.1. The maximum depth of the tunnel is approximately 400 m, with most of the alignment being at depths in excess of 100 m.

The route for the tunnel was identified as having intrinsic hazards including:

- potentially high water pressures, up to 40 bars (4 MPa) pressure
- fault zones, possibly associated with significant groundwater inflows
- highly sheared and closely jointed rocks
- some rocks with high strength and abrasivity
- possible cavernous limestone with groundwater
- old mine workings (coal).

Halcrow, with Nick Swannell as Project Manager, was commissioned to advise a consortium of contractors tendering for the construction of the tunnel and as part of the brief carried out a thorough hazard and risk assessment of the route for both TBM and drill and blast options. Several state-of-the art reports by Dr Graham Garrard of Halcrow were based on existing ground investigations along the route, mapping and further GI together with an in-depth review of tunnel case histories in similar terrain throughout the world by Dr. Laurie Richards.

A risk assessment method was developed to make a quantitative and objective assessment of the construction methods of the tunnel. The risks associated with tunnel excavation are dependent on the hazards encountered and were defined for this project with respect to

Figure AE-1.1 The Young Dong (now Solan) Mountain Loop Tunnel Project, South Korea, after Daewoo Corporation. Figure modified from Kim *et al.* (2001).

programme (rather than other issues such as safety or cost as might have been done). The likelihood of a hazard occurring was assigned one of three levels and consequence of each hazard assumed to be at one of five levels, as set out in Table AE-1.1 below.

The level of risk for each hazard can be determined by finding its likelihood of occurrence and considering its consequence. The level of risk associated with the hazard is then established conventionally as follows:

Level of Risk = Likelihood × Consequence

Once the level of risk has been ascertained, it can be compared with Table AE-1.2 below to identify the action that should be taken to mitigate the risk.

Having made an assessment of the risk associated with each hazard, appropriate mitigation measures are considered. The residual risk remaining after mitigation is then assessed in the same way to determine acceptability or otherwise.

Table AE-1.1

LIKELIHOOD			CONSEQUENCE		
TITLE	**DESCRIPTION**	**SCALE**	**TITLE**	**DESCRIPTION**	**SCALE**
Probable	Likely to occur during the construction of the tunnel, possibly on more than one occasion	3	Catastrophic	Total loss of a section of tunnel	5
			Critical	Major damage or delay to tunnel or major environmental impact affecting programme	4
Occasional	Likely to occur at least once during construction of the tunnel	2	Serious	Some damage or delay to tunnel or some environmental impact affecting programme	3
			Marginal	A routine maintenance repair to tunnel or minor hindrance	2
Remote	Unlikely to occur during construction of the tunnel	1	Negligible	Of little consequence to programme	1

Risk assessment

The assessment of risks associated with the use of a shielded TBM was carried out separately from that for drill and blast excavation.

Table AE-1.3 provides a brief summary of the hazards identified for a drill and blast option with level of probability of occurrence and likely consequence and hence risk. Possible ways of mitigating the risk are identified and the residual likely risk identified. The original reports go into far more detail of the nature of hazards, ways to mitigate them and the potential ways those mitigation measures might prove unsuccessful leaving residual risks. An example of a summary hazard and risk register for this project (TBM option) is given at Appendix E-2.

Conclusions

It was found that the number of hazards and residual risks associated with a shielded TBM construction would be greater than for the drill and blast method. The principal reasons were:

- the relative inflexibility of mechanised excavation and lining systems to deal with conditions for which they may not have been specifically designed.

– the dependence of the tunnel progress entirely on the performance and reliability of a single item of mechanical plant, which would require a high level of technological input for its successful operation and maintenance.

Further details are given by Kim *et al.* (2001).

Table AE-1.2

Consequence / Likelihood	Catastrophic	Critical	Serious	Marginal	Negligible
Probable	15	12	9	6	3
Occasional	10	8	6	4	2
Remote	5	4	3	2	1

Score	
10–15	Very high risk – not acceptable for tunnel construction – need to apply mitigation measures to eliminate or reduce risk
6–9	High risk – apply mitigation measures to eliminate or reduce risk. Residual risk at this level indicates need for active management control and response plans to be well developed with well trained personnel, materials and plant readily available
1–5	Low risk – may be accepted if mitigating measures are in place under active management control

Table AE-1.3 Programme risk assessment for excavation by drill and blast.

(L = likelihood, C = consequence, R = L × C = risk). Residual risk is likely outcome *after* application of mitigation measures.

NO	HAZARD	RISK	RISK LEVEL			MITIGATION MEASURES	RESIDUAL RISK LEVEL		
			L	C	R		L	C	R
1	Highly jointed rock mass (possibly in association with high-pressure water). See Hazard 3 for water ingress specifically.	Ravelling ground, roof falls and sidewall and/or face instability with high amount of primary support.	3	4	12	1. Reduce length of excavation advance; face support and/or buttressing and/or partial face advance. 2. Reduce powder factor to lessen blast damage. 3. Increase rock support and install rock support in the form of rock bolts and steel fibre reinforced shotcrete without delay. 4. Probing and pre-injection.	2	2	4
2	Fault zones	Soft ground or mixed face conditions with potential roof falls and sidewalls instability requiring a high degree of primary support.	3	4	12	1. Reduce length of excavation advance; face support and/or buttressing and/or partial face advance. 2. Reduce powder factor. 3. Increase rock support and install rock bolts, steel fibre reinforced shotcrete, lattice girders and spilling bars without delay. 4. Provision of probe drilling to identify these features ahead of the excavation face. 5. Provision of Tunnel Seismic Prediction (TSP) to identify fault zones ahead of the excavation face. 6. Provision of instrumentation to monitor movement to optimise support.	3	2	6

Table AE-1.3 (continued) Programme risk assessment for excavation by drill and blast.

NO	HAZARD	RISK	RISK LEVEL			MITIGATION MEASURES	RESIDUAL RISK LEVEL		
			L	*C*	*R*		*L*	*C*	*R*
3	Water ingress, possibly under high pressure up to 40 bar (4 MPa)	Water in cavities, joints and fissures in the rock mass entering excavation and causing instability of ground. Difficulties with shotcrete application.	3	4	12	1. Tunnel drive to be up-grade to allow water to drain. (Not possible with all drives.) 2. Provision of pumps to cope with high flows and backup systems to deal with pumps and power failures. 3. Provision of probe drilling to identify areas of high water flows and to carry out pre-injection grouting to stem the flow. 4. Excavation equipment systems to be rated to IP68 or equivalent. 5. Use drainage channels to control inflows prior to shotcreting.	2	2	4
4	Cavities in the rock mass (including mine workings) possibly associated with water inflow	Instability of tunnel face, roof fall and side wall instability. Flooding. Need for major structural work or infilling.	3	4	12	1. Provision of TSP to identify cavities in advance of excavation. 2. Provision of probe drilling to determine extent of cavities and provide means for grouting or other advance stabilisation measures. 3. Reduce length of excavation advance.	2	3	6

continued overleaf

Table AE-1.3 (continued) Programme risk assessment for excavation by drill and blast.

NO	HAZARD	RISK	RISK LEVEL			MITIGATION MEASURES	RESIDUAL RISK LEVEL		
			L	C	R		L	C	R
5	Tunnel atmosphere and ventilation including accumulation of explosive and noxious gases	Explosion risk. Possible accumulation of explosive and or noxious gas. Methane, associated with coal or other sources is a flammable gas, lighter than air and can give rise to explosion. In large quantities it can also cause asphyxiation. Other gases such as carbon dioxide, carbon monoxide, sulphur dioxide and hydrogen ulphide are noxious.	3	5	15	1. Provision of adequate fresh air from the portal to the excavation face. 2. Provision of adequate and suitable atmospheric monitoring system. 3. Avoid the use of dry shotcrete mix. 4. Use explosive appropriate to tunnels prone to fire risk. 5. Standby generators to power fans.	1	4	4
6	Mechanical breakdown	Failure of key item of plant.	3	3	9	1. Planned maintenance strategy. 2. Maintain spare plant items. 3. Maintain stocks of spares.	3	1	3
7	Use of explosives	Premature detonation or uncontrolled explosion.	2	5	10	1. Employ qualified staff. 2. Comply with safety regulations. 3. Use proper storage and transport facilities. 4. Use non-electric detonators.	1	5	5

Appendix E-2 Example of hazard and risk prediction table

Example table of predicted hazards at option/design stage for TBM with potential intersection of old mine workings, karst, collapsed and weak zones; this is part of a more detailed report prepared for a specific project and was supported by similar tables for drill and blast options, detailed text, recommended methods of working, international case studies and discussion.

Number	Hazard			Possible mitigation measures prior to and during construction to reduce likelihood or consequence of hazard. Notes * indicates technique more likely than others to be useful	Impact of mitigation measures and risk			Summary of main residual risks
	Title	Example of circumstances	Consequence of event on tunnel		Likelihood (L)	Consequence (C)	Risk (L × C)	
1	Inrush of water or unconsolidated ground	cavity, fault, area of collapsed ground, shaft, adit	H	1. GI[1] boreholes and packer tests 2. pre-construction geological desk study and risk assessment* 3. surface grouting ahead of tunnel 4. probe drilling ahead of the face through TBM[2] head. Monitoring of water volume and pressure* 5. inspection of probe drill cuttings for weak or unconsolidated material, monitoring of strata colour and odour for unanticipated changes* 6. TSP[3] to detect likely geological structures* 7. monitoring of water make from face and significant changes in volume or pressure* 8. ground probing radar to detect cavities, clay or high density of fractures 9. geological mapping to detect changes in structure, strata distress	L	H	H	1. failure to identify inrush conditions 2. blockage of probe holes with debris, or unconsolidated material resulting in underestimation of water pressure and volume 3. failure of grouting programme 4. failure of EPBM[4] bulkhead 5. injury to workforce 6. partial or complete loss of tunnel drive

continued overleaf

1. GI = Ground investigation
2. TBM = Tunnel boring machine
3. TSP = Tunnel Seismic Profiling or Prediction
4. EPBM = Earth pressure balance tunnelling machine

(continued)

| Number | Hazard | | Consequence of event on tunnel | Possible mitigation measures prior to and during construction to reduce likelihood or consequence of hazard. Notes * indicates technique more likely than others to be useful | Impact of mitigation measures and risk | | | Summary of main residual risks |
	Title	Example of circumstances			Likelihood(L)	Consequence(C)	Risk (L x C)	
				10. grouting ahead of tunnel face* 11. consider also likelihood of hazards: 2, 3, 5, 6, 7, 8, 9, 10, 11, 16 and 17				
2	Inrush of flammable gas	as 1 above	M	1. GI boreholes 2. pre-construction geological desk study and risk assessment* 3. probe drilling ahead of and around the face through TBM head. Monitoring of character of flush return* 4. monitoring of tunnel atmosphere (gas monitoring at top of face) to detect significant changes in character or composition* 5. inspection of probe drill cuttings to identify weak or unconsolidated material, monitoring of strata colour and odour for unanticipated changes* 6. TSP to detect likely geological structures* 7. monitoring of tunnel atmosphere for significant changes in character or composition* 8. GPR to detect cavities, clay or high density of fractures 9. geological mapping to detect changes in structure, strata distress	L	L	L	1. failure to identify presence of gas 2. explosion during probe drilling because of flush failure at the drill bit 3. escape of gas and explosion in the driveage 4. injury to workforce 5. temporary evacuation of drive during remedial works or 6. temporary loss of drive

(continued)

Number	Hazard – Title	Example of circumstances	Consequence of event on tunnel	Possible mitigation measures prior to and during construction to reduce likelihood or consequence of hazard. Notes * indicates technique more likely than others to be useful	Impact of mitigation measures and risk – Likelihood(L)	Consequence(C)	Risk (L x C)	Summary of main residual risks
				10. provision of adequate ventilation (and if necessary methane drainage system)* 11. adoption of flammable gas working practices (no smoking policy, safety light policies, intrinsically safe plant and equipment 12. consider also likelihood of hazards: 1, 3, 4, 5, 7, 8, 9, 10, 11, 12 and 17				
3	Noxious gases (eg CO_2, SO_2, H_2S	as 1 above	M	1. GI boreholes 2. pre-construction geological desk study and risk assessment* 3. geological mapping of face cuttings to detect changes in structure and strata, presence of organic material, sulphides, etc* 4. monitoring of tunnel atmosphere (gas monitoring at top and base of face) to detect significant changes in character or composition* 5. probe drilling ahead of and around the face through TBM head. Monitoring of character of flush return 6. provision of adequate ventilation* 7. consider also presence of hazard 2	L	L	L	1. failure to identify gases 2. injury to workforce 3. temporary evacuation of drive during remedial works

continued overleaf

Number	Hazard		Consequence of event on tunnel	Possible mitigation measures prior to and during construction to reduce likelihood or consequence of hazard Notes * indicates technique more likely than others to be useful	Impact of mitigation measures and risk			Summary of main residual risks
	Title	Example of circumstances			Likelihood(L)	Consequence(C)	Risk (L x C)	
4	Radon	as 1 above	L	1. pre-construction geological desk study and risk assessment* 2. monitoring of tunnel atmosphere for significant changes in character or composition 3. provision of adequate ventilation* 4. consider also hazard 3	L	L	L	1. failure to identify radon at elevated levels 2. long term affects on workforce health 3. temporary evacuation of drive during remedial works
5	Excessive volume of groundwater	as 1 above plus permeable strata and fractured ground associated with 1 above	H	1. hydrological testing in GI boreholes* 2. pre-construction geological desk study and risk assessment, tunnel depth* 3. probe drilling ahead of face through control valve grouted into rock ahead of TBM cutters and monitoring of water volume and pressure* 4. packer testing to measure permeability ahead of face* 5. TSP to detect likely geological structures* 6. monitoring of water make from face and significant changes in volume or pressure* 7. ground probing radar to detect cavities, clay or high density of fractures	M	H	H	1. failure of grouting programme ground freezing 2. underestimation of inflow 3. failure of EPBM bulkhead 4. failure of pumps 5. temporary evacuation of drive or 6. temporary loss of drive

(continued)

Number	Hazard Title	Example of circumstances	Consequence of event on tunnel	Possible mitigation measures prior to and during construction to reduce likelihood or consequence of hazard Notes * indicates technique more likely than others to be useful	Impact of mitigation measures and risk — Likelihood(L)	Impact of mitigation measures and risk — Consequence(C)	Impact of mitigation measures and risk — Risk (L x C)	Summary of main residual risks
				8. geological mapping to anticipate permeable or fractured structure, strata distress etc 9. grouting ahead of face* 10. advanced dewatering ahead of tunnel from the surface 11. ground freezing ahead of face 12. consider also hazard 1				
6	Excessive groundwater pressure	as 5 above	M	1. hydrological testing in GI boreholes 2. pre-construction geological desk study and risk assessment, tunnel depth* 3. probe drilling ahead of face through control valve grouted into rock ahead of TBM cutters and monitoring of water volume and pressure* 4. monitoring of water make from face and significant changes in volume or pressure* 5. TSP to detect likely geological structures 6. ground probing radar to detect cavities, clay or high density of fractures	M	H	H	1. failure of grouting programme ground freezing 2. underestimation of groundwater pressure 3. failure of EPBM bulkhead 4. failure of pumps 5. temporary evacuation of drive or 6. temporary or permanent loss of drive

continued overleaf

(continued)

Number	Hazard: Title	Example of circumstances	Consequence of event on tunnel	Possible mitigation measures prior to and during construction to reduce likelihood or consequence of hazard — Notes * indicates technique more likely than others to be useful	Likelihood(L)	Consequence(C)	Risk (L x C)	Summary of main residual risks
				7. geological mapping to anticipate permeable or fractured structure, strata distress 8. grouting ahead of face* 9. ground freezing ahead of face 10. advanced dewatering ahead of tunnel from the surface 11. use permeable liner to reduce pressure on liner 12. consider also hazard 1 and 2				
7	Collapsed ground	karst areas (boulder chokes), shafts, areas of collapsed mine workings	M	1. GI boreholes 2. pre-construction geological desk study and risk assessment 3. probe drilling ahead of and around face through TBM head. Monitoring of rates of drilling and character of drill cuttings (presence of clay, infill materials etc)* 4. monitoring of water make from face 5. TSP to detect change in reflection characteristic of strata 6. geological mapping to identify disturbed or fractured structure and strata distress	L	M	M	1. failure of grouting programem or strata control foams 2. stability problems and damage to cutter head 3. higher than anticipated loading on tunnel linings leading to premature failure 4. higher post construction maintenance 5. temporary or permanent loss of drive

(continued)

Number	Hazard Title	Example of circumstances	Consequence of event on tunnel	Possible mitigation measures prior to and during construction to reduce likelihood or consequence of hazard Notes * indicates technique more likely than others to be useful	Impact of mitigation measures and risk Likelihood(L)	Consequence(C)	Risk (L x C)	Summary of main residual risks
				7. monitoring relative volume of spoil produced and driveage distance to identify caving ground 8. use of strata control (eg bentonite) foams ahead of face 9. grouting ahead of face 10. grouting behind liner to fill cavities 11. consider also hazard 1, 2, 8, 9, 10, 11, 12, 13, 14, 15, 16, 17				
8	Cavity << tunnel diameter	karst, old mine workings, possibly faults	M	1. GI boreholes 2. pre-construction geological desk study and risk assessment 3. probe drilling ahead of and around face through TBM head. Monitoring of rates of drilling and character of drill cuttings (presence of clay, infill materials etc)* 4. TSP to detect air or water filled cavity (unlikely to detect small cavities) 5. geological mapping to identify disturbed or fractured structure and strata distress 6. consider also hazard 1, 2, 9, 10, 11, 12, 13, 14, 15, 16, 17	L	L	L	1. failure of grouting progmeram 2. difficulty in maintaining tunnel alignment 3. higher risk of water or gas inrush (see above)

continued overleaf

(continued)

Number	Hazard			Possible mitigation measures prior to and during construction to reduce likelihood or consequence of hazard Notes * indicates technique more likely than others to be useful	Impact of mitigation measures and risk			Summary of main residual risks
	Title	Example of circumstances	Consequence of event on tunnel		Likelihood(L)	Consequence(C)	Risk (L x C)	
9	Cavity ~ tunnel diameter	karst, old partial extraction mine workings	M	1. GI boreholes 2. pre-construction geological desk study and risk assessment 3. probe drilling ahead of and around face through TBM head. Monitoring of rates of drilling and character of drill cuttings (presence of clay, infill materials etc)* 4. TSP to detect cavity 5. geological mapping to identify disturbed or fractured structure and strata distress 6. grouting ahead of face to fill cavities to create plug to improve driving conditions and help minimise TBM deviation* 7. grouting behind lining to fill cavities* 8. consider also hazard 1, 2, 8, 11, 12, 13, 14, 15, 16, 17	L	L	L	1. failure to detect cavity 2. failure of grouting programme 3. difficulty in maintaining tunnel alignment 4. higher risk of water or gas inrush (see above)

Number	Hazard		Consequence of event on tunnel	Possible mitigation measures prior to and during construction to reduce likelihood or consequence of hazard Notes * indicates technique more likely than others to be useful	Impact of mitigation measures and risk			Summary of main residual risks
	Title	Example of circumstances			Likelihood(L)	Consequence(C)	Risk (L x C)	
10	Cavity >> tunnel diameter	karst, old partial extraction mine workings	H	1. GI boreholes 2. pre-construction geological desk study and risk assessment 3. probe drilling ahead of and around through TBM head. Monitoring of rates of drilling and character of drill cuttings (presence of clay, infill materials, etc)* 4. TSP to detect cavity 5. geological mapping to identify disturbed, fractured or distressed strata 6. grout cavities ahead of tunnel face (may be very difficult)* 7. drain cavity, take TBM off line and construct dummy tunnel rings on tunnel line, grout remaining cavities around lining then advance through dummy tunnel. Note: this option may not be possible due to water make* 8. consider also hazard 2, 3, 4, 8, 9, 10, 11, 12, 14, 15, 17 and particularly hazards 1 and 16 during construction	L	H	H	1. failure to detect cavity systems 2. failure of grouting programme 3. unable to drain cavities 4. significant delay to programme 5. significantly increased risk if water and gas inrush and settlement (hazards 1, 2, 3 and 16)

continued overleaf

(continued)

| Number | Hazard | | Consequence of event on tunnel | Possible mitigation measures prior to and during construction to reduce likelihood or consequence of hazard Notes * indicates technique more likely than others to be useful | Impact of mitigation measures and risk | | | Summary of main residual risks |
	Title	Example of circumstances			Likelihood(L)	Consequence(C)	Risk (L x C)	
11	Variable face conditions	karst boulder beds and cave/mine working infill, fault zones and highly folded/ disturbed areas	M	1. GI boreholes 2. pre-construction geological desk study and risk assessment 3. probe drilling ahead of face through TBM head and monitoring rates of drilling and character of drill cuttings* 4. geological mapping to detect changes in structure* 5. TSP to detect likely geological structures which could be associated with variable face conditions 6. ground probing radar to detect clay or high densities of fractures 7. use TBM face with ability to deal with small amounts of soft material (tunnel alignment may be compromised*) 8. consider also likelihood of hazards: 2, 3, 5, 6, 7, 8, 9, 10, 11, 16 and 17	M	L	M	1. failure to anticipate conditions 2. damage to cutters 3. stability problems and damage to cutter head 4. delays to programme 5. increased risk of water or gas inrush

Number	Hazard		Consequence of event on tunnel	Possible mitigation measures prior to and during construction to reduce likelihood or consequence of hazard. Notes * indicates technique more likely than others to be useful	Impact of mitigation measures and risk			Summary of main residual risks
	Title	Example of circumstances			Likelihood(L)	Consequence(C)	Risk (L x C)	
12	Metal obstructions	old mine workings, surface GI boreholes (lost drill strings) and lost bits from advance drilling	H	1. pre-construction geological desk study and risk assessment (presence of old mine workings) 2. maintenance of good GI drilling records and records of advanced drilling programme* 3. monitoring of face production rates* 4. stop machine draw back and use cutting or explosives to remove obstacles* 5. Consider also likelihood of 1, particularly 2, 7, 8, 9, 10, 11, 13, 14, 15, 16, 17	M	H	H	1. failure to identify possibility of metal in face 2. stability problems and damage to cutter head 3. unable to cut or remove metal from ahead of TBM due to face instability or water make 4. delays to programme 5. difficulty in maintaining tunnel alignment
13	Subsidence along tunnel line	karst and old mine workings beneath tunnel line	H	1. GI boreholes, identification of karst or mineworkings 2. pre-construction geological desk study and risk assessment* 3. probe drilling ahead of face and beneath tunnel to detect areas of weak ground. Monitoring rates of drilling and character of drill cuttings* 4. identification of karstic or mining features*	L	H	H	1. failure to identify unconsolidated ground 2. failure of grouting programme 3. workings/karst too deep to consolidate from tunnel 4. settlement of TBM, delays to programme and loss of line 5. failure of tunnel lining post construction

continued overleaf

Number	Hazard		Consequence of event on tunnel	Possible mitigation measures prior to and during construction to reduce likelihood or consequence of hazard Notes * indicates technique more likely than others to be useful	Impact of mitigation measures and risk			Summary of main residual risks
	Title	Example of circumstances			Likelihood(L)	Consequence(C)	Risk (L x C)	
				5. geological mapping to detect disturbed or distressed strata* 6. grouting ahead and below tunnel to consolidate strata* 7. post construction microgravity survey (gradiometry) along line of tunnel to identify cavities and areas of untreated ground or to confirm consolidation undertaken satisfactorily* 8. consider also likelihood of hazards: 1, 2, 5, 7, 8, 9, 10, 11 and 15				
14	Ground heave	proximity to faults, shear zones, old mine workings, folded ground and places where stress conditions atypical	M	1. GI boreholes structural interpretation 2. pre-construction geological desk study and risk assessment 3. probe drilling ahead of face and beneath tunnel to detect areas of weak ground. Monitoring rates of drilling and character of drill cuttings* 4. identification of mining features 5. geological mapping to identify weak or disturbed strata* 6. use copy cutters in weak ground to increase tunnel diameter* 7. consider also likelihood of hazards: 15	M	M	M	1. failure to identify ground conditions 2. difficulties in moving TBM forward 3. difficulties in placing lining 4. delays in programme 5. incorrect lining selection 6. post construction lining failure 7. increased maintenance costs

Number	Hazard Title	Example of circumstances	Consequence of event on tunnel	Possible mitigation measures prior to and during construction to reduce likelihood or consequence of hazard. Notes * indicates technique more likely than others to be useful	Impact of mitigation measures and risk Likelihood(L)	Consequence(C)	Risk (L x C)	Summary of main residual risks
15	Squeezing ground	as 14 above	M	1. GI boreholes structural interpretation 2. pre-construction geological desk study and risk assessment 3. probe drilling ahead of face and beneath tunnel to detect areas of weak ground. Monitoring rates of drilling and character of drill cuttings* 4. identification of mining features 5. geological mapping to identify weak or disturbed strata* 6. use of copy cutters in weak ground to increase tunnel diameter* 7. consider also likelihood of hazards: 14	M	M	M	1. failure to identify ground conditions 2. difficulties in moving TBM forward 3. difficulties in placing lining 4. delays in programme 5. incorrect lining selection 6. post construction failure of tunnel lining 7. increased maintenance costs
16	Shock loading of tunnel lining	collapse of karst, old mine workings above, below or adjacent to tunnel lining. Earthquake, fault rupture	H	1. GI boreholes proving of extensive karst features or open mine workings 2. pre-construction geological desk study and risk assessment* 3. seismic hazard assessment 4. detection of cavities >> tunnel diameter from probe drilling ahead of and around the face through TBM head* 5. TSP to detect cavities in vicinity of tunnel bore *	L	H	H	1. failure to detect cavities 2. failure of grouting programme 3. inability to grout workings due to depth topography 4. tunnel collapse and temporary loss of drive 5. selection of incorrect tunnel lining

continued overleaf

Number	Hazard Title	Example of circumstances	Consequence of event on tunnel	Possible mitigation measures prior to and during construction to reduce likelihood or consequence of hazard — Notes * indicates technique more likely than others to be useful	Impact of mitigation measures and risk — Likelihood(L)	Consequence(C)	Risk (L x C)	Summary of main residual risks
				6. use concrete rather than spheroidal steel lining 7. post construction microgravity survey (gradiometry) along line of tunnel to locate and determine size of cavities in vicinity (30m radius) of tunnel 8. consider also likelihood of hazards: 1, 2, 3, 5, 6, 7, 8, 9, 10, 11, 16 and 17				6. post construction damage to tunnel lining 7. higher risk of water inrush through damaged lining
17	Contaminated ground	industrial and domestic effluents disposed of down mine shafts	L	1. GI boreholes proving contamination 2. GI investigation identifying post mining uses of mine shafts and adits* 3. pre-construction geological desk study and risk assessment* 4. monitoring of spoil properties during tunnelling* 5. monitoring of tunnel atmosphere 6. consider also likelihood of hazards: 1, 2, 3, 7, 8, 9, 10, 11, 12, and 16	L	L	L	1. failure to identify contaminants 2. higher risk of hazard 2 from petroleum wastes 3. injury (short and long-term) to workforce 4. selection of incorrect tunnel lining 5. corrosion of tunnel lining

Appendix E-3   Example risk register

Table of typical tunnelling hazards and possible mitigation measures to be considered during design and construction (modified from Brown, 1999 with reference to Channel Tunnel Rail Link).

Hazard (alphabetical)	Aspect	Mitigations/actions
Access/egress	To site and work areas	Safe routes and methods
Biological health hazards	Burial sites, etc.	Site investigation Liaison with authorities Avoid disturbance/minimise Appropriate disposal
Confined spaces	Asphyxiation, explosion, flooding, heat, humidity: a) Existing confined spaces b) Confined spaces to be constructed/ maintained	Minimise need for working in confined spaces in design Reduce need to enter confined spaces Safe working practices
Contaminated land	Contaminated ground Ground gas	Site investigation to identify Avoid disturbance Ventilation and monitoring Appropriate disposal
Demolition and site clearance	To existing structures	Survey structures and condition Consider stability Plan and phase work to minimise disturbance Fencing and security
	New structures	Design to ensure practical and safe sequence Communication
Earthworks	Ground movements	Site investigation Minimise earthworks Consider effect on existing structures Adequate information to Contractor
Excavation	Collapse, or falls associated with ground movements	GI to identify design constraints Minimise deep excavation Consider existing structures Adequate information to Contractor
	Areas prone to flooding (cavernous limestone, old mine workings, etc.)	Site investigation Liaise with authorities re flood potential Safety plan
	Contaminated ground, ground gas, old mine workings	See contaminated land above Probing ahead/geophysics

Hazard (alphabetical)	Aspect	Mitigations/actions
Fire/explosion	General construction Confined spaces and tunnels	Use non-flammable and benign materials Appropriate plant and machinery Emergency systems including liaison with emergency services
Mechanical lifting operations	Impacts and loads	Design works to minimise interaction with existing structures Design detailing and documentation
Maintenance	Access/egress to working environment	Incorporate into design High durability materials and details Allow for access Address residual risks
Manual handling		Minimise Design to ensure legislation compliance
Noise and vibration	General construction	Specification of methods and techniques Set limits and ensure compliance with health and safety requirements
Public safety		Minimise disruption and interaction by phasing works and design Security fencing and control Minimise influence of subsequent maintenance works
Services	Overhead and underground services	Agree with utility providers Avoid diversions Avoid work near services Provide information on services and signage
Site plant/traffic		Ensure adequate site size Traffic control measures Phase works Maximise separation between plant and personnel
Substances hazardous to health		Identify hazards Eliminate or substitute Reduce exposure Proper handling
Temporary stability of structures	Existing structures affected by works	Consider existing structure stability Provide info to Contractor Condition surveys
Unexploded ordnance (bombs)		Site investigation including desk study Survey and use specialists Emergency plan

References

AASHTO 2007. *LRFD Bridge Design Specification*, 4th edition. American Association of State Highway and Transportation Officials, Washington DC.

Abrahamson, M.W. 1992. Abbeystead – a legal view. *Proceedings of the ICE, Civil Engineering*, 92, 98–99.

Al-Harthi, A.A. & Hencher, S.R. 1993. Modelling stability of the Pen-Y-Clip tunnel, UK. *Safety and Environmental Issues in Rock Engineering, Eurock '93*, Lisbon: Balkema, 451–459.

Al-Sanad, H.A., Shaqour, F.M., Hencher, S.R. & Lumsden, A.C. 1990. The influence of changing groundwater levels on the geotechnical behaviour of desert sands. *Quarterly Journal of Engineering Geology*, 23, 357–364.

Allum, J.A.E. 1966. *Photogeology and Regional Mapping*. Oxford: Pergamon Press, 107p.

Ambraseys, N.N. & Hendron, A.J. Jr. 1968. Dynamic behaviour of rock masses. Chapter 7 in *Rock Mechanics in Engineering Practice*, Stagg, K.G. & Zienkiewicz, O.C. (eds.), New York: Wiley, 203–236.

Ambraseys, N.N., Simpson, K.A. & Bommer J.J. 1996. Prediction of horizontal response spectra in Europe. *Earthquake Engineering & Structural Dynamics*, 25, 371–400.

Ambraseys, N.N. & Srbulov, M. 1995. Earthquake-induced displacements of slopes. *Soil Dynamics & Earthquake Engineering*, 14, 59–72.

Ameen, M.S. 1995. Fractography: fracture topography as a tool in fracture mechanics and stress analysis. An introduction. In *Fractography: Fracture Topography as a Tool in Fracture Mechanics and Stress Analysis*, Ameen, M.S. (ed.), London: Geological Society London Special Publication No. 92, 1–10.

American Society for Testing and Materials 1999. *Standard Practice for Rock Core Drilling and Sampling of Rock for Site Investigation*, ASTM, D 2113–99, 20p.

Anderson, A.E., Weiler, M., Alila, Y. & Hudson, R.O. 2008. Dye staining and excavation of a lateral preferential flow network. *Hydrology and Earth System Sciences Discussions*, 5, 1043–1065.

Anderson, E.M. 1951. *The Dynamics of Faulting*. Edinburgh: Oliver and Boyd, 206p.

Anon 1970. The logging of rock cores for engineering purposes. *Quarterly Journal of Engineering Geology*, 3, 1–14.

Anon 1972. The preparation of maps and plans in terms of engineering geology. *Quarterly Journal of Engineering Geology*, 5, 293–382.

Anon 1985. Liquefied petroleum gas caverns at South Killingholme. Photographic feature. *Quarterly Journal of Engineering Geology*, 18, 1, ii–iv.

Anon 1995. The description and classification of weathered rocks for engineering purposes. Geological society engineering group working party report. *Quarterly Journal of Engineering Geology*, 28, 207–242.

Archambault, G., Verreault, N., Riss, J. & Gentier, S. 1999. Revisiting Fecker-Rengers-Barton's methods to characterize joint shear dilatancy. In *Rock Mechanics for Industry*, Amadei, Kranz, Scott & Smeallie (eds.), Rotterdam: Balkema, 423–430.

Association of Geotechnical & Geoenvironmental Specialists 2006. AGS *Guidelines for Good Practice in Site Investigation*, 2p.

Atkinson, J.H. 2000. Non-linear soil stiffness in routine design. *Géotechnique*, 50, 487–508.

Atkinson, J.H. 2007. *The Mechanics of Soils and Foundations*, 2nd edition. London: Taylor & Francis, 442p.

Australian Standards 1993. *Geotechnical site investigations*. AS–1726.

Baecher, G.B. & Christian, J.T. 2003. *Reliability and Statistics in Geotechnical Engineering*. London and New York: Wiley, 605p.

Bahat, D., Grossenbacher, K. & Karasaki, K. 1999. Mechanism of exfoliation joint formation in granitic rocks, Yosemite National Park. *Journal of Structural Geology*, 21, 85–96.

Bandis, S.C. 1990. Scale effects in the strength and deformability of rocks and rock joints. In *Scale Effects in Rock Masses*, Pinto da Cunha (ed.), Rotterdam: Balkema, 59–76.

Bandis, S.C., Lumsden, A.C. & Barton, N.R. 1981. Experimental studies of scale effects on the shear behaviour of rock joints. *International Journal of Rock Mechanics and Mining Sciences and Geomechanics Abstracts*, 18, 1–21.

Banyard, J.K, Coxon, R.E. & Johnston, T.A. 1992. Carsington Reservoir – reconstruction of the dam. *Proceedings of the ICE – Civil Engineering*, 92, 106–115.

Barla, G., Barla, M., Camusso, M. & Martinotti, M.E. 2007. Setting up a new direct shear testing apparatus. *Proceedings of the 11th Congress of the International Society for Rock Mechanics*, 1, 415–418.

Barla, G. & Jarre, P. 1993. Tunnelling under an industrial waste landfill in Italy: Environmental controls and excavation procedures. *Safety and Environmental Issues in Rock Engineering, Eurock '93*, Lisbon: Balkema, 259–266.

Barla, G. & Pelizza, S. 2000. TBM Tunnelling in difficult conditions. *Proceedings of GeoEng 2000*, Melbourne, Australia, 329–354.

Barton, N.R. 1973. Review of a new shear-strength criterion for rock joints. *Engineering Geology*, 7, 287–332.

Barton, N.R. 1990. Scale effects or sampling bias? *Proceedings of 1st International Workshop on Scale Effects in Rock Masses*, Loen, Norway, 31–55.

Barton, N.R. 2000. *TBM Tunnelling in Jointed and Faulted Rock*. Rotterdam: Balkema, 173p.

Barton, N. 2002. Some new Q-value correlations to assist in site characterization and tunnel design. *International Journal of Rock Mechanics and Mining Science*, 39, 185–216.

Barton, N.R. 2003. TBM or drill-and-blast. *Tunnels & Tunnelling International*, June, 20–22.

Barton, N.R. 2005. Comments on 'A critique of QTBM'. *Tunnels & Tunnelling International*, July, 16–19.

Barton, N.R. & Bandis, S.C. 1990. Review of predictive capabilities of JRC-JCS model in engineering practice. In *Rock Joints, Proceedings International Symposium on Rock Joints*, Loen, Norway, Barton, N. & Stephansson, O. (eds.), Rotterdam: Balkema, 603–610.

Bates, A.D. 1981. Profit or loss pivot on pre-dredging surveys. *Dredging + Port Construction*, April 1981, Intec Press, UK. [Referenced in PIANC (2000) Site investigation requirements for dredging works, Report of Working Group No. 23, 32p.]

Baynes, F.J. 2007. Sources of geotechnical risk. *Engineering Geology in Geotechnical Risk Management*, Hong Kong Regional Group of the Geological Society of London, 1–12.

Beer, A.J., Stead, D. & Coggan, J.S. 2002. Estimation of the Joint Roughness Coefficient (JRC) by visual comparison. *Rock Mechanics & Rock Engineering*, 35, 65–74.

Bell, D.K. & Davidson, B.J. 1996. Response identification of Pacoima Dam for the 1994 Northridge Earthquake. *Eleventh World Conference on Earthquake Engineering*, Oxford, Disc 2, Paper No. 774.

Bell, F.G. 1999. *Geological Hazards*. London and New York: Taylor & Francis.

Benson, R. 1989. Design of unlined and lined pressure tunnels. *Tunnelling and Underground Space Technology*, 4, 2, 155–170.

Bieniawski, Z.T. 1976. Rock mass classification in rock engineering. In *Exploration for Rock Engineering, Proceedings of Symposium*, Bieniawski, Z.T. (ed.), 1, 97–106.

Bienawski, Z.T. 1989. *Engineering Rock Mass Classifications*. New York: Wiley, 251p.

Bird, M.I., Chang, C.H., Shirlaw, J.N., Tan, T.S. & Teh, T.S. 2003. The age and origin of the quaternary sediments of Singapore with emphasis on the Marine Clay. *Proceedings of Underground Singapore 2003*, Singapore, 428–440.

Bisci, C., Dramis, F. & Sorriso-Valvo, M. 1996. Rock Flow (Sackung). Chapter 7.2 in *Landslide Recognition*, Dikau, R., Brunsden, D., Schrott, L. & Ibsen, M.L. (eds.), Chichester: Wiley, 150–160.

Black, J.H. 2010. The practical reasons why slug tests (including falling and rising head tests) often yield the wrong value of hydraulic conductivity. *Quarterly Journal of Engineering Geology*, 43, 345–358.

Black, J.H., Barker, J.A. & Woodman, N.D. 2007. *An Investigation of 'Sparse Channel Networks'*. Report No. R-07–35. Svensk Kärnbränslehantering AB (SKB), 123p.

Blake, J.R., Renaud, J.-P. Anderson, M.G. & Hencher, S.R. 2003. Retaining wall impact on slope hydrology using a high-resolution finite element model. *Computers and Geotechnics*, 30, 6, 431–442.

Blockley, D.I. 2011. Engineering safety. *Proceedings of the Institution of Civil Engineers: Forensic Engineering*, 164, Issue FE1, 7–13.

Blyth, F.G.H. & de Freitas, M.H. 1984. *A Geology for Engineers*, 7th edition. London: Edward Arnold, 325p.

Bo, M.W., Na, Y.M. & Arulrajah, A. 2009. Densification of granular soil by dynamic compaction. *Proceedings of the Institution of Civil Engineers: Ground Improvement*, 162, 121–132.

Bock, H. 2006. Common ground in engineering geology, soil mechanics and rock mechanics: past, present and future. *Bulletin Engineering Geology & Environment*, 65, 209–216.

Bommer, J.J. & Martinez-Pereira, A. 1999. The effective duration of earthquake strong motion. *Journal of Earthquake Engineering*, 3, 127–172.

Bond, A.J. & Harris, A. 2008. *Decoding Eurocode 7*. London: Taylor and Francis, 598p.

Bond, A.J. & Simpson, B. 2010. Pile design to Eurocode 7 and the UK National Annex. Part 2: UK National Annex. *Ground Engineering*, January, 28–31.

Bowden, F.P. & Tabor, D. 1950. *The Friction and Lubrication of Solids. Part I*. Oxford: Clarendon Press.

Bowden, F.P. & Tabor, D. 1964. *The Friction and Lubrication of Solids. Part II*. Oxford: Clarendon Press.

Bowles, J.E. 1996. *Foundation Analysis and Design*, 5th edition. New York: McGraw Hill, 1175p.

Brady, B.H.G. & Brown, E.T. 2004. *Rock Mechanics for Underground Mining*, 3rd edition. Berlin and New York: Springer.

Brambati, A., Carbognin, L., Quaia, T., Teatini, P. & Tosi, L., 2003. The Lagoon of Venice: geological setting, evolution and land subsidence. *Episodes: Journal of International Geoscience*, 26, 3, 264–268.

Brand E.W., Hencher, S.R. & Youdan, J.D. 1983. Rock slope engineering in Hong Kong. *Proceedings of the 5th International Rock Mechanics Congress*, Melbourne, 17–24.

Bray, R.N., Bates, A.D. & Land, J.M. 1997. *Dredging. A Handbook for Engineers*, 2nd edition. Oxford: Butterworth-Heinemann, 434p.

Bridges, M.C. 1990. Identification and characterisation of sets of fractures and faults in rock. *Proceedings of the International Symposium on Rock Joints*, Loen, Norway, 19–25.

British Standards Institution 1964. BS3618:1964, Sec. 5. *Glossary of Mining Terms, Geology*, 18p.

British Standards Institution 1981a. BS 6031:1981 *Code of Practice for Earthworks*.

British Standards Institution 1981b. BS 5930:1981 *Code of Practice for Site Investigations* (Formerly CP 2001), 147p.

British Standards Institution 1986. BS 8004:1986 *Code of Practice for Foundations*.

British Standards Institution 1989. BS 8081:1989 *Code of Practice for Ground Anchorages*.

British Standards Institution 1990. BS 1377:1990 *Methods of Test for Soils for Civil Engineering Purposes (9 parts)*.

British Standards Institution 1994. BS 8002:1994 *Code of Practice for Earth Retaining Structures*.

British Standards Institution 1999. BS 5930:1999 *Code of Practice for Site Investigation*. 206p.

British Standards Institution 2000. BS EN 1537:2000 *Execution of Special Geotechnical Work – Ground Anchors*.

British Standards Institution 2001. BS 10175:2001 *Code of Practice for Investigation of Potentially Contaminated Sites*.

British Standards Institution 2002. EN ISO 14688–1. *Geotechnical Investigation and Testing – Identification and Classification of Soil – Part 1: Identification and Description*.

British Standards Institution 2003. EN ISO 14689–1. *Geotechnical Investigation and Testing – Identification and Classification of Rock – Part 1: Identification and Description*.

British Standards Institution 2004. EN 1997–1:2004. *Eurocode 7: Geotechnical Design. Part 1: General Rules*. Code of Practice.

British Standards Institution 2007. EN 1997–2:2007. *Eurocode 7: Geotechnical Design. Part 2: Ground Investigation and testing*. Code of Practice.

British Tunnelling Society 2003. *The Joint Code of Practice for Risk Management of Tunnel Works in the UK*, 18p.

British Tunnelling Society 2005. *Closed-Face Tunnelling Machines and Ground Stability*. London: Thomas Telford.

Broch, E., Myrvang, A.M. &. Stjern, G. 1996. Support of large rock caverns in Norway. *Tunnelling and Underground Space Technology*, 11, 11–19.

Broms B.B. & Lai P.H. 1995. The Republic Plaza in Singapore – Foundation Design, *Bengt B. Broms Symposium on Geotechnical Engineering*, Singapore, 3–24.

Brook, N. 1993. The measurement and estimation of basic rock strength. In *Comprehensive Rock Engineering*, Volume 3, Hudson, J. (ed.), Oxford: Pergamon Press, 41–66.

Brown, E.T. 2008. Estimating the mechanical properties of rock masses. In *Proceedings, 1st Southern Hemisphere International Rock Mechanics*

Symposium, Perth, 16–19 September, Potvin, Y., Carter, J., Dyskin, A. & Jeffrey, R. (eds.), **1**, 3–22.

Brown, R.H.A. 1999. The management of risk in the design and construction of tunnels. *Korean Geotechnical Society, Tunnel Committee Seminar, 21st Seminar*, 21 September, 22p.

Brunsden, D. 2002. Geomorphological roulette for engineers and planners: some insights into an old game. *Quarterly Journal of Engineering Geology & Hydrogeology*, 35, 101–142.

Buckingham, R.J. 2003. *Problems Associated with Water Ingress into Hard Rock Tunnels*. MSc (Applied Geosciences) dissertation, Hong Kong University, unpublished, 39p.

Bulletin of Atomic Scientists 1976. *Preface to the Royal Commission on Environmental Pollution*. Sixth Report, Nuclear Power and the Environment.

Bunce, C.M., Cruden, D.M. & Morgenstern, N.R. 1997. Assessment of the hazard from rockfall on a highway. *Canadian Geotechnical Journal*, 34, 344–356.

Burland, J. 2007. Terzhagi: back to the future. *Bulletin Engineering Geology and Environment*, 66, 29–33.

Burland, J.B. 1990. On the compressibility and shear strength of natural clays. *Géotechnique*, 40, 329–378.

Burland, J.B. 2008. Interlocking, and peak and design strengths. Discussion of Schofield, A.N., 2006, *Géotechnique*, 58, 527.

Burland, J.B. & Burbidge, M.C. 1985, Settlement of foundations on sand and gravel. *Proceedings of the Institution of Civil Engineers*, 78, 1325–1381.

Burnett, A.D., Brand, E.W. & Styles, K.A. 1985. Terrain classification mapping for a landslide inventory in Hong Kong. *Proceedings IVth International Conference and Field Workshop on Landslides*, Tokyo, 63–68.

Buttling, S. 1986. Remedial works to a major rock slope in Hong Kong. *Proceedings of the Conference on Rock Engineering and Excavation in an Urban Environment*, Hong Kong, 41–48.

Byerlee, J.D. 1978. Friction of rocks. *Pure Applied Geophysics*, 116, 615–626.

CALTRANS 2010. *Soil and Rock Logging, Classification, and Presentation Manual*, 81p. http://www.dot.ca.gov/hq/esc/geotech/sr_logging_manual/page/logging_manual_2010.pdf

Carter, T.G., de Graaf, P., Booth, P., Barrett, S. & Pine, R. 2002. Integration of detailed field investigations and innovative design key factors to the successful widening of the Tuen Mun Highway. *Proceedings of the 22nd Annual Seminar*, organised by the Geotechnical Division of the Hong Kong Institution of Engineers, 8 May 2002, 187–201.

Carter, T.G., Diederichs, M.S. & Carvalho, J.L. 2008. Application of modified Hoek-Brown transition relationships for assessing strength and post-yield behaviour at both ends of the rock competence scale. *Proc. SAIMM*, 108, 325–338.

Carvalho, J.L., Carter, T.G. & Diederichs, M.S. 2007. An approach for prediction of strength and post yield behaviour for rock masses of low intact strength. *Proceedings 1st Canadian-US Rock Mechanics Symposium. Meeting Society's Challenges and Demands*, Vancouver, 249–257.

Casagrande, A. 1965. Classification and identification of soils. *Transactions American Society Civil Engineers*, 113, 901–992.

Cederstrom, M., Thorsall P.-E., Hildenwall, B. & Westberg, S.-B. 2005. Incident with loss of seven post-tensioned 72 ton anchors in a dam. *Dam Safety 2005 Proceedings*, Association of State Dam Safety Officials.

Challinor, J.A. 1964. *A Dictionary of Geology*, 2nd edition. Cardiff: University of Wales Press, 289p.

Chandler, R.J. 1969. The effects of weathering on the shear strength properties of Keuper Marl. *Géotechnique*, 19, 321–334.

Chandler, R.J. 2000. Clay sediments in depositional basins: the geotechnical cycle. *Quarterly Journal Engineering Geology and Hydrogeology*, 33, 7–39.

Chaplow, R. 1996. The geology and hydrogeology of Sellafield: an overview. *Quarterly Journal of Engineering Geology*, 29, S1–S12.

Charles, J.A. 2002. Ground improvement: the interaction of engineering science and experience based technology. *Géotechnique*, 52, 527–532.

Chen, T.-C., Lin, M.-L. & Hung, J.-J. 2003. Pseudostatic analysis of Tsao-Ling rockslide caused by Chi-Chi earthquake. *Engineering Geology*, 71, 31–47.

Cheng, Y. & Liu, S.C. 1993. Power caverns of Mingtan Pumper Storage Project, Taiwan. In *Comprehensive Rock Engineering*, Volume 5, Surface and Underground Project Case Histories (Editor in Chief, J.A. Hudson; Volume Editor, E. Hoek), Oxford: Pergamon Press, 111–132.

Chigira, M., Wang, W-N, Furya, T. & Kamai, T. 2003. Geological causes and geomorphological precursors of the Tsaoling landslide triggered by the 1999 Chi-Chi earthquake, Taiwan. *Engineering Geology*, 68, 259–273.

Chigira, M. & Yahi, H. 2006. Geological and geomorphological characteristics of landslides triggered by the 2004 Mid Niigata prefecture earthquake in Japan. *Engineering Geology*, 82, 202–221.

China Ministry of Construction 2009. *Code of Investigation of Geotechnical Engineering*. National Standard of the People's Republic of China, GB50021–2001 (2009 edition), 269p.

Cho, S-M., Ahn, S-S., Lee, Y-K. & Oh, S-T. 2009a. Incheon Bridge: facts and the state of the art. *Proceedings of the International Commemorative Symposium for the Incheon Bridge*, 18–28.

Cho, S-M., Kim, J-H., Lee, H-G., Kim, Z-C., Shin, S-H., Song, M-J., Kim, D-J. & Park, Y-H. 2009b. Pile load tests on the Incheon Bridge Project. *Proceedings of the International Commemorative Symposium for the Incheon Bridge*, 488–499.

Choot, E.B. 1983. *Landslide Studies 1982 Case Study No. 9 Tsing Yi (2)*. Special Project Report No. SPR 10/83, Geotechnical Control Office, Hong Kong, 44p.

Chung, S.G., Kim, G.J., Kim, M.S. & Ryu, C.K. 2007. Undrained shear strength from field vane test on Busan clay. *Marine Georesources & Geotechnology*, 25, 167–179.

Cole, K.W. 1988. *Foundations*. ICE Works Construction Guides, London: Thomas Telford Ltd., 82p.

Combault, J., Morand, P. & Pecker, A. 2000. Structural response of the Rion Antirion Bridge. *Proceedings of the 12th World Conference on Earthquake Engineering*, Auckland, Australia.

Construction Industry Research and Information Association (CIRIA) 1995. *Remedial Treatment for Contaminated Land. Volume III: Site Investigation and Assessment.*

Construction Industry Research and Information Association (CIRIA) 2005. *The Rock Manual. The Use of Rock in Hydraulic Engineering*, 2nd edition. C683, CIRIA, London.

Clayton, C.R.I. 1995. *The Standard Penetration Test (SPT): Methods and Use*. CIRIA Report 143, 144p.

Clayton, C.R.I. 2001. *Managing Geotechnical Risk: Improving Productivity in UK Building and Construction*. London: Thomas Telford, 80p.

Clayton, C.R.I. 2011. Stiffness at small strain: research and practice. *Géotechnique*, 61, 5–37.

Clayton, C.R.I, Simons, N.E. & Mathews, M.C. 1995. *Site Investigation. A Handbook for Engineers*, 2nd edition. London: Granada, 424p.

Clover, A.W. 1986. Slope stability on a site in the volcanic rocks of Hong Kong. *Proceedings of the Conference on Rock Engineering and Excavation in an Urban Environment, Hong Kong*, 121–134.

Coburn, A. & Spence, R. 1992. *Earthquake Protection*. Chichester: Wiley, 355p.

Combault, J., Morand, P. & Pecker, A. 2000. Structural response of the Rion-Antirion Bridge. 12th World Conference Earthquake Engineering, Paper 1609, 7p.

Cooper, A.H. & Waltham, A.C. 1999. Photographic Feature: Subsidence caused by gypsum dissolution at Ripon, North Yorkshire. *Quarterly Journal of Engineering Geology*, 32, 4, 305–310.

Coulson, J.H. 1971. Shear strength of flat surfaces of rock. *Proceedings of the 13th US Symposium on Rock Mechanics*, Urbana, Illinois, 77–105.

Craig, R.F. 1992. *Soil Mechanics*, 5th edition. Chapman & Hall, 427p.

Crawford, A.M. & Curran, J.H. 1982. The influence of rate and displacement dependant, shear resistance on the response of rock slopes to seismic loads. International Journal of Rock Mechanics & Mining Sciences and Geomechanics Abstracts, 19, 1–8.

Culshaw, M.G. 2005. From concept towards reality: developing the attributed 3D geological model of the shallow subsurface. *Quarterly Journal of Engineering Geology and Hydrogeology*, 38, 231–284.

Cunha, A.P. 1990. Scale effects in rock mechanics. In *Proceedings 1st International Workshop on Scale Effects in Rock Masses*, Pinto da Cunba (ed.), Loen, Norway, 3–27.

Darracott, B.W. & McCann, D.M. 1986. Planning engineering geophysical surveys. In *S I Practice: Assessing BS5930, Geological Society, Engineering Geology Special Publication No. 2*, A.B. Hawkins (ed.), 85–90.

Das, B.M. & Sivakugan, N. 2007. Settlements of shallow foundations on granular soil – an overview. *International Journal of Geotechnical Engineering*, 1, 19–29.

Davies, R.J., Mathioas, S.A., Swarbrick, R.E. & Tingay, M.J. 2011. Probabilistic longevity estimate for the LUSI mud volcano, East Java. *Journal of the Geological Society, London*, 168, 517–523.

Davis, E.H. & Poulos, H.G. 1967. Laboratory investigations of the effects of sampling. *Civil Engineering Transactions (Australia)*, CE9, 86–94.

Davis, G.H. & Reynolds, S.J. 1996. *Structural Geology of Rocks and Regions*, 2nd edition. John Wiley & Sons, 776p.

Dearman, W.R. & Coffey, J.R. 1981. Effects of evaporite removal on the mass properties of limestone. *Bulletin of the International Association of Engineering Geology*, 24, 91–96.

Dearman, W.R. & Fookes, P.G. 1974. Engineering geological mapping for civil engineering practice in the United Kingdom. *Quarterly Journal of Engineering Geology and Hydrogeology*, 7, 223–256.

Deere, D.U. 1968. Geological considerations. Chapter 1 in *Rock Mechanics in Engineering Practice*, Stagg, K.G. & Zienkiewicz, O.C. (eds.), New York: Wiley, 1–20.

Deere, D.U. 1971. The foliation shear zone – an adverse engineering geologic feature of metamorphosed rocks. *Journal of the Boston Society of Civil Engineers*, 60, 4, 1163–1176.

Deere, D.U. & Deere, D.W. 1988. The RQD index in practice. *Proceedings Symposium on Rock Classification for Engineering Purposes*, ASTM Special Technical Publications 984, Philadelphia, 91–101.

de Freitas, M.H. 2009. Geology; its principles, practice and potential for Geotechnics. *Quarterly Journal of Engineering Geology and Hydrogeology*, 42, 397–441.

DeGraff, J.M. & Aydin, A. 1987. Surface morphology of columnar joints and its significance to mechanics and direction of joint growth. *Geological Society of America Bulletin*, 99, 605–617.

Derbyshire, E. 2001. Geological hazards in loess terrain, with particular reference to the loess regions of China. *Earth-Science Reviews*, 54, 231–260.

Dering, C. 2003. Discussion. *Proceedings 14th SE Asian Geotechnical Conference*, Hong Kong, 3, 41–43.

Dershowitz, W. & LaPointe, P. 1994. Discrete fracture approaches for oil and gas applications. In *Proceedings 1st North American Rock Mechanics Symposium, Rock Mechanics*, Nelson & Laubach (eds.), Rotterdam: Balkema, 19–30.

Dershowitz, W., Lee, G., Geier, J., Foxford, T., LaPointe, P. & Thomas, A. 1996. *FracMan Interactive Discrete Feature Data Analysis, Geometric Modeling, and Exploration Simulation, User Documentation, Version 2.5*. Redmond, Washington: Golder Associates.

Devonald, D.M., Thompson, J.A., Hencher, S.R. & Sun, H.W. 2009. Geomorphological landslide models for hazard assessment – a case study at Cloudy Hill. *Quarterly Journal Engineering Geology*, 42, 473–486.

Diederichs, M.S. 2003. Rock fracture and collapse under low confinement conditions. *Rock Mechanics and Rock Engineering*, 36, 339–381.

Dobie, M.J.D. 1987. Slope instability in a profile of weathered norite. *Quarterly Journal of Engineering Geology and Hydrogeology*, 20, 279–286.

Döse, C. Stråhle, A., Rauséus, G., Samuelsson, E. & Olsson, O. 2008. *Revision of BIPS-Orientations for Geological Objects in Boreholes from Forsmark and Laxemar*. Svensk Kärnbränslehantering AB (SKB Report No. P-08–37), 97p.

Douglas, P.M. & Voight, B. 1969. Anisotropy of granites: a reflection of microscopic fabric, *Géotechnique*, 19, 376–398.

Dowding, C.H. 1985. *Blast Vibration Monitoring and Control*. Saddle River, NJ: Prentice-Hall Inc., 297p.

Dowrick, D.J. 1988. *Earthquake Resistant Design for Engineers and Architects*, 2nd edition. Chichester: John Wiley & Sons.

Dumbleton, M.J. & West, G. 1970 *Air Photograph Interpretation for Road Engineers in Britain*. Ministry of Transport, RRL Report, LR369, Road Research Laboratory, Crowthorne.

Dunnicliff, J. 1993. *Geotechnical Instrumentation for Monitoring Field Performance*. New York: Wiley, 577p.

Dusseault, M.B. & Morgenstern, N.R. 1979. Locked sands. *Quarterly Journal of Engineering Geology*, 12, 117–131.

Early, K.R. & Skempton, A.W. 1972. Investigations of the landslide at Walton's Wood, Staffordshire. *Quarterly Journal of Engineering Geology*, 5, 19–41.

Eberhardt, E., Stead, D. & Coggan, J.S. 2004. Numerical analysis of initiation and progressive failure in natural rock slopes – the 1991 Randa rockslide. *International Journal of Rock Mechanics and Mining Sciences*, 41, 69–87.

Eberl, D.D. 1984. Clay formation and transformation in rocks and soils. *Philosophical Transactions, Royal Society, London*, A311, 241–257.

Ebuk, E.J. 1991. *The Influence of Fabric on the Shear Strength Characteristics of Weathered Granites*. Unpublished PhD thesis, the University of Leeds, 486p.

Ebuk, E.J., Hencher, S.R. & Lumsden, A.C. 1993. The influence of structure on the shearing mechanism of weakly bonded soils. *Proceedings 26th Annual Conference of the Engineering Group of the Geological Society, The Engineering Geology of Weak Rock*, Leeds, 207–215.

Edwards, R.J.G. 1971. The practical application of rock bolting – Jeffrey's Mount rock cut on the M6. *Journal Institution Highway Engineers*, 17, 12, 21–27.

Egan, D. 2008. The ground: clients remain exposed to unnecessary risk. Proceedings of the Institution of Civil Engineers, *Geotechnical Engineering*, 161, 189–195.

Eisma, D. (ed.) 2006. *Dredging in Coastal Waters*. Balkema: Taylor & Francis, 244p.

Engelder, T. 1985. Loading paths to joint propagation during a tectonic cycle: an example from the Appalachian Plateau, USA. *Journal of Structural Geology*, 7, 459–476.

Engelder, T. 1999. Transitional-tensile fracture propagation: a status report. *Journal of Structural Geology*, 21, 1049–1055.

Engelder, T. & Peacock, D. 2001. Joint development normal to regional compression during flexural-slow folding: the Listock buttress anticline, Somerset, England. *Journal of Structural Geology*, 23, 259–277.

Fecker, E. & Rengers, N. 1971. Measurement of large scale roughness of rock planes by means of profilograph and geological compass. *Proceedings Symposium on Rock Fracture*, Nancy, France, Paper 1–18.

Fell, R., Ho, K.K.S., Lacasse, S. & Leroi, E. 2005. A framework for landslide risk assessment and management. *Landslide Risk Management*, Hung, Fell, Couture & Eberhardt (eds.), 3–25.

Fletcher, C.J.N. 2004. *Geology of Site Investigation Boreholes from Hong Kong.* Applied Geoscience Centre, Hong Kong Construction Association and AGS (Hong Kong), 132p.

Fletcher, C.J.N., Wightman, N.R. & Goodwin, C.R. 2000. Karst related deposits beneath Tung Chung New Town: implications for deep foundations. *Proceedings Conference on Engineering Geology HK 2000*, Institution of Mining and Metallurgy, Hong Kong, 139–149.

Fogg, G.E., Noyes, C.D. & Carle, S.F. 1998. Geologically based model of heterogeneous hydraulic conductivity in an alluvial setting. *Hydrogeology Journal*, 6, 1, 131–143.

Fookes P.G. 1997. Geology for engineers: the geological model, prediction and performance. *Quarterly Journal of Engineering Geology*, 30, 293–431.

Fookes, P.G., Baynes, F.J. & Hutchinson, J.M. 2000. Total geological history: a model approach to the anticipation, observation and understanding of ground conditions. *GeoEng 2000*, Melbourne, 1, 370–460.

Fookes, P.G. & Hawkins, A.B. 1988. Limestone weathering: its engineering significance and a proposed classification scheme. *Quarterly Journal of Engineering Geology*, 21, 7–31.

Fookes, P.G. & Parrish, D.G. 1969. Observations on small-scale structural discontinuities in the London Clay and their relationship to regional geology. *Quarterly Journal of Engineering Geology*, 1, 217–240.

Fookes, P.G., Sweeney, M., Manby, C.N.D. & Martin, R.P. 1985. Geological and geotechnical engineering aspects of low-cost roads in mountainous terrain. *Engineering Geology*, 21, 1–13, 17–53, 57–65, 69–85, 91–99, 102, 105–109, 115 –131, 133, 135–152.

Fry, N. 1984. *The Field Description of Metamorphic Rocks*. Milton Keynes, UK: The Open University Press, 110p.

Fujii, Y., Takemura, T., Takahashi, M. & Lin, W. 2007. Surface features of uniaxial tensile fractures and their relation to rock anisotropy in Inada granite. *International Journal of Rock Mechanics and Mining Science*, 44, 98–107.

Note – all the following GCO and GEO publications are available to download from: http://www.cedd.gov.hk/eng/publications/http://www.cedd.gov.hk/eng/publications/

Geotechnical Control Office 1979. *Geotechnical Manual for Slopes*, 1st edition. Geotechnical Control Office, Hong Kong, 242p.

Geotechnical Control Office 1984a. *Geotechnical Manual for Slopes*, 2nd edition. Geotechnical Control Office, Hong Kong, 295p.

Geotechnical Control Office 1984b. *Mid-Levels Study: Report on Geology, Hydrology and Soil Properties*. Geotechnical Control Office, Public Works Department, Hong Kong.

Geotechnical Control Office 1987. *Guide to Site Investigation (Geoguide 2)*. Hong Kong Government, 359p.

Geotechnical Control Office 1988. *Guide to Rock and Soil Descriptions (Geoguide 3)*. Geotechnical Engineering Office, Hong Kong, 189p.

Geotechnical Control Office 1990. *Review of Design Methods for Excavations*. GCO Publication No. 1/90, 192p.

Geotechnical Engineering Office 1992. *Guide to Cavern Engineering (Geoguide 4)*. Geotechnical Engineering Office, Hong Kong, 156p.

Geotechnical Engineering Office 1993. *Guide to Retaining Wall Design (Geoguide 1)*, 2nd edition. Geotechnical Engineering Office, Hong Kong.

Geotechnical Engineering Office 2000a. *Highway Slope Manual*. Geotechnical Engineering Office, Hong Kong, 114p.

Geotechnical Engineering Office 2000b. *Technical Guidelines on Landscape Treatment and Bio-engineering of Man-made Slopes and Retaining Walls*. GEO Publication No. 1/2000, Geotechnical Engineering Office, Hong Kong, 146p.

Geotechnical Engineering Office 2006. *Foundation Design and Construction*. Geotechnical Engineering Office, Hong Kong, 348p.

Geotechnical Engineering Office 2007. *Engineering Geological Practice in Hong Kong*, GEO Publication No. 1/2007, 278p.

Geotechnical Engineering Office 2009. *Catalogue of Notable Tunnel Failure Case Histories (up to December 2008)*. Geotechnical Engineering Office, Hong Kong.

Gioda, G. & Sakurai, S. 2005. Back analysis procedures for the interpretation of field measurements in geomechanics. *International Journal for Numerical and Analytical Methods in Geomechanics*, 11, 555–583.

Glossop, R. 1968. The rise of geotechnology and its influence on engineering practice. Eighth Rankine Lecture. *Géotechnique*, 18, 107–150.

Gomes, J.P., Batista A.L. & Oliveira, S. 2009. Damage-Chemo-Viscoelastic Model on the Analysis of Concrete Dams under Swelling Processes. *Proceedings International Conference on Long Term Behaviour of Dams (LTBD09)*, Graz.

Goodman, R.E. 1980. *Introduction to Rock Mechanics*. New York: John Wiley & Sons.

Goodman, R.E. 1993. *Engineering Geology. Rock in Engineering Construction*. New York: Wiley, 412p.

Goodman, R.E. 2002. Karl Terzaghi's legacy in geotechnical engineering. *Geo-strata*, ASCE, October.

Goodman, R.E. 2003. Karl Terzhagi and engineering geology. In *Geotechnical Engineering*, Ho, K.K.S. & Li, K.S. (eds.), Rotterdam: Balkema, 115–122.

Gordon, T., Scott, M.J. & Statham, I. 1996. The identification of bedding shears and their implications for road cutting design and construction. *Prediction and Performance in Rock Mechanics and Rock Engineering, Eurock '96*, Torino, Italy, 597–603.

Graham, R.H. 1981. Gravity sliding in the Maritime Alps. *Geological Society, London, Special Publications*, 9, 335–352.

Green, P. & Western, R. 1994. *Time to Face the Inevitable*. A Submission from Friends of the Earth Ltd. to the UK Department of the Environment's Review of Radioactive Waste Management Policy, 72p.

Green, R.G. & Hawkins, A.B. 2005. Rock cut on the A5 at Glyn Bends, North Wales, UK. *Bulletin Engineering Geology and Environment*, 64, 95–109.

Grose, W.J. & Benton, L. (2005). Hull wastewater flow transfer tunnel: tunnel collapse and causation investigation. *Geotechnical Engineering*, 158, 4, 179–185.

Hack, R. 1998. *Slope Stability Probability Classification, SSPC*, 2nd edition. International Institute for Aerospace Survey and Earth Sciences (ITC), Publication No. 43, 258p.

Haimson, B.C. 1992. Designing pre-excavation stress measurements for meaningful rock characterization. In *Rock Characterization, Eurock'92*, Hudson, J.A. (ed.), London: Thomas Telford, 221–226.

Halcrow Asia Partnership 1998a. *Report on the Ching Cheung Road Landslide of 3rd August 1997*. Geotechnical Engineering Office, Hong Kong, GEO Report No. 78, 47 p.

Halcrow Asia Partnership Ltd. 1998b. *Report on the Landslide at Ten Thousand Buddha's Monastery of 2 July 1997*. GEO Report No. 77, Geotechnical Engineering Office, Hong Kong, 96p. http://www.cedd.gov.hk/eng/publications/geo_reports/geo_rpt077.htm

Hancock, P.L. 1985. Brittle microtectonics: principles and practice. *Journal of Structural Geology*, 7, 437–457.

Hancock, P.L. 1991. Determining contemporary stress directions from neotectonic joint systems. *Philosophical Transactions Of the Royal Society Land A*, 337, 29–40.

Haneberg, W.C. 2008. Using close range terrestrial digital photogrammetry for 3-D rock slope modelling and discontinuity mapping in the United States. *Bulletin of Engineering Geology and the Environment*, 4, 457–469.

Hardy, W.B. & Hardy, J.K. 1919. Note on static friction and on the lubricating properties of certain chemical substances. *The Philosophical Magazine*, 38, 32–48.

Harris, D.I., Mair, R.J., Love, J.P., Taylor, R.N. & Henderson, T.O. 1994. Observations of ground and structure movements for compensation grouting during tunnel construction at Waterloo station. *Géotechnique*, 44, 691–713.

Hartwell, D.J. 2006. Discussion: Hull wastewater flow transfer tunnel: tunnel collapse and causation investigation. *Geotechnical Engineering*, 2, 125–126.

Hashash, Y.M.A., Hook, J.J., Schmidt, B. & Yao, J.I.-C. 2001. Seismic design and analysis of underground structures. *Tunnelling and Underground Space Technology*, 16, 247–293.

Haszeldine, R.S. & Smythe, D.K. (eds.) 1996. *Radioactive Waste Disposal at Sellafield, UK. Site Selection, Geological and Engineering Problems*. University of Glasgow, 520p.

Hawkins, A.B. (ed.) 1986. *Site Investigation Practice: Assessing BS 5930*. Geological Society of London, Engineering Geology Special Publication No. 2, 423p.

Hawkins, A.B., Larnach, W.J., Lloyd, I.M. & Nash, D.F.T. 1989. Selecting the location, and the initial investigation of the SERC soft clay test bed site. *Quarterly Journal of Engineering Geology and Hydrogeology*, 22, 281–316.

Hawkins, A.B. & Pinches, G.M. 1987. Cause and significance of heave at Llandough Hospital, Cardiff – a case history of ground floor heave due to gypsum growth. *Quarterly Journal of Engineering Geology*, 20, 41–47.

Head, J.M. & Jardine, F.M. 1992. Ground-borne vibrations arising from piling. *Technical note 142*, Construction Industry Research and Information Association (CIRIA), UK.

Heald, M.T. & Larese, R.E. 1974. Influence of coatings on quartz cementation. *Journal Sedimentary Petrology*, 44, 1269–1274.

Health and Safety Executive 1985. *The Abbeystead Explosion*. A report of the investigation by the Health and Safety Executive into the explosion on 23 May 1984 at the valve house of the Lune/Wyre Water Transfer Scheme at Abbeystead.

Health and Safety Executive 2000. *The collapse of NATM tunnels at Heathrow Airport*. 110p.

Heathcote, J.A., Jones, M.A. & Herbert, A.W. 1996. Modelling groundwater flow in the Sellafield area. *Quarterly Journal of Engineering Geology*, 29, S59–S81.

Hencher, S.R. 1976. Correspondence: A simple sliding apparatus for the measurement of rock friction. *Géotechnique*, 26, 4, 641–644.

Hencher, S.R. 1977. *The Effect of Vibration on the Friction between Planar Rock Surfaces*. Unpublished PhD thesis, Imperial College Science and Technology, London University.

Hencher, S.R. 1981a. Friction parameters for the design of rock slopes to withstand earthquake loading. *Proceedings of Conference on Dams and Earthquake, ICE*, London, 79–87.

Hencher, S.R. 1981b. *Report on Slope Failure at Yip Kan Street (11SW-D/C86) Aberdeen on 12th July 1981.* Geotechnical Control Office Report No. GCO 16/81, Geotechnical Control Office, Hong Kong, 26p.

Hencher, S.R. 1983a. *Landslide Studies 1982 Case Study No. 4 South Bay Close.* Special Project Report No. SPR 5/83, Geotechnical Control Office, Hong Kong, 38p.

Hencher, S.R. 1983b. *Landslide Studies 1982 Case Study No. 5 Tuen Mun Highway.* Special Project Report No. SPR 6/83, Geotechnical Control Office, Hong Kong, 42p.

Hencher, S.R. 1983c. *Landslide Studies 1982 Case Study No. 10 Ching Cheung Road.* Special Project Report No. SPR 11/83, Geotechnical Control Office, Hong Kong, 38p.

Hencher, S.R. 1983d. *Landslide Studies 1982 Case Study No. 1 Chai Wan Road.* Special Project Report No. SPR 2/83, Geotechnical Control Office, Hong Kong, 34p.

Hencher, S.R. 1983e. *Summary Report on Ten Major Landslides in 1982.* Special Project Report No. SPR 1/83, Geotechnical Engineering Office, Hong Kong, 28p.

Hencher, S.R. 1983f. *Landslide Studies 1982 Case Study No. 8 Tsing Yi (1).* Special Project Report No. SPR 9/83, Geotechnical Control Office, Hong Kong, 36p.

Hencher, S.R. 1985. Limitations of stereographic projections for rock slope stability analysis. *Hong Kong Engineer*, 13, 7, 37–41.

Hencher, S.R. 1987. The implications of joints and structures for slope stability. In *Slope Stability – Geotechnical Engineering and Geomorphology*, Anderson, M.G. & Richards, K.S. (eds.), UK: John Wiley & Sons Ltd, 145–186.

Hencher, S.R. 1995. Interpretation of direct shear tests on rock joints. In *Proceedings 35th US Symposium on Rock Mechanics*, Deaman & Schultz (eds.), Lake Tahoe, 99–106.

Hencher, S.R. 1996a. Fracture flow modelling: Proof of evidence. In *Radioactive Waste Disposal at Sellafield, UK: Site Selection, Geological and Engineering Problems*, Haszeldine, R.S. & Smythe, D.K. (eds.), published by Department of Geology and Applied Geology, University of Glasgow, 349–358. http://www.foe.co.uk/archive/nirex/sfoe6.html

Hencher, S.R. 2000. Engineering geological aspects of landslides. Keynote paper. Proceedings Conference on Engineering Geology HK 2000, Institution of Mining and Metallurgy, Hong Kong, 93–116.

Hencher, S.R. 2006. Weathering and erosion processes in rock – implications for geotechnical engineering. *Proceedings Symposium on Hong Kong Soils and Rocks*, March 2004, Institution of Mining, Metallurgy and Materials and Geological Society of London, 29–79.

Hencher, S.R. 2007. Hazardous ground conditions – reducing the risks. Keynote Paper. *Engineering Geology in Geotechnical Risk Management*, Hong Kong Regional Group of the Geological Society of London, 32–40.

Hencher, S.R. 2008. The 'new' British and European standard guidance on rock description. A critique by Steve Hencher. *Ground Engineering*, 41, 7, 17–21.

Hencher, S.R. 2010. Preferential flow paths through soil and rock and their association with landslides. *Hydrological Processes*, 24, 1610–1630.

Hencher, S.R. & Acar, I.A. 1995. The Erzincan Earthquake, Friday, 13 March 1992. *Quarterly Journal of Engineering Geology*, 28, 313–316.

Hencher, S.R., Anderson, M.G. & Martin R.P. 2006. Hydrogeology of landslides. *Proceedings of International Conference on Slopes*, Malaysia, 463–474.

Hencher, S.R. & Daughton, G. 2000. Anticipating geological problems. *The Urban Geology of Hong Kong, Geological Society of Hong Kong Bulletin*, 6, 43–62.

Hencher, S.R. & Knipe, R.J. 2007. Development of rock joints with time and consequences for engineering. *Proceedings of the 11th Congress of the International Society for Rock Mechanics*, 1, 223–226.

Hencher, S.R., Lee, S.G., Carter, T.G. & Richards, L.R. 2011. Sheeting joints – Characterisation, Shear Strength and Engineering. *Rock Mechanics and Rock Engineering*, 44, 1–22.

Hencher, S.R., Liao, Q.H. & Monighan, B. 1996. Modelling slope behaviour for open pits. *Transactions of the Institution of Mining & Metallurgy*, 105, A37–A47.

Hencher, S.R. & Mallard, D.J. 1989. On the effects of sand grading on driven pile performance. *Proceedings of the International Conference on Piling and Deep Foundations*, London, 255–264.

Hencher, S.R. & Martin, R.P. 1984. The failure of a cut slope on the Tuen Mun road in Hong Kong. *Proceedings of the International Conference on Case Histories in Geotechnical Engineering*, St Louis, Missouri, 2, 683–688.

Hencher, S.R., Massey, J.B. & Brand, E.W. 1985. Application of back analysis to some Hong Kong landslides. *Proceedings of the 4th International Symposium on Landslides*, Toronto, 1, 631–638.

Hencher, S.R. & McNicholl, D.P. 1995. Engineering in weathered rock. *Quarterly Journal of Engineering Geology*, 28, 253–266.

Hencher, S.R. & Richards, L.R. 1989. Laboratory direct shear testing of rock discontinuities. *Ground Engineering*, 22, 2, 24–31.

Hencher, S.R., Sun, H.W. & Ho, K.K.S. 2008. The investigation of underground streams in a weathered granite terrain in Hong Kong. In *Geotechnical and Geophysical Site Characterization*, Huang & Mayne (eds.), London: Taylor & Francis, 601–607.

Hencher, S.R., Toy, J.P. & Lumsden, A.C. 1993. Scale dependent shear strength of rock joints. In *Scale Effects in Rock Masses 93*, Pinto da Cunha (ed.), Rotterdam: Balkema, 233–240.

Hencher, S.R., Tyson, J.T. & Hutchinson, P. 2005. Investigating substandard piles in Hong Kong. *Proceedings 3rd International Conference on Forensic Engineering*, Institution of Civil Engineers, London, 107–118.

Heuer, R.E. 1974. Important ground parameters in soft ground tunnelling. *Proceedings of Speciality Conference on Subsurface Exploration for Underground Excavation and Heavy Construction*, ASCE, 41–45.

Hight, D. 2009. Nicoll Highway, Singapore: the post collapse investigations. Meeting Report by Cabarkapa, Z. *Ground Engineering*, October, 9–13.

Hill, S.J. & Wallace, M.I. 2001. Mass modulus of rock for use in the design of deep foundations. In *Proceedings 14th Southeast Asian Geotechnical Conference*, Hong Kong, Ho, K.K.S. & Li, K.S. (eds.), 1, 333–338.

Ho, H.Y., King, J.P. & Wallace, M.I. 2006. *A Basic Guide to Air Photo Interpretation in Hong Kong*. Applied Geoscience Centre, University of Hong Kong, 115p.

Ho, K.K.S., Sun H.W. & Hui, T.H.H. 2003. *Enhancing the Reliability and Robustness of Engineered Slopes*. Geotechnical Engineering Office, Hong Kong, GEO Report No. 139.

Hoek, E. 1968. Brittle failure of rock. In *Rock Mechanics in Engineering Practice*. Stagg & Zienkiewicz (eds.), London: Wiley and Sons, 99–124.

Hoek, E. 1999. Putting numbers to geology – an engineer's viewpoint. The 2nd Glossop Lecture. *Quarterly Journal of Engineering Geology*, 32, 1–20.

Hoek, E. 2000. Big tunnels in bad rock. 2000 Terzhagi Lecture. *ASCE Journal of Geotechnical and Geoenvironmental Engineering*, 127, 726–740.

Hoek, E. 2009. A slope stability problem in Hong Kong. Published online at http://www.rocscience.com/hoek/pdf/7_A_slope_stability_problem_in_Hong_Kong.pdf

Hoek, E. & Bray, J.W. 1974. *Rock Slope Engineering*. London: Institution of Mining and Metallurgy, 309p.

Hoek, E. & Brown, E.T. 1980. *Underground Excavations in Rock*. Institution of Mining and Metallurgy, London, 527p.

Hoek, E. & Brown, E.T. 1997. Practical estimates of rock mass strength. *International Journal of Rock Mechanics & Mining Sciences*, 34, 1165–1186.

Hoek, E. & Diederichs, M.S. 2006. Empirical estimation of rock mass modulus. *International Journal of Rock Mechanics & Mining Sciences*, 43, 203–215.

Hoek, E., Kaiser, P.K. & Bawden, W.F. 1995. *Support of Underground Excavations in Rock*. Rotterdam: A.A. Balkema, 215p.

Hoek E. & Marinos, P. 2000. Predicting Tunnel Squeezing. *Tunnels and Tunnelling International*, Part 1, 32/11, 45–51, November 2000; Part 2, 32/12, 33–36, December 2000.

Hoek, E. & Moy, D. 1993. Design of large powerhouse caverns in weak rock. In *Comprehensive Rock Engineering*, Volume 5, Surface and Underground Project Case Histories (Editor in Chief, J.A.Hudson; Volume Editor, E. Hoek), Oxford: Pergamon Press, 85–110.

Holmes, A. 1965. *Principles of Physical Geology*, 2nd edition. London: Nelson & Sons.

Holzhausen, G.R. 1989. Origin of sheet structure, 1. Morphology and boundary conditions. *Engineering Geology*, 27, 225–278.

Horn, H.M. & Deere, D.U. 1962. Frictional characteristics of minerals. *Géotechnique*, 12, 319–335.

Hoshino, K. 1993. Geological evolution from the soil to the rock: Mechanism of lithification and change of mechanical properties. In *Geotechnical Engineering of Hard Soils – Soft Rocks*, Anagnostopoulos *et al.* (eds.), Rotterdam: Balkema, 131–138.

Houghten, D.A. & Wong, C.M. 1990. Implications of the karst marble at Yuen Long for foundation investigation and design. *Hong Kong Engineer*, June, 19–27.

Hudson, J.A. 1989. *Rock Mechanics Principles in Engineering Practice*. CIRIA Ground Engineering Report: Underground Construction, London: Butterworths, 72p.

Hudson, J.A. 1992. *Rock Engineering Systems*. New York: Ellis Horwood, 185p.

Hudson, J.A. & Harrison, J.P. 1992. A new approach to studying complete rock engineering problems. *Quarterly Journal of Engineering Geology and Hydrogeology*, 25, 93–105.

Hudson, J.A. & Harrison, J.P. 1997. *Engineering Rock Mechanics. An Introduction to the Principles*. Oxford: Pergamon Press, 444p.

Hudson, R.R. & Hencher, S.R. 1984. The delayed failure of a large cut slope in Hong Kong. *Proceedings of the International Conference on Case Histories in Geotechnical Engineering*, St Louis, Missouri, 679–682.

Hungr, O., Corominas, J. & Eberhardt, E. 2005a. Estimating landslide motion mechanism, travel distance and velocity. In *Landslide Risk Management*, Hungr, Fell, Couture & Eberhardt (eds.), London: Taylor & Francis Group, 99–128.

Hungr, O., Fell, R., Couture, R. & Eberhardt, E. (eds.) 2005b. *Landslide Risk Management*. London: Taylor & Francis Group, 764p.

Hunt, R.E. 2005. *Geotechnical Engineering Investigation Handbook*, 2nd edition. Boca Raton, FL: Taylor & Francis, 1066p.

Hutchinson, J.N. 2001. Reading the ground: morphology and geology in site appraisal. *Quarterly Journal of Engineering Geology and Hydrogeology*, 334, 7–50.

Institution Civil Engineers 1996. *Sprayed Concrete Linings (NATM) for Tunnels in Soft Ground*. London: Thomas Telford, 88p.

Institution Civil Engineers 2005. *NEC3 Engineering and Construction Contract*. London: Thomas Telford.

Institute of Geological Sciences 1971. *British Regional Geology: Northern England*, 4th edition. HMSO, 125p.

International Society for Rock Mechanics 1974. *Suggested Methods for Determining Shear Strength*. Final Draft, February 1974.

International Society for Rock Mechanics 1978. Suggested methods for the quantitative description of discontinuities in rock masses. *International Journal of Rock Mechanics and Mining Sciences & Geomechanics Abstracts*, 15, 319–368.

International Society for Rock Mechanics. 1981. Basic geotechnical description of rock masses. *International Journal of Rock Mechanics Mining Sciences and Geomechanics Abstracts*, 18, 85–110.

Irfan, T.Y. 1996. Mineralogy, fabric properties and classification of weathered granites in Hong Kong. *Quarterly Journal of Engineering Geology*, 29, 5–35.

Irfan, T.Y. & Tang, K.Y. 1993. *Effect of the Coarse Fractions on the Shear Strength of Colluvium*. Geotechnical Engineering Office, GEO Report No. 23, Hong Kong, 232p.

Itasca 2004. *UDEC Version 4.0 User's Guide*, 2nd edition.

Iverson, R.M. 2000. Landslide triggering by rain infiltration. *Water Resources Research*, 36, 1897–1910.

Jahns, R.H. 1943. Sheet structure in granites, its origin and use as a measure of glacial erosion in New England. *Journal of Geology*, 51, 71–98.

Jaksa, M.B., Goldsworthy, J.S., Fenton, G.A., Kaggwa, W.S., Griffiths, D.V., Kuo, Y.L. & Poulos, H.G. 2005. Towards reliable and effective site investigations. *Géotechnique*, 55, 2, 109–121.

Janbu, N. 1973. Slope stability computations. *Embankment-Dam Engineering*, New York: John Wiley & Sons, Inc., 47–86.

Jardine, R.J., Symes, M.J. & Burland, J.B. 1984. The measurement of soil stiffness in the triaxial apparatus. *Géotechnique*, 34, 323–340.

Jiao, J.J., Ding, G.P. & Leung, C.M. 2006. Confined groundwater near the rockhead in igneous rocks in the Mid-Levels area, Hong Kong, China. *Engineering Geology*, 84, 207–219.

Jiao, J.J. & Malone, A.W. 2000. An hypothesis concerning a confined groundwater zone in slopes of weathered igneous rocks. *Symposium on Slope Hazards and their Prevention*, Hong Kong, 165–170.

Jiao, J.J., Wang, X-S. & Nandy, S. 2005. Confined groundwater zone and slope instability in weathered igneous rocks in Hong Kong. *Engineering Geology*, 80, 71–92.

Jones, A. 1971. Soil piping and stream channel initiation. *Water Resources Research*, 7, 3, 602–609.

Jones, L.D. 2006. Monitoring landslides in hazardous terrain using terrestrial LIDAR: an example from Montserrat. *Quarterly Journal of Engineering Geology and Hydrogeology*, 39, 371–373.

Kamewada, S., Gi, H.S., Taniguchi, S. & Yoneda, H. 1990 Application of borehole image processing system to survey of tunnel. In *Proceedings of the International Symposium on Rock Joints*, Barton, N.R. & Stephansson, O., (eds.), Loen, Norway, Rotterdam: Balkema, 51–58.

Karakuş, M.M. & Fowell, R.J. 2004. An insight into the New Austrian Tunnelling Method (NATM). *Rockmec'2004-VIIth Regional Rock Mechanics Symposium*, Sivas, Turkey, 14p.

Karrow, P.F. & White, O.L. 2002. A history of neotectonic studies in Ontario. *Tectonophysics*, 353, 3–15.

Katsura, S., Kosugi, K., Mizutani, T., Okunaka, S. & Mizuyama, T. 2008. Effects of bedrock groundwater on spatial and temporal variations in soil mantle ground-water in a steep granitic headwater catchment. *Water Resources Research*, 44, W09430.

Keefer, D.K. 1984. Landslides Caused by Earthquakes, *Geological Society of America Bulletin*, 95, 406–421.

Keefer, D.K. 2002. Investigating landslides caused by earthquakes – a historical review. *Surveys in Geophysics*, 23, 473–510.

Kikuchi, K. & Mito, Y. 1993. Characteristics of seepage flow through the actual rock joints. *Proceedings 2nd International Workshop on Scale Effects in Rock Masses*, Lisbon, 305–312.

Kim Y.I., Hencher, S.R., Yoon, Y.H. & Cho, S.G. 2001. Determination of the construction method for the Young dong tunnel by risk assessment. *Korea Society of Civil Engineers, Seminar*, Seoul, 200–206.

Knill, J.L. 1976. *Cow Green revisited*. Inaugural Lecture, Imperial College of Science and Technology, University of London.

Knill, J.L. 1978. Geology and Foundations. Chapter 2 in *Foundation Engineering in Difficult Ground*, Bell, F.G. (ed.), London: Newnes-Butterworths, 116–131.

Knill, J.L. 2002. Core values: the first Hans Cloos Lecture. *Proceedings 9th AEG Congress*, Durban, 1–45.

Kovacevic, N., Hight, D.W. & Potts, D.M. 2007. Predicting the stand-up time of temporary London Clay slopes at Terminal 5, Heathrow Airport. *Géotechnique*, 57, 63–74.

Krauskopf, K.B. 1988. *Radioactive Waste Disposal and Geology. Topics in the Earth Sciences*, Volume 1, London: Chapman & Hall, 145p.

Krynine, D.P. & Judd, W.R. 1957. *Principles of Engineering Geology and Geotechnics*. New York: McGraw-Hill, 730p.

Kwong, A.K.L. 2005. Drawdown and settlement measured at 1.5 km away from SSDS Stage 1 Tunnel C (a perspective from hydro-geological modelling). *Proceedings of K.W.Lo Symposium*, London, Ontario, Canada, Session D, 1–24.

Kulander, B.R. & Dean, S.L. 1995. Observations on fractography with laboratory experiments for geologists. In *Fractography: Fracture Topography as a Tool in Fracture Mechanics and Stress Analysis*, Ameen, M.S. (ed.), Geological Society London Special Publication, 92, 59–82.

Kulander, B.R., Dean, S.L. & Ward, B.J. 1990. *Fractured Core Analysis. Interpretation, Logging and Use of Natural and Induced Fractures in Core*. AAPG Methods in Exploration Series, No. 8, 88p.

Kulatilake, P.H.S.W., Shou, G., Huang, T.H. & Morgan, R.M. 1995. New peak shear strength criteria for anisotropic rock joints. *International Journal of Rock Mechanics and Mining Science & Geomechanics Abstracts*, 32, 673–669.

Kveldsvik, V., Nilsen, B., Einstein, H.H. & Nadim, F. 2008. Alternative approaches for analyses of a 100,000m^3 rock slide based on Barton-Bandis shear strength criterion. *Landslides*, 5, 161–176.

Lambe, T.W. & Whitman, R.V. 1979. *Soil Mechanics*. New York: Wiley, 553p.

Leach, B. & Herbert, R. 1982. The genesis of a numerical model for the study of the hydrogeology of a steep hillside in Hong Kong. *Quarterly Journal of Engineering Geology*, 15, 243–259.

Lee, K.C., Jeng, F.S., Huang, T.S., Hsieh, Y.M. & Orense, R.P. 2010. Can tilt test provide insight regarding frictional behaviour of sandstone under seismic excitation? Proceedings New Zealand Society for Earthquake Engineering 2010 Conference, Paper 19, 8p.

Lee, S.G. & Hencher, S.R. 2009. The repeated collapse of a slope despite continuous reassessment and remedial works. *Engineering Geology*, 107, 16–41.

Leeder, M. 1999. *Sedimentology and Sedimentary Basins*. Oxford: Blackwell Publishing, 592p.

Legislative Council Panel on Environmental Affairs 2000. Paper for discussion on 25 October 2000.

Lerouiel, S. & Tavernas, F. 1981. Pitfalls of back-analysis. *Proceedings 10th ICSMFE*, Stockholm, 1, 185–190.

Li, Z.H., Huang, H.W., Xue, Y.D. & Yin, J. 2009. Risk assessment of rockfall hazards on highways. *Georisk: Assessment and Management of Risk for Engineered Systems and Geohazards*, 3, 14–154.

Lindquist, E.S. & Goodman, R.E. 1994.Strength and deformation properties of a physical model melange. *Proceedings 1st North American Rock Mechanics Symposium*, Austin, Texas, 843–858.

Lumb, P. 1962. Effect of rain storms on slope stability. *Proceedings of the Symposium on Hong Kong Soils*, Hong Kong, 73–87.

Lumb, P. 1976. Discussion on the assessment of landslide potential with recommendations for future research by A.A. Beattie & E.P.Y. Chau. *Hong Kong Engineer*, 4, 2, 55.

Lumsdaine, R.W. & Tang, K.Y. 1982. A comparison of slopes stability calculations. *Proceedings of the Seventh Southeast Asian Geotechnical Conference*, Hong Kong, 31–38.

Maidl, B., Schmid, L., Ritz W. & Herrenknecht, M. 2008. *Hardrock Tunnel Boring Machines*. Berlin: Ernst and Young, 343p.

Malone, A.W. 1998. Slope movement and failure: evidence from field observations of landslides associated with hillside cuttings in saprolites in Hong Kong. *Proceedings 13th Southeast Asian Geotechnical Conference*, Taipei, 2, 81–90.

Magnus R., Teh C.I., Lau J.M. 2005. *Report of the Committee of Inquiry into the Incident at the MRT Circle Line Worksite that Led to the Collapse of Nicoll Highway on 20 April 2005*.

Malone, A.W., Hansen, A., Hencher, S.R. & Fletcher, C.J.N. 2008. Post-failure movements of a large slow rock slide in schist near Pos Selim, Malaysia. *The 10th International Symposium on Landslides and Engineered Slopes*, Xian, 457–461.

Mandl, G. 2005. *Rock Joints. The Mechanical Genesis*. Berlin and New York: Springer, 221p.

Marcusson, W.F. III, Hynes, M.E. & Franklin, A.G. 1990. Evaluation and use of residual strength in seismic safety analysis of embankments. *Earthquake Spectra*, 6, 529–572.

Marinos, P. & Hoek, E. 2000. GSI – a geologically friendly tool for rock mass strength estimation. *Proceedings of GeoEng 2000*, Melbourne, Australia, 1422–1442.

Marshall, R.H. & Flanagan, R.F. 2007. Singapore's Deep Tunnel Sewerage System – Experiences and Challenges. *RETC 2007*, Toronto, 1308–1319.

Martel, S.J. 2006. Effect of topographic curvature on near-surface stresses and application to sheeting joints. *Geophysical Research Letters*, 33, LO1308.

Martin, D.C. 1994. Quantifying drilling-induced damage in samples of Lac du Bonnet granite. *Proceedings 1st North American Rock Mechanics Symposium*, Austin, Texas, 419–427.

Martin, R.P. & Hencher, S.R. 1986. Principles for description and classification of weathered rocks for engineering purposes. In *S I Practice: Assessing BS5930, Geological Society, Engineering Geology Special Publication No. 2*, Hawkins, A.B. (ed.), 299–308.

Masset, O. & Loew, S. 2010. Hydraulic conductivity distribution in crystalline rocks, derived from inflows to tunnels and galleries in the Central Alps, Switzerland. *Hydrogeology Journal*, 18, 863–891.

Mathews, M.C., Clayton, C.R.I. & Owen, Y. 2000. The use of field geophysical techniques to determine ground stiffness profiles. *Proceedings Institution Civil Engineers, Geotechnical Engineering*, January, 31–42.

Matsuo, S. 1986. Tackling floods beneath the sea. *Tunnels & Tunnelling*, March, 42–45.

Maunsell Consultants Asia Ltd. 2000. *Investigation of unusual settlement in Tsuen Kwan O Town Centre*. Final Report, November 2000, Territory Development Department, Hong Kong Government.

Maury, V. 1993. An overview of tunnel, underground excavation and boreholes collapse mechanisms. In *Comprehensive Rock Engineering*, Volume 4, Editor in Chief (J. Hudson), Oxford: Pergamon Press, 369–412.

Mayne, P.W., Christopher, B.R. & deJong, J. 2001. *Manual on Subsurface Investigations*. National Highway Institute, Publication No. FHWA NHI-01-031 Federal Highway Administration, Washington, DC, 305p.

MacGregor, F., Fell, R., Mostyn, G.R., Hocking, G. & McNally, G. 1994. The estimation of rock rippability. *Quarterly Journal of Engineering Geology*, 27, 123–144.

McCabe, B.A., Nimmons G.J. & Egan, D. 2009. A review of field performance of stone columns in soft soils. *Proceedings of the Institution of Civil Engineers: Geotechnical Engineering*, 162, 323–334.

McGown, A., Marsland, A., Radwan, A.M. & Gabr, A.W.A. 1980. Recording and interpreting soil macrofabric data. *Géotechnique*, 30, 4, 417–447.

McLearie, D.D., Forekman, W., Hansmire, W.H. & Tong, E.K.H. 2001. Hong Kong Strategic Sewage Disposal Scheme Stage 1 deep tunnels. *2001 RETC Proceedings*, 487–498.

McMahon, B.K. 1985. Geotechnical design in the face of uncertainty. *Journal of the Australian Geomechanics Society*, 10, 7–19.

McNicholl, D.P., Pump, W.L. & Cho, G.W.F. 1986. Groundwater control in large scale slope excavations – five case histories from Hong Kong. In *Groundwater in Engineering Geology*, Cripps, J.C. *et al.* (eds.), Geological Society, Engineering Geology Special Publication no. 3, 561–576.

Mead, L. & Austin, G.S. 2005. Dimension stone. *Industrial Minerals and Rocks*, 7th edition. Littleton CO: AIME-Society of Mining Engineers, 907–923.

Megaw, T.M. & Bartlett, J.V. 1981. *Tunnels: Planning, Design, Construction, Volume 1*. New York: Ellis Horwood Limited, 284p.

Menard, L. & Broise, Y. 1975. Theoretical and practical aspects of dynamic consolidation. *Géotechnique*, 25, 3–18.

Michie, U. 1996. The geological framework of the Sellafield area and its relationship to hydrogeology. *Quarterly Journal of Engineering Geology*, 29, S13–S28.

Mikkelsen, P.E. & Green, G.E. 2003. Piezometers in fully grouted boreholes. *Proceedings Symposium on Field Measurements in Geomechanics*, Oslo, 1–10.

Montgomery, D.R., Dietrich, W.E. & Heffner, J.T. 2002. Piezometric response in shallow bedrock at CB1: Implications for runoff generation and landsliding. *Water Resource Research*, 38, 12, 1274, 10-1–10-18.

Moore, R. & Brunsden, D. 1996. A physico-chemical mechanism of seasonal mudsliding. *Géotechnique*, 46, 259–278.

Moore, R., Hencher, S.R. & Evans, N.C. 2001. An approach for area and site-specific natural terrain hazard and risk assessment, Hong Kong. *Proc. 14th SEA Geotech. Conference*, Hong Kong, 155–60.

Morgenstern, N.R. 2000. Common ground. *Proceedings of GeoEng 2000*, Melbourne, Australia, 1, 1–30.

Morgenstern, N.R. & Price, V.E. 1965. The analysis of the stability of general slip surfaces. *Géotechnique*, 15, 79–93.

Moye, D.G. 1955. Engineering Geology for the Snowy Mountains scheme. *Journal of the Institution of Engineers*, Australia, 27, 287–298.

Muir, T.R.C., Smethurst, B.K. & Finn, R.P. 1986. Design and construction of high rock cuts for the Kornhill Development, Hong Kong. *Proceedings of Conference on Rock Engineering and Excavation in an Urban Environment*, Hong Kong, Institution of Mining and Metallurgy, 309–329.

Muir Wood, A. 2000. *Tunnelling: Management by Design*. London: E & FN Spon, 320p.

Murray, A.D. & Gray, A.M. 1997. The Pergau Hydroelectric Project Part 3: civil engineering design. *Proceedings institution Civil Engineers Water, Maritime & Energy*, 124, 173–188.

Newmark, N.M. 1965. Effects of earthquakes on dams and embankments. *Géotechnique*, 2, 139–160.

New Zealand Geotechnical Society 2005. *Field Description of Soil and Rock*. 36p.

Nichols, T.C. Jr. 1980. Rebound, its nature and effect on engineering works. *Quarterly Journal of Engineering Geology*, 13, 133–152.

Nicholson, D.T., Lumsden, A.C. & Hencher, S.R. (2000). Excavation-induced deterioration of rock slopes. *Proceedings of Conference on Landslides in Research, Theory and Practice*, Cardiff: Thomas Telford, Volume 3, 1105–1110.

Nirex 2007. *Geosphere Characterisation Report: Status Report*. October 2006, 198p.

Nixon, P.J. 1978. Floor heave in buildings due to the use of pyritic shales as fill material. *Chemistry and Industry*, 160–164.

Norbury, D. 2010. *Soil and Rock Description in Engineering Practice*. CRC Press, 288p.

North Wales Geology Association 2006. *Newsletter*, 46, July, p. 1.

Nowson, W.J.R. 1954. The history and construction of the foundations of the Asia Insurance Building, Singapore. *Proceedings Institution Civil Engineers*, Pt 1, 3, 407–443.

Oldroyd, D.R. 2002. Nirex and the great denouement. Chapter 20 in: *Earth, Water, Ice and Fire: Two Hundred Years of Geological Research in the English Lake District*, Geological Society Memoir No. 25. xvi, London, Bath: Geological Society of London, 328p.

Ollier, C.D. 1975. *Weathering*, 2nd impression. London: Longman Group Limited, 304p.

Ollier, C.D. 2010. Very deep weathering and related landslides. In *Weathering as a Predisposing Factor to Slope Movements*, Calceterra, D. & Parise, M. (eds.), Geological Society, London: Engineering Geology Special Publications, 23, 5–14.

Olsson, O. & Gale, J.E. 1995. Site assessment and characterization for high-level nuclear waste disposal: results from the Stripa Project, Sweden. *Quarterly Journal of Engineering Geology and Hydrogeology*, 28, Supplement 1, S17–S30.

Olsson, R. & Barton, N. 2001. An improved model for hydromechanical coupling during shearing of rock joints. *International Journal of Rock Mechanics & Mining Sciences*, 38, 317–329.

Orr, W.E., Muir-Wood, A., Beaver, J.L., Ireland, R.J. & Beagley, D. 1991. Abbeystead Outfall Works: background to repairs and modifications – and the lessons learned. *Journal of the Institution of Water and Environmental Management*, 5, 7–20.

Osipov, V.I. 1975. Structural bonds and the properties of clays. *Bulletin of the International Association of Engineering Geology*, 12, 13–20.

Palmström, A. 1982. The volumetric joint count – a useful and simple measure of the degree of rock jointing. *Proceedings 4th Congress International Association Engineering Geologists*, Delhi, 221–228.

Papaliangas, T., Lumsden, A.C., Manolopoulou, S. & Hencher, S.R. 1990. Shear strength of modelled filled rock joints. *Proceedings of the International Symposium on Rock Joints*, Loen, Norway, 275–282.

Papaliangas, T.T., Hencher, S.R. & Lumsden, A.C. 1994. Scale independent shear strength of rock joints. *Proceedings IV CSMR/ Integral Approach to Applied Rock Mechanics*, Santiago, Chile, 123–133.

Papaliangas, T.T., Hencher, S.R. & Lumsden, A.C. 1995. A comprehensive peak strength criterion for rock joints. *Proceedings 8th International Congress on Rock Mechanics*, Tokyo, 1, 359–366.

Parks, C.D. 1991. A review of the mechanisms of cambering and valley bulging. In *Quarternary Engineering Geology*, Forster, A., Culshaw, M.G., Cripps, J.C., Little, J.A. & Moon, C. (eds.), Geological Society Engineering Geology Special Publication No.7, 381–388.

Parry, S., Campbell, S.D.G. & Fletcher, C.J.N. 2000. Kaolin in Hong Kong saprolites – genesis and distribution. *Proceedings Conference Engineering Geology HK 2000*, IMM HK Branch, 63–70.

Patton, F.D. 1966. Multiple modes of shear failure through rock. *Proceedings 1st International Congress International Society Rock Mechanics*, Lisbon, 1, 509–513.

Peck, R.B., Hanson, W.E. & Thornburn, T.H. 1994. *Foundation Engineering*, 2nd edition. New York: John Wiley and Sons, 514p.

Petley, D. 2011 (ongoing). Dave's Landslide Blog. http://daveslandslideblog.blogspot.com/

Pettifer, G.S. & Fookes, P.G. 1994. A revision of the graphical method for assessing the excavatability of rock. *Quarterly Journal of Engineering Geology*, 27, 145–164.

Phillips, F.C. 1973. *The Use of Stereographic Projection in Structural Geology*, 3rd edition. London: Edward Arnold, 90p.

Phillipson, H.B. & Chipp, P.N. 1982. Air foam sampling of residual soils in Hong Kong. *Proceedings of the ASCE Specialty Conference on Engineering and Construction in Tropical and Residual Soils*, Honolulu, 339–356.

PIANC 2000. *Site Investigation Requirements for Dredging Works*. Report of Working Group No. 23 of the Permanent Technical Committee II, International Navigation Association, 32p.

Picarelli, L. & Di Maio, C. 2010. Deterioration processes of hard clays and clay shales. In *Weathering as a Predisposing Factor to Slope Movements*, Calceterra, D. & Parise, M. (eds.), Geological Society, London: Engineering Geology Special Publications, 23, 15–32.

Pierson, T.C. 1983. Soil pipes and slope stability. *Quarterly Journal of Engineering Geology*, 16, 1–15.

Pine, R.J. & Roberds, W.J. 2005. A risk-based approach for the design of rock slopes subject to multiple failure modes – illustrated by a case study in Hong Kong. *International Journal of Rock Mechanics & Mining Sciences*, 42, 261–275.

Pinho, A., Rodrigues-Carvalho, J.A., Gomes, C.F. & Duarte, I.M. 2006. Overview of the evaluation of the state of rock weathering by visual inspection. 10th IAEG Congress, Engineering Geology for Tomorrow's Cities, Nottingham, Paper 260.

Pirazzoli, P.A. 1996. *Sea-Level Changes. The Last 20000 Years*. Chichester: John Wiley & Sons, 211p.

Piteau, D.R. 1973. Characterising and extrapolating rock joint properties in engineering practice. *Rock Mechanics*, Supplement 2, 2–31.

Pollard, D.D. & Aydin, A. 1988. Progress in understanding jointing over the past century. *Geological Society America Bulletin*, 100, 1181–1204.

Pope, R.G., Weeks, R.C. & Chipp, P.N. 1982. Automatic recording of standpipe piezometers. *Proceedings of the 7th Southeast Asian Geotechnical Conference*, Hong Kong, 1, 77–89.

Pöschl, I. & Kleberger, J. 2004 Geotechnical risk in rock mass characterisation – a concept. *Course on Geotechnical Risks in Rock Tunnelling*, Aveiro, Portugal, April, 11p.

Poulos, H.G. 2005. Pile behavior – consequences of geological and construction imperfections. 40th Terzaghi Lecture, *Journal Geotechnical & Geoenvironmental Engineering*, ASCE, 131, 538–563.

Powderham, A.J. 1994. An overview of the observational method: development in cut and cover and bored tunnelling projects. *Géotechnique*, 44, 619–636.

Power, C.M. 1998. *Mechanics of Modelled Rock Joints under True Stress Conditions Determined by Electrical Measurements of Contact Area*. Unpublished PhD thesis, University of Leeds, 293p.

Power, C.M. & Hencher, S.R. 1996. A new experimental method for the study of real area of contact between joint walls during shear. *Rock Mechanics, Proceedings 2nd North American Rock Mechanics Symposium*, Montreal, 1217–1222.

Power, M.S., Rosidi, D., Kaneshiro, J.Y. 1998. Seismic vulnerability of tunnels and underground structures revisited. In *North American Tunneling*, Ozdemir, L. (ed.), Rotterdam: Balkema, 243–250.

Preene, M. & Brassington, F.C. 2003. Potential groundwater impacts from civil engineering works. *Water & Environment Journal*, 17, 59–64.

Priest, S.D. 1993. *Discontinuity Analysis for Rock Engineering*. London: Chapman & Hall, 473p.

Priest S.D. & Brown E.T. 1983 Probabilistic stability analysis of variable rock slopes. *Transactions Institution Mining & Metallurgy*, 92, A1–12.

Price, N.J. 1959. Mechanics of jointing in rocks. *Geological Magazine*, 96, 149–167.

Price, N.J. & Cosgrove, J. 1990. *Analysis of Geological Structures*. Cambridge: Cambridge University Press.

Puller, M. 2003. *Deep Excavations: A Practical Manual*, 2nd edition. London: Thomas Telford Ltd., 584p.

Pye, K. & Miller, J.A. (1990). Chemical and biochemical weathering of pyritic mudrocks in a shale embankment. *Quarterly Journal of Engineering Geology*, 23, 365–382.

Quinn J.D., Philip L.K. & Murphy W. 2009. Understanding the recession of the Holderness Coast, East Yorkshire, UK: a new presentation of temporal and spatial patterns. *Quarterly Journal of Engineering Geology and Hydrogeology*, 42, 165–178.

Rawnsley, K.D. 1990. *The Influence of Joint Origin on Engineering Properties*. Unpublished PhD thesis, University of Leeds, 388p.

Rawnsley, K.D., Hencher, S.R. & Lumsden, A.C. 1990. Joint origin as a predictive tool for the estimation of geotechnical properties. *Proceedings of the International Symposium on Rock Joints*, Loen, Norway, 91–96.

Rawnsley, K.D., Rives, T., Petit, J.-P., Hencher, S.R. & Lumsden, A.C. 1992. Joint development in perturbed stress fields near faults. *Journal of Structural Geology*, 14, 939–951.

Reason, J. 1990. *Human Error*. New York: Cambridge University Press.

Richards, K.S. & Reddy, K.R., 2007. Critical appraisal of piping phenomena in earth dams. *Bulletin Engineering Geology Environment*, 66, 381–402.

Richards, L.R. & Cowland, J.W. 1982. The effect of surface roughness on the field shear strength of sheeting joints in Hong Kong granite. *Hong Kong Engineer*, 10, 39–43.

Richards, L.R. & Cowland, J.W. 1986. Stability evaluation of some urban rock slopes in a transient groundwater regime. *Proceedings Conference on Rock Engineering and Excavation in an Urban Environment*, IMM, Hong Kong, 357–363, (Discussion 501–6).

Richards, L.R. & Read, S.A.L. 2007. New Zealand greywacke characteristics and influences on rock mass behaviour. *Proceedings 11th Congress of the International Society of Rock Mechanics*, Lisbon, 1, 359–364.

Richey, J.E. 1948. *Scotland. The Tertiary Volcanic Districts*, 2nd edition revised. British Regional Geology, Her Majesty's Stationery Office, 105p.

Rico, M., Benito, G., Salgueiro, A.R., Díez-Herrero, A. & Pereira, H.G. 2008. Reported tailings dam failures. A review of the European incidents in the worldwide context. *Journal of Hazardous Materials*, 152, 846–852.

RILEM 2003. Recommended Test Method AAR-I, Detection of Potential Alkali-Reactivity of Aggregates – Petrographic Method, *Materials & Structures*, 36, 480–496.

Rives, T., Rawnsley, K.D. & Petit, J.-P. 1994. Analogue simulation of natural orthogonal joint set formation in brittle varnish. *Journal of Structural Geology*, 16, 419–429.

Robertshaw, C. & Tam, T.K. 1999. Tai Po to Butterfly Valley Treated Water Transfer. *World Tunnelling*, 12, 8, 383–385.

Rodriguez, C.E. 2001. Hazard *Assessment of Earthquake-induced Landslides on Natural Slopes*. Unpublished PhD thesis, Imperial College of Science and Technology, University of London, 367p.

Rogers, C.M. & Engelder, T. 2004. The feedback between joint-zone development and downward erosion of regularly spaced canyons in the Navajo Sandstone, Zion National Park, Utah. In *The Initiation, Propagation and Arrest of Joints and Other*

Fractures, Cosgrove, J.W. & Engelder, T. (eds.), Geological Society Publication, 231, 49–71.

Roorda, J., Thompson, J.C. & White, O.L., 1982. The analysis and prediction of lateral instability in highly stressed, near-surface rock strata. *Canadian Geotechnical Journal*, 19, 4, 451–462.

Romana, M. 1991. SMR Classification. *Proceedings 7th Congress Rock Mechanics*, ISRM. Aachen, Germany, 2, 955–960.

Roscoe, K.H., Schofield, A.N. & Wroth, C.P. 1958. On the yielding of soils. *Géotechnique*, 8, 22–53.

Ruxton, B.P. & Berry, L. 1957 The weathering of granite and associated erosional features in Hong Kong. *Bulletin of the Geological Society of America*, 68, 1263–1292.

Sadiq, A.M. & Nasir, S.J. 2002. Middle Pleistocene karst evolution in the State of Qatar, Arabian Gulf. *Journal of Cave and Karst Studies*, 64, 132–139.

Salehy, M.R., Money, M.S. & Dearman, W.R. 1977. The occurrence and engineering properties of intra-formational shears in Carboniferous rocks. In *Proceedings Conference on Rock Engineering*, Newcastle, Attewell, P.B. (ed.), Newcastle: London, 311–328.

Santamarina, J.C. & Cho, G.C. (2004). Soil behaviour: The role of particle shape. In *Advances in Geotechnical Engineering: The Skempton Conference*, Jardine, R.J. et al. (eds.), 29–31 March 2004, London: Thomas Telford 1, 604–617.

Sarma, S.K. 1975. Seismic stability of earth dams and embankments. *Géotechnique*, 25, 743–761.

Sausse, J. & Genter, A. 2005. Types of permeable fractures in granite. *Geological Society, London, Special Publications* 2005, 240, 1–14.

Schmidt, B. 1966. Discussion on earth pressures at rest related to stress history. *Canadian Geotechnical Journal*, 3, 239–242.

Schneider, H.J. 1976. The friction and deformation behaviour of rock joints. *Rock Mechanics*, 8, 169–184.

Schofield, A.N. 2006. Interlocking, and peak and design strengths. *Géotechnique*, 56, 357–358.

Scholz, C.H. 1990. *The Mechanics of Earthquakes and Faulting*. Cambridge: Cambridge University Press, 439p.

Seed, H.B., Idriss, I.M. & Kiefer, F.W. 1968. *Characteristics of Rock Motions during Earthquakes*, EERC 68–5, University of California, Berkeley, September.

Seed, H.B. & Idriss, I.M. 1982. *Ground Motions and Soil Liquefaction during Earthquakes*, Monograph No. 5, Earthquake Engineering Research Institute, Berkeley, California, 134p.

Selby, M.J. 1993. *Hillslope Materials and Processes*, 2nd edition. Oxford: Oxford University Press, 451p.

Seol, H. & Jeong, S. 2009. Load-settlement behavior of rock-socketed drilled shafts using Osterberg-Cell tests. *Computers and Geotechnics*, 36, 1134–1141.

Sewell, R.J., Campbell, S.D.G., Fletcher, C.J.N., Lai, K.W. & Kirk, P.A. 2000. *The Pre-Quaternary Geology of Hong Kong*. Geotechnical Engineering Office, Civil Engineering Department, Hong Kong SAR Government, 181p plus 4 maps.

Shaw, R. & Owen, R.B. 2000. Geomorphology and ground investigation planning. *Proceedings Conference on Engineering Geology HK 2000*, Institution of Mining and Metallurgy, Hong Kong, 151–163.

Shibata, T. & Taparaska, W. 1988. Evaluation of liquefaction potential of soil using cone penetration tests. *Soils and Foundations*, 28, 49–60.

Shirlaw, J.N. 2006. Discussion: Hull wastewater flow transfer tunnel: tunnel collapse and causation investigation. *Geotechnical Engineering*, 2, 126–127.

Shirlaw, J.N., Broome, P.B., Chandrasegaran, S., Daley, J., Orihara, K., Raju, G.V.R., Tang, S.K., Wong, K.S. & Yu, K. 2003. The Fort Canning Boulder Bed. *Underground Singapore 2003. Engineering Geology Workshop*, 388–407.

Shirlaw, J. N., Hencher S.R. & Zhao, J. 2000. Design and construction issues for excavation and tunnelling in some tropically weathered rocks and soils. *Proceedings of GeoEng 2000*, Melbourne, Australia, Volume 1: Invited Papers, 1286–1329.

Shirlaw, N., Ong, J.C.W., Rosser, H.B., Tan, C.G., Osborne, N.H. & Heslop, P.E. 2003. Local settlements and sinkholes due to EPB tunneling, *Proceedings ICE, Geotech. Engineering*, 156 (GE4), 193–211.

Simons, N., Menzies, B. & Mathews, M. 2001. *A Short Course in Soil and Rock Slope Engineering*. London: Thomas Telford, 432p.

Singapore Standards 2003, *Code of Practice for Foundations*. CP4:2003, Spring, Singapore, 253p.

Skempton, A.W. 1953. The colloidal 'activity' of clays. *Proceedings 3rd International Conference of Soil Mechanics and Foundation Engineering*, 1, 57–61.

Skempton, A.W. 1970. The consolidation of clays by gravitational compaction. *Quarterly Journal of Engineering Geology*, 125, 373–411.

Skempton, A.W., Leadbetter, A.D. & Chandler, R.J. 1989. The Mam Tor landslide, North Derbyshire. *Philosophical Transactions Royal Society London*, A, 329, 503–547.

Skempton, A.W. & Macdonald, D.H. 1956. The allowable settlement of buildings. *Proceedings of the Institution of Civil Engineers*, Part 3, 5, 727–784.

Skempton, A.W., Norbury, D., Petley, D.J. & Spink, T.W. 1991. Solifluction shears at Carsington, Derbyshire. Geological Society, London, *Engineering Geology Special Publications*, 7, 381–387.

Skempton, A.W., Schuster, R.L. & Petley, D.J. 1969. Joints and fissures in the London Clay at Wraysbury and Edgeware. *Géotechnique*, 19, 205–217.

Skempton, A.W. & Vaughan, P.R. 1993. The failure of Carsington Dam. *Géotechnique*, 43, 1, 151–173.

Slob, S. 2010. *Automated Rock Mass Characterisation Using 3-D Terrestrial Laser Scanner*. Unpublished PhD thesis, Technical University of Delft, 287p.

Smart, P.L. 1985. Applications of fluorescent dye tracers in the planning and hydrological appraisal of sanitary landfills. *Quarterly Journal of Engineering Geology*, 18, 275–286.

Smith, M.R. & Collis, L. (eds.) 2001. *Aggregates*. Geological Society, London, Engineering Geology Special Publications No. 17, 339p.

Spang, K. & Egger, P. 1990. Action of fully-grouted bolts in jointed rock and factors of influence, *Rock Mechanics and Rock Engineering*, 23, 201–229.

Spink, T. 1991. Periglacial discontinuities in Eocene clays near Denham, Buckinghamshire. In *Quarternary Engineering Geology*, Forster, A., Culshaw, M.G., Cripps, J.C., Little, J.A. & Moon, C. (eds.), Geological Society Engineering Geology Special Publication No. 7, 389–396.

Starfield, A.M. & Cundall, P.A. 1988. Towards a methodology for rock mechanics modelling. *International Journal of Rock Mechanics and Mining Sciences & Geomechanics Abstracts*, 25, 99–106.

Starr, D., Dissanayake, A., Marks, D., Clements, J. & Wijeyakulasuriya, V. 2010. South West Transport Corridor Landslide. *Queensland Roads*, 8, 57–73.

Statham, I. & Baker, M. 1986. Foundation problems on limestone: A case history from the Carboniferous Limestone at Chepstow, Gwent. *Quarterly Journal of Engineering Geology*, 19, 191–201.

Sterling, R.L. 1993. The expanding role of rock engineering in developing national and local infrastructures. In *Comprehensive Rock Engineering*, Volume 5. Surface and Underground Project Case Histories, (Editor in Chief, J.A. Hudson; Volume Editor, E. Hoek), Oxford: Pergamon Press, 1–27.

Stille, H. & Palmström, A. 2003. Classification as a tool in rock engineering. *Tunnelling and Underground Space Technology*, 18, 4, 331–345.

Stow, D.A.V. 2005. *Sedimentary Rocks in the Field. A Colour Guide.* Manson Publishing Ltd., 320p.

Streckeisen, A. 1974. Classification and nomenclature of plutonic rocks: IUGS Subcommission on the Systematics of Igneous Rocks. *Geologische Rundschau*, 63, 773–786.

Streckeisen, A. 1980. Classification and nomenclature of volcanic rocks, lamprophyres, carbonatites and melilitic rocks: IUGS Subcommission on the Systematics of Igneous Rocks. *Geologische Rundschau*, 69, 1, 194–207.

Stringer, C. 2006. *Homo Britannicus*. London: Allen Lane, 319p.

Styles, K.A. & Hansen, A. 1989. *Geotechnical Area Studies Programme: Territory of Hong Kong* (GASP Report No. XII). Geotechnical Control Office, Hong Kong, 356p.

Swannell, N.G. & Hencher, S.R. Cavern design using modern software. *Proceedings of 10th Australian Tunnelling Conference*, Melbourne, March 1999.

Sweeney, D.J. & Robertson, P.K. 1979. A fundamental approach to slope stability problems in Hong Kong. *Hong Kong Engineer*, 7, 10, 35–44.

Tada, R. & Siever, R. 1989. Pressure solution during diagenesis. *Annual Reviews Earth Planetary Sciences*, 17, 89–118.

Talbot, D.K., Hodkin, D.L. & Ball, T.K. (1997) Radon investigations for tunnelling projects: a case study from St. Helier, Jersey. *Quarterly Journal of Engineering Geology*, 30, 2, 115–122.

Tam, A. 2010. Lofty aspirations at destination West Kowloon. *Hong Kong Engineer*, May, 8–11.

Tanner, K. & Walsh, P. 1984. *Hallsands: A Pictorial History*. Kingsbridge: Tanner & Walsh, 32p.

Tapley, M.J., West, B.W., Yamamoto, S. & Sham, R. 2006. Challenges in construction of Stonecutters Bridge and progress update. *International Conference on Bridge Engineering*, Hong Kong.

Tarkoy, P.J. 1991. Appropriate support selection, *Tunnels and Tunneling*, 23 (October), 42–45.

Terzhagi, K. 1925. The physical causes of proportionality between pressure and frictional resistance. From *Erdmechanic*, translated by A. Casagrande in *From Theory to Practice in Soil Mechanics*, J. Wiley.

Terzhagi, K. 1929. *The Effect of Minor Geological Details on the Stability of Dams*. American Institute of Mining and Metallurgical Engineers, Technical Publication 215, 31–44.

Thomas, A.L. & La Pointe, P.R. 1995. Conductive fracture identification using neural networks. *Proceedings 35th US Symposium on Rock Mechanics*, Lake Tahoe, 627–632.

Thorpe, R.S. & Brown, G.C. 1985. *The Field Description of Igneous Rocks*. Milton Keynes, UK: The Open University Press, 154p.

Thrush, P.W. & the Staff of the US Bureau of Mines 1968. *A Dictionary of Mining, Mineral and Related Terms*. US Department of the Interior, 1269p.

Tika, Th., Sarma, S.K. & Ambraseys, N. 1990. Earthquake induced displacements on pre-existing shear surfaces in cohesive soils. *Proc. 9th Eur. Conference on Earthquake Engineering*, 4A, 16–25.

Todd, D.K. 1980. *Groundwater Hydrology*, 2nd edition. New York: Wiley, 535p.

Tomlinson, M.J. 2001. *Foundation Design and Construction*, 7th edition. England: Prentice Hall, 569p.

Tottergill, B.W. 2006. *FIDIC Users' Guide: A Practical Guide To The 1999 Red And Yellow Books: Incorporating Changes And Additions To The 2005 MDB Harmonised Edition*. London: Thomas Telford.

Trenter, N. 2003. Understanding and containing geotechnical risk. *Proceedings of the Institution of Civil Engineers: Civil Engineering*, 156, 42–48.

Trotter, F.M., Hollingworth, S.E., Eastwood, T. & Rose, W.C.C. 1937. Gosforth District (One-inch Geological Sheet 37 New Series). In *Memoirs of the Geological Survey of Great Britain*, Department of Scientific and Industrial Research.

Trurnit, P. 1968. Pressure solution phenomena in detrital rocks. *Sedimentary Geology*, 2, 89–114.

Tsuji, H., Sawada, T. & Takizawa, M. 1996. Extraordinary inundation accidents in the Seikan undersea tunnel. *Proceedings Institution Civil Engineers, Geotechnical Engineering*, 119, 1–14.

Tucker, M.E. 1982. *The Field Description of Sedimentary Rocks*. Milton Keynes, UK: The Open University Press, 112p.

Turner, A.K. 2006. Challenges and trends for geological modelling and visualisation. *Bulletin Engineering Geology and Environment*, 65, 109–127.

Twidale, C.R. 1973. On the origin of sheet jointing. *Rock Mechanics*, 5, 163–187.

Uchida, T., Kosugi, K. & Mizuyama, T. 2001. Effects of pipeflow on hydrological process and its relation to landslide: a review of pipeflow studies in forested headwater catchments. *Hydrological Processes*, 15, 2151–2174.

Ulusay, R., Hudson, J.A. (eds.) 2006. *The Complete ISRM Suggested Methods for Rock Characterisation, Testing and Monitoring*.

van der Berg J.P., Clayton C.R.I., Powell D.B. 2003. Displacements ahead of an advancing NATM tunnel in the London clay. *Géotechnique*, 53, 767–784.

Varley, P.M. & Warren, C.D. 1996. History of the geological investigations for the Channel Tunnel. Chapter 2 in *Engineering Geology of the Channel Tunnel*, Harris, C.S., Hart, M.B., Varley, P.M. & Warren, C. (eds.), London: Thomas Telford, 5–18.

Vaughan, P.R. 1969. A note on sealing piezometers in boreholes. *Géotechnique*, 19, 405–413.

Vervoort, A. & De Wit, K. 1997. Correlation between dredgeability and mechanical properties of rock. *Engineering Geology*, 47, 259–267.

Vidal Romani, J.R. & Twidale, C.R. 1999. Sheet fractures, other stress forms and some engineering implications. *Geomorphology*, 31, 13–27.

Wakasa, S., Matsuzaki, H., Tanaka, Y. & Matskura, Y. 2006. Estimation of episodic exfoliation rates of rock sheets on a granite dome in Korea from cosmogenic nuclide analysis. *Earth Surface Processes and Landforms*, 31, 1246–1256.

Wallis, K. 2000. *Breakthroughs Made on Hong Kong Sewer*. T & T International, December, 2000.

Wallis, S. 1999. Heathrow failures highlight NATM (abuse?) misunderstandings. *Tunnel*, 3/99, 66–72.

Waltham, A.C. 2002. *Foundations of Engineering Geology*, 2nd edition. London: Spon Press, 92p.

Waltham, A.C. 2008. Lötschberg tunnel disaster, 100 years ago. *Quarterly Journal of Engineering Geology & Hydrogeology*, 41,131–136.

Walton, J.G., 2007. *Unforeseen Ground Conditions and Allocation of Risk: Before the Roof Caved in*. Paper delivered at the first of two sessions organised by the Society of Construction Law on 1 May 2007.

Warner, J. 2004. *Practical Handbook of Grouting – Soil, Rock and Structures*. New York: John Wiley & Sons.

Weltman, A.J. & Healy, P.R. 1978. *Piling in 'Boulder Clay' and other Glacial Tills*. DOE and CIRIA Piling Development Group, Report PG5, 78p.

West, G., Carter, P.G., Dumbleton, M.J. & Lake, L.M. 1981. Site investigation for tunnels, *International Journal Rock Mechanics Mining Science*, 18, 345–367.

West, L.J., Hencher, S.R. & Cousens, T.C. 1992. Assessing the stability of slopes in heterogeneously graded soils. *Proceedings 6th International Symposium on Landslides*, Christchurch, New Zealand, 1, 591–595.

Whitten, D.G.A. & Brooks, J.R.V. 1972. *The Penguin Dictionary of Geology*. Penguin Books, 495p plus appendices.

Wong, H.N. 2005. Landslide risk assessment for individual facilities. In *Landslide Risk Management*, Hungr, Fell, Couture & Eberhardt, (eds.), London: Taylor & Francis Group, 237–296.

Woodworth, J.B. 1896. On the fracture system of joints, with remarks on certain great fractures. *Proceedings of the Boston Society of Natural History*, 27, 163–184.

Wyllie, D.C. 1999. *Foundations on Rock: Engineering Practice*, 2nd edition. London: E & FN Spon.

Wyllie, D.C. & Mah, C.W. 2004. *Rock Slope Engineering*, 4th edition. London: E & FN Spon, 431p.

Yin, Y., Wang, F. & Sun, P. 2008. Landslide hazards triggered by the 12 May 2008 Wenchuan Earthquake, Sichuan, China, 2008, *Proceedings First World Landslide Forum*, Tokyo, 1–17.

Young, G.M. 2008. Origin of enigmatic structures: field and geochemical investigation of columnar joints in sandstones, Island of Bute, Scotland. *The Journal of Geology*, 116, 527–536.

Younger, P.L. & Manning, D.A.C. 2010. Hyper-permeable granite: lessons from test pumping in the Eastgate Geothermal Borehole, Weardale, UK. *Quarterly Journal of Engineering Geology & Hydrogeology*, 43, 5–10.

Yu, Y.F., Siu, C.K. & Pun, W.K. 2005. *Guidelines on the Use of Prescriptive Measures for Rock Slopes*. GEO Report 161, 34p.

Yu, Y.S. & Coates, D.F. 1970. *Analysis of Rock Slopes Using the Finite Element Method*. Department of Energy, Mines and Resources Mines Branch, Mining Research Centre, Research Report, R229 (Ottawa).

Zare, S. & Bruland, A. 2006. Comparison of tunnel blast design models. *Tunnelling and Underground Space Technology*, 21, 533–541.

Zhang, L & Einstein, H.H. 2010. The planar shape of rock joints. *Rock Mechanics & Rock Engineering*, 43, 55–68.

Index

Bold numbers depict figures, tables or boxes.